Hacking Cybersecurity Principles

Empowering You to Navigate Core Cyber Security Concepts

Alec Sklepic
CISSP, IRAP Assessor
Cybersecurity Advisor

Published by CyOps Consulting ABN: 21 484 733 463

North Adelaide, SA, 5006, Australia

info@cyops.com.au

www.cyops.com.au

ISBN Paperback: 978-1-7642544-0-3

NATIONAL LIBRARY OF AUSTRALIA

A catalogue record for this book is available from the National Library of Australia

Disclaimer

To my dearest PJ, whose unwavering belief in me came long before my own and whose love transforms the impossible into reality.

A Gift for Readers

Don't Miss This Essential Resource

Cyber incidents aren't a matter of *if*—they're a matter of *when*. That's why I've created a practical Incident Response Checklist to help you respond quickly and effectively when it matters most.

As a reader of *Hacking Cybersecurity Principles*, you can receive the checklist by visiting cyops.com.au and leaving a message. It will then be sent straight to your inbox.

Don't leave your organisation exposed—equip yourself with a clear, actionable plan before an incident strikes.

Table of Contents

Who is This Book For?

This book is for practitioners and professionals who already understand the basics of cybersecurity and want to strengthen their command of the principles that underpin effective security practice. Rather than focusing on tools or transient techniques, it concentrates on clarifying foundational concepts and showing how they apply in real-world, operational contexts.

Hacking Cybersecurity Principles is designed for professionals looking to deepen their thinking and sharpen their judgement. Early- to mid-career practitioners will gain a clearer mental model of how core principles influence architecture, risk decisions, governance, and operational trade-offs. Experienced professionals transitioning into security roles will find it bridges theory and practice, helping translate existing knowledge into a security-first mindset.

The book is equally relevant to senior practitioners, consultants, and leaders who need a principled lens for decision-making, strategy, and communication with stakeholders. If you are responsible for managing risk, advising others, or shaping security outcomes, this guide reinforces the concepts that matter long after specific technologies change. It meets you where you are, and helps you think more clearly about why cybersecurity works the way it does.

What You'll Get:

- Clear explanations of core cybersecurity concepts
- Step-by-step guidance through fundamental principles
- Real-world examples that make complex ideas easier to understand
- Practical tips or hacks for implementing basic security measures

- A solid foundation for further learning in cybersecurity.

Who is This Book *Not* For?

While *Hacking Cybersecurity Principles* provides a solid foundation, it may not align with your goals if you seek advanced technical training or specific tool tutorials. Those pursuing mastery in ethical hacking and penetration testing or seeking detailed guidance on security tools like Nessus or Wireshark, will need more specialised resources. This guide intentionally stays broad rather than deep—you won't find complex security configurations, implementations or a comprehensive history of cybersecurity here, as this publication focuses on practical fundamentals. However, general knowledge of technology and Information and Communication Technology (ICT) will assist you on this journey.

Introduction

All technology, whether software, hardware or mechanical, regardless of age, contains flaws that can be exploited if an attacker is properly motivated and resourced. Cybersecurity is about managing the risk of exploitation through prevention and response.

Imagine a world where a single software patch could bring global systems to their knees. The stark reality is that this world, once confined to science fiction, is one we already live in, as became evident in 2024 when a faulty update by a major security provider crippled an estimated 8.5 million computers, disrupting essential services worldwide.[1] Banks, airports, government agencies, hospitals and supermarkets found themselves paralysed, their operations grinding to a halt. This incident serves as a sobering reminder of the fragility of our IT ecosystem.

As cyber threats constantly evolve, we must recognise that relying solely on technological solutions is no longer sufficient. While cutting-edge tools and their applications remain key to our defensive strategy, their limitations are becoming increasingly apparent. The time has come to shift our focus from a purely technological approach to one rooted in fundamental cybersecurity principles.

Why focus on principles rather than tools? Consider this: learning to use a cybersecurity tool, such as a vulnerability scanner, may be helpful now, but that knowledge will quickly become outdated. However, understanding cybersecurity principles will enable you to adapt to new technologies and emerging threats rapidly. It's akin to the adage about teaching someone to fish rather than simply giving them one.

This approach forms the basis of *Hacking Cybersecurity Principles*. This book stems from the pressing need to develop security strategies that encompass not just the *what* of cybersecurity but also the *why*, *when* and *where*. By understanding core principles, you can maximise your security investments and address the most severe and imminent threats facing you and/or your organisation today.

In recent years the number of people affected by data breaches has soared from millions to over a billion[2]. An alarming rise that exemplifies the growing scale and reach of cyber threats. These numbers aren't just abstract data points; they represent real people whose lives have been disrupted and who faced potential identity theft, financial loss or privacy violation. The message is clear: cybersecurity is no longer a niche concern but a fundamental necessity in our interconnected world. It's a shared responsibility that extends from individuals to multinational corporations and governments.

Throughout my nearly three-decade career in IT and cybersecurity, I've encountered many professionals who feel overwhelmed by the rapid pace of technological change, the ever-expanding threat environment and the myriad tools available to combat these challenges. It's too easy to become overwhelmed by the minutiae, losing sight of the bigger picture. This is where the power of core cybersecurity principles comes into play. Grounding yourself in these fundamental principles is the most effective way to navigate this complexity. They act as a compass, guiding you through the noise and helping you make informed, strategic decisions.

I'm frequently asked, "How do you get started in this field?" I once posed this question myself to a professor lecturing on cybersecurity at a prominent university. When I sought guidance, his advice was disappointingly vague: "Do something big", he said. To this day, I'm not sure if he was serious. Nonetheless, my answer to this question, while simple, is hopefully more specific: entering cybersecurity is like building a house—you need a solid foundation. This book will guide you in constructing that base, preparing you for a lasting career. It won't make you an instant professional, but it will set you on the right path, priming you to become a cybersecurity expert.

Regardless of your starting point, what matters most is developing a solid understanding of core cybersecurity principles. They provide a stable foundation that remains relevant even as specific technologies evolve, allowing you to adapt confidently to new challenges.

By mastering these fundamental concepts, you'll be equipped to:

- Navigate the complex world of cybersecurity with clarity and purpose—Understanding core principles helps you see the big picture, allowing you to make sense of the often-confusing cybersecurity field.

- Make informed, strategic decisions about security investments and priorities—With a solid grasp of core principles, you can better evaluate different solutions and strategies and ensure your resources are used effectively.

- Adapt quickly to new technologies and emerging threats—The principles provide a framework for understanding and addressing new challenges, even when faced with unfamiliar technologies or attack vectors.

- Bridge knowledge gaps and develop a comprehensive understanding of essential concepts—Cybersecurity touches on many different areas of IT and business. Core cybersecurity principles help you connect these dots, giving you a holistic view.

- Advance your career in cybersecurity and potentially earn more—Employers value professionals who understand not just the *how* but the *why* of cybersecurity. A strong foundation in these principles makes you a more versatile and valuable team member.

Driven by expertise and purpose, I wrote this book to address critical knowledge gaps for both curious beginners and seasoned professionals.

Drawing from years of hands-on experience, I've created a comprehensive resource that transforms complex security concepts into actionable insights. This book delivers practical strategies, empowering individuals and organisations to not only understand modern cyber threats but to build resilient defence mechanisms against them. My aim is to spark the passion of a new generation of cybersecurity practitioners while providing current professionals with the knowledge they need to excel in this dynamic field.

Hacking Cybersecurity Principles is organised into three main sections, each building upon the previous to create a cohesive and holistic understanding of the field, as follows:

Part I: Foundations of Cybersecurity:

This section lays the groundwork for our cybersecurity journey. We'll start by exploring the fundamentals and tracing the evolution of cybersecurity from its early days to the present. We'll dive into understanding cyber threats and vulnerabilities and tackle the all-important topic of risk management.

You'll learn about the CIA Triad, not the secret government organisation, but the pillars of Confidentiality, Integrity and Availability in information security. We'll unpack each concept, demonstrating how they form the bedrock of any security strategy.

We'll also briefly cover key industry frameworks and standards, such as the National Institute of Standards and Technology (NIST) Cybersecurity Framework (CSF) and ISO 27001. These frameworks provide structured approaches to implementing cybersecurity measures and are widely used across various industries.

Finally, we'll explore the critical role of threat intelligence. You'll learn how it is gathered, analysed and used so you can stay one step ahead of potential attackers.

Part II: Implementing Cybersecurity Measures:

This is where we roll up our sleeves and build upon our foundational knowledge. We'll explore implementing cybersecurity measures, starting with governance (a fancy word for 'who's in charge of what').

You'll then walk through the entire cybersecurity lifecycle: govern, identify, protect, detect, respond and recover. We'll also examine asset management and vulnerability identification, discussing how to keep track of your digital assets and identify potential weak points in your defences.

You'll learn about various protection strategies, from access control and basic encryption to secure software development practices. We'll cover detection capabilities, including intrusion detection systems and Security Information and Event Management (SIEM) tools.

We'll also discuss incident response processes, teaching you how to prepare for, respond to and recover from security incidents. You'll learn about the importance of having a well-crafted Incident Response Plan (IRP) and how to conduct effective post-incident reviews.

Part III: Human Factors:

In this final section, we focus on the crucial human side of cybersecurity.

You'll navigate the tricky waters of legal and ethical considerations in cybersecurity, including discussions on privacy laws, ethical hacking and the moral dilemmas cybersecurity professionals often face.

You'll also discover how to measure the effectiveness of cybersecurity through metrics and performance evaluation. After all, if you can't measure it, you can't secure it. We'll explore different types of security metrics and how to use them to drive improvement in your security posture.

Finally, we'll look at the vital topic of cultivating a security-conscious culture. This includes security awareness training and strategies for getting buy-in where needed. Remember, the strongest firewall in the world is useless if someone leaves the door open.

As you read *Hacking Cybersecurity Principles*, you are encouraged to approach the material with an open mind and a willingness to engage deeply with its concepts. It is imperative to recognise that cybersecurity is not merely a set of technical skills but a mindset—a way of thinking critically about the digital world.

This book blends theoretical foundations with practical application to deepen your understanding of cybersecurity. These elements are conveyed through several types of content presented throughout the book. *Strategic Briefings* ground key concepts in real-world scenarios, while *Tactical Briefings* offer hands-on applications to build your skills. Interactive elements, like *Revision Questions* and *Cybersecurity Hacks*, reinforce learning through engagement. You'll explore essential definitions, dissect major cyber incidents and navigate industry-standard frameworks, emerging technologies and evolving threats.

Note that we use 'cybersecurity' as a single word throughout this publication. While 'cyber security' (two words) is common in some regions, particularly Australia and the United Kingdom, I have chosen the single-word form for consistency and simplicity. Both forms are correct and essentially interchangeable.

As we embark on this journey together, we begin the process of hacking cybersecurity principles—not to exploit them but to master, implement and use them to create a safer digital world for all. The challenges ahead are significant, but with the knowledge gained here, you will be well-equipped to move from feeling overwhelmed by cyber threats and technology to feeling empowered and prepared.

Part I

Foundations of Cybersecurity

So, are you ready to take control of your cyberse
potential victim to knowledgeable defender? Turn
frontlines and let's break down the principles behin

Chapter 1: Foundations of Cybersecurity

– Start Here –

Objectives

By the end of this chapter, you will be able to:

- Define cybersecurity and explain its critical role in the modern digital landscape
- Identify and describe the three pillars of the CIA Triad (Confidentiality, Integrity and Availability)
- Distinguish between threats, vulnerabilities and risks in cybersecurity
- Understand the basic principles of cyber defence and security controls
- Recognise different types of threat actors and their motivations.

Cybersecurity, in its essence, is the practice of safeguarding systems, networks and data from malicious attacks designed to compromise, steal or disrupt our digital lives. As technology becomes more deeply interwoven into every aspect of our personal and professional lives, understanding and implementing robust cybersecurity measures is no longer optional—it's essential.

The journey of the field of cybersecurity reflects the rapid technological evolution of our digital ecosystem. In the early decades of computing, security was predominantly a physical concern, involving controlling access to computer rooms and hardware. The 1960s and 1970s saw minimal digital security practices, with most efforts focused on protecting tangible computing infrastructure. As networks emerged and computers became interconnected, security threats shifted dramatically. The proliferation of the internet in the 1990s marked a pivotal moment, expanding potential vulnerabilities and introducing new digital risks that extended far beyond physical boundaries.

Today's cybersecurity landscape is characterised by unprecedented complexity. Cyber threats have evolved from simplistic virus infections to sophisticated, multi-layered attacks orchestrated by increasingly advanced threat actors. Nation-states, criminal organisations and individual hackers now deploy complex strategies leveraging Artificial Intelligence (AI), social engineering and intricate technological vulnerabilities to breach our cyber defences.

Contemporary digital infrastructure now faces extraordinary cybersecurity challenges, where the implications of security breaches extend far beyond traditional risk management paradigms. A single security breach can result in catastrophic consequences, financial losses, operational disruption, reputational damage and potential legal ramifications. Organisations and individuals alike must recognise that cybersecurity is not just a technical challenge but a fundamental business and personal survival strategy.

Welcome to the foundational principles of cybersecurity. This chapter establishes the foundation for understanding cybersecurity fundamentals. By exploring core concepts, fundamental principles and the dynamic threat environment, you will be equipped with the key knowledge needed to navigate the complex digital landscape. From understanding the CIA Triad (Confidentiality, Integrity and Availability) to exploring the nuanced relationship between threats, vulnerabilities and risks, this chapter serves as a launching point for your cybersecurity journey.

The Security Imperative

Every 39 seconds, a cyberattack occurs. As digital technology integrates into every aspect of life, the risks grow just as rapidly. Our modern society operates through a vast network of interconnected digital systems that deliver unprecedented convenience. However, this reliance also exposes us to evolving threats that require robust security measures.

Modern Digital Landscape

The internet is now a global digital ecosystem where over five billion people interact daily through cloud services, social media and digital commerce platforms. Your organisation operates in this hyperconnected environment, where traditional network boundaries have dissolved, replaced by complex interactions spanning multiple countries and jurisdictions. As organisations embrace these innovations, however, security risks escalate.

Digital transformation is reshaping industries at an accelerating pace as organisations adopt cloud computing, Internet of Things (IoT) devices and AI. The mainstream adoption of remote work means employees now access sensitive corporate resources from diverse locations and devices. This transformation significantly expands your potential attack surface (the total points where an attacker might breach your systems), creating urgent security challenges that demand attention.

At the same time, the threat landscape is shifting in parallel with these changes. Cybercriminals employ increasingly sophisticated techniques, leveraging AI to automate and scale attacks. Previously rare threats, such as zero-day exploits (attacks targeting undiscovered software vulnerabilities), ransomware and supply chain attacks, have now become common risks. Meanwhile, nation-state actors target critical infrastructure and valuable intellectual property, escalating the stakes for cybersecurity professionals.

The Need for Proactive Security

Addressing these threats requires a proactive approach, strong cybersecurity strategies, continuous monitoring and investment in advanced defence mechanisms. Organisations must recognise that cybersecurity is no longer an IT concern but a core business imperative. By taking decisive action, businesses can safeguard their data, protect their operations and maintain trust in an increasingly volatile digital landscape.

Why Cybersecurity Matters

Digital asset protection is now a central concern as organisations expand the storage of highly valuable assets in digital form. This includes intellectual property, customer data, financial records and operational systems. The compromise of these assets has devastating consequences, from financial losses to operational disruption. Effective cybersecurity measures are necessary to protect these critical assets from unauthorised access, theft or manipulation.

Business continuity now depends heavily on a cyber ecosystem. Even brief disruptions to critical systems can result in significant financial losses and customer dissatisfaction. Organisations must ensure their digital infrastructure remains operational and resilient against various threats, from technical failures to targeted attacks. This requires security measures that protect systems and data while enabling rapid incident recovery.

Reputation management has taken on new importance. Security breaches can severely damage an organisation's reputation, eroding customer trust and brand value. The impact of a major security incident often extends far beyond immediate financial losses, potentially affecting customer relationships, partner confidence and market value for years to come. Just look at how long it took for Optus to recover from its 2022 data breach.[3] Organisations must demonstrate their commitment to security to maintain stakeholder trust.

Regulatory compliance is another area of increasing complexity as governments respond to the growing and ever-evolving cyber threat environment. Organisations must navigate a maze of data protection regulations, industry standards and security requirements. From *The Australian Privacy Principles* (APPs) and *The General Data Protection Regulation* (GDPR) to sector-specific regulations, compliance mandates comprehensive security controls and documentation. Failure to meet these requirements results in significant fines and legal consequences.

The imperative for effective cybersecurity has never been stronger. As organisations continue to transform their operations and expand their digital footprint, they must adopt security measures that address both current and emerging threats. This requires a strategic approach that balances security requirements with business objectives while ensuring compliance with changing regulatory requirements.

Strategic Briefing: The Australian National University Data Breach

In 2019, the Australian National University (ANU) learned a hard lesson when hackers stole 19 years of sensitive personal data from staff, students and visitors. The attack began in November 2018 and remained undetected for six months, causing significant damage. This breach is a reminder that even well-established institutions must vigilantly guard their digital assets.

The Incident

The ANU attack revealed just how vulnerable sensitive data can be. The hackers stole a treasure trove of personal information: from basic details like names and addresses to highly sensitive data including tax file numbers, bank account details and academic records.[4] Using sophisticated Advanced Persistent Threat (APT) techniques, they operated like silent thieves, moving undetected through the university's systems for months while steadily collecting information.

Impact and Response

The breach's impact was profound, affecting approximately 200,000 individuals. The financial implications were substantial and included costs such as investigating the breach, enhancing security measures and notifying affected individuals. The reputational damage was also considerable, as the ANU had to reassure its community and the public about steps taken to secure its systems and prevent future incidents.[5]

Following the breach, the university took decisive action to rebuild its cyber defences with a comprehensive review and overhaul of its cybersecurity protocols. They worked with the Australian Signals Directorate (ASD) to strengthen their security system and ensure best practices were followed.[6]

They also rolled out stronger safeguards, including improved monitoring and detection systems, enhanced access controls and increased cybersecurity awareness training for staff and students.

Lessons Learned

The ANU data breach offers several key lessons in cybersecurity fundamentals:

- **Importance of Detection and Monitoring**—The prolonged period during which attackers remained undetected emphasises the need for effective monitoring and intrusion detection systems.
- **Advanced Persistent Threats (APTs)** —The sophistication of the attack highlights the growing threat of APTs and the necessity for advanced security measures to counter such threats.
- **Comprehensive Cybersecurity Strategy**—This incident emphasises the necessity of a holistic cybersecurity approach incorporating people, processes and technology. Regular security audits, employee training and collaboration with cybersecurity experts are fundamental components of this strategy.

Core Cybersecurity Concepts

Cybersecurity relies on core principles that enable individuals to understand, assess and safeguard digital assets. Central to these principles is the CIA Triad—confidentiality, integrity and availability. Given their significance, these concepts will be explored in depth in later chapters.

Core Components of Cybersecurity

The CIA Triad is the foundation of information security. These three principles guide the development of security policies, strategies, architectures and controls worldwide.

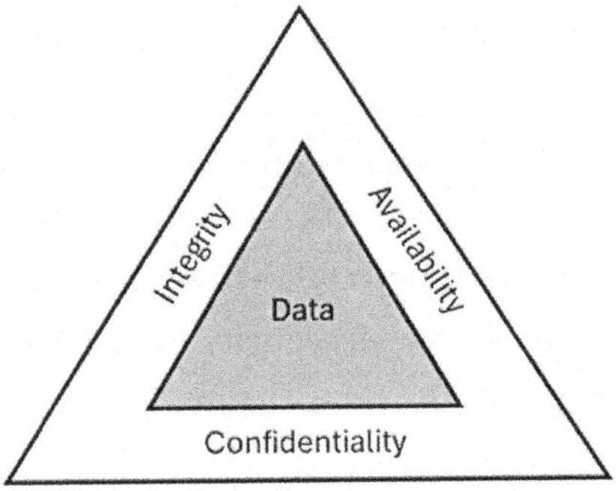

The CIA Triad

Confidentiality: This principle ensures that sensitive information is accessible only to those authorised to view it. Confidentiality is achieved through mechanisms such as encryption, access controls and authentication protocols.

Integrity: Integrity involves maintaining the accuracy and completeness of data. This means that data cannot be altered in an unauthorised manner, ensuring information is reliable and trustworthy.

Availability: Availability ensures information and resources are accessible to authorised users when needed. This is critical for business operations, as downtime can lead to significant losses.

Threat vs. Vulnerability vs. Risk

To effectively manage cybersecurity, you must understand how threats, vulnerabilities and risks interact. These distinct but interconnected elements require specific mitigation strategies.

A threat represents any potential source of harm to information systems, whether natural (floods, earthquakes), human (malicious actors, negligent staff) or environmental (power failures, fire). Vulnerabilities are exploitable weaknesses in your systems, arising from factors such as inadequate physical security, software flaws, misconfigurations or poor security practices. When threats exploit vulnerabilities, they create risk—the measurable potential for loss or damage.

The relationship between these elements is often described by the formula:

$$\text{Risk} = \text{Vulnerabilities} \times \text{Threats}.$$

This approach reflects how security professionals evaluate potential losses and prioritise controls (more on this in Chapter 4). By systematically analysing threats and vulnerabilities, you can accurately assess risk levels and implement proportionate security measures.

Risks vs. Controls

We now know the relationship between threats, vulnerabilities and risk. Equally important is the risk management process. The goal of risk management is twofold: first, to reduce the likelihood of a risk occurring, and second, to minimise its potential impact if it does materialise. These goals drive the implementation of controls.

Organisations protect their assets by deploying controls, specific measures and procedures designed to address identified threats and vulnerabilities. These controls must achieve a delicate balance: providing robust protection while maintaining operational efficiency. For example, while strict password policies reduce unauthorised access risks, overly complex requirements might lead to staff writing passwords down, creating new vulnerabilities.

The key challenge in risk management lies in developing cost-effective controls that protect assets without impeding business operations. Effective security strategies implement proportional controls that address specific risks while preserving organisational agility. This approach ensures that security measures protect your systems while supporting, rather than hindering, operational goals.

Examples of mitigation strategies and controls:

- *Strategies to Mitigate Cyber Security Incidents:* https://www.cyber.gov.au/resources-business-and-government/essential-cyber-security/essential-eight
- *Information Security Manual (ISM):* https://www.cyber.gov.au/resources-business-and-government/essential-cyber-security/ism
- *NIST SP 800-53 Rev. 5 Security and Privacy Controls for Information Systems and Organizations:* https://csrc.nist.gov/pubs/sp/800/53/r5/upd1/final

Assets and Attack Surfaces

Your organisation's cybersecurity strategy must begin with a clear understanding of what you're protecting (assets) and how attackers might reach them (attack surfaces). Assets represent everything valuable within your organisation, from physical systems and data to intellectual property and operational capabilities. Each asset requires protection proportional to its business impact and sensitivity level, making asset classification a critical first step in security planning.

Attack surfaces encompass every potential entry point into your systems, including network interfaces, software vulnerabilities, user access credentials and physical access points. Digital transformation continuously expands these attack surfaces, as cloud services, mobile devices and IoT deployments create new opportunities for exploitation. For example, a single cloud-based application might expose multiple entry points through its web interface, Application Programming Interface (API) connections and user authentication systems, while remote work environments further expand potential vulnerability points.

The relationship between assets and attack surfaces creates complex security challenges that require careful attention. Each business-critical asset typically connects to multiple attack vectors, requiring comprehensive protection strategies. To manage this complexity effectively, you must systematically identify and categorise assets by their criticality, map all potential attack surfaces across both digital and physical domains, implement targeted protection mechanisms for high-value assets, and continuously monitor and reduce potential entry points. This systematic approach ensures no critical assets remain exposed through overlooked attack vectors.

By methodically reducing attack surfaces while focusing on asset protection, you cost-effectively strengthen your security posture. This strategic approach ensures that your most valuable resources receive appropriate protection while managing overall organisational risk exposure. Regular assessment and adjustment of these protection mechanisms help maintain security effectiveness as your technology landscape evolves and new threats emerge.

Common Cybersecurity Terms

Understanding cybersecurity begins with knowing its key terms. These essential terms are crucial for grasping cybersecurity and defence strategies.

At the core of cybersecurity threats lies **malware**, encompassing various forms of malicious software designed to compromise information systems. This includes **viruses** that self-replicate across systems, **worms** that propagate through network connections, **trojans** that deceive users with seemingly legitimate functions and **ransomware** that encrypts data for extortion purposes. Each variant poses distinct risks to system integrity and data security.

Phishing is a significant human-focused threat that employs social engineering techniques to compromise security through deception. Attackers systematically target individuals through various communication channels, primarily email but increasingly SMS and social media platforms, to obtain sensitive credentials and financial information. This threat vector specifically exploits human trust mechanisms rather than technical vulnerabilities.

Organisations implement multiple layers of defence to protect against these threats. **Encryption** is a fundamental security control, converting sensitive data into protected formats to ensure confidentiality during transmission and storage. This cryptographic protection forms the backbone of secure data management practices.

Firewalls are critical network security components, establishing controlled boundaries between trusted internal networks and external environments. Through systematic traffic monitoring and rule enforcement, firewalls enable access control and **network segmentation**, essential for a **defence-in-depth** strategy. More on this later.

Zero-day exploits present a particular challenge in the security environment, targeting previously unknown vulnerabilities before patches become available. These threats emphasise the importance of security measures beyond simple patch management.

Multi-Factor Authentication (MFA) is key for grasping cybersecurity access control measures. It is a security process that strengthens access control by requiring users to verify their identity through multiple methods. These typically include:

- Something you know (like a password or PIN)
- Something you have (like a smartphone or security token)
- Something you are (like a fingerprint or facial recognition).

By requiring at least two different types of verification, MFA provides significantly stronger protection than traditional single-password systems.

Intrusion Detection Systems (IDS) provide continuous monitoring capabilities, analysing network and system activities to identify potential security incidents. This automated surveillance enables rapid detection and response to security violations, supporting proactive security management.

An organisation's **security posture** encompasses the totality of its security capabilities, including technical controls, human resources, policies and response procedures (**people, process and technology**). This approach to security management enables organisations to maintain defences while adapting to emerging threats.

While these terms cover the basic concepts and will assist you with the material in this book, cybersecurity encompasses many more. For a more comprehensive understanding, consult the following additional resources:

- *ASD-ACSC. Glossary*: https://www.cyber.gov.au/learn-basics/view-resources/glossary
- *NIST. Glossary*: https://csrc.nist.gov/glossary
- *SANS Institute. Glossary of Cyber Security Terms*: https://www.sans.org/security-resources/glossary-of-terms/

Basic Principles of Cyber Defence

While the CIA Triad is the foundation of information security, there are several other concepts upon which effective cybersecurity hinges. These are the basic principles of cyber defence. They include preventive and detective security controls, the concept of defence-in-depth and the critical importance of patch management.

Preventive, Detective and Corrective Security Controls

Preventative security controls are designed to thwart potential cyber threats before they can cause harm. They aim to reduce the attack surface and prevent unauthorised access to systems and data. Examples of preventive controls include firewalls, which act as barriers to block unauthorised traffic, and encryption, which secures data by making it unreadable without the correct decryption key. Access controls, such as MFA, also serve as preventive measures by ensuring that only authorised users can access sensitive information even if passwords are compromised.

Detective security controls identify and respond to cybersecurity incidents that have already occurred. They play a crucial role in recognising security breaches and initiating appropriate responses. Examples of detective controls include IDS, which monitors network traffic; and Security Information and Event Management (SIEM) systems, which analyse logs and alerts to detect potential security incidents and use threat hunting capabilities to search for hidden threats. Regular security audits and vulnerability assessments also fall under detective controls, as they help identify weaknesses and describe a security posture.

Corrective security controls are strategic mechanisms designed to mitigate damage and restore system functionality after a security incident or breach. These controls focus on minimising the impact of a cybersecurity event, addressing vulnerabilities and returning systems to a secure operational state. Examples of corrective controls include IRPs outlining specific steps to contain security breaches and system restoration procedures that rebuild compromised systems. Malware removal tools, system rollback capabilities and automated recovery scripts are also corrective controls that neutralise threats and minimise potential long-term damage quickly.

By integrating preventative, detective and corrective security controls, we can develop a multi-layered approach to cybersecurity, ensuring strong protection and rapid response capabilities.

Defence-in-Depth

This multi-layered approach to cybersecurity centres on creating multiple lines of defence, thereby significantly increasing the difficulty for attackers to breach an organisation. This embodies the defence-in-depth principle, which recognises that no single measure is entirely foolproof. By layering complementary and overlapping controls, organisations can enhance resilience, ensuring that if one barrier is bypassed, others remain to protect critical systems and data.

The defence-in-depth approach involves deploying multiple layers of security controls at different system levels, such as network, application and data. For example, an organisation might use firewalls, Intrusion Prevention Systems (IPS) and antivirus software to protect its network while also implementing access controls and encryption at the application level. Additionally, endpoint security measures like antivirus and anti-malware software provide another layer of protection. The goal of defence-in-depth is to ensure that if one control fails or a vulnerability is exploited, other layers of the onion-like model remain to mitigate or contain the threat.

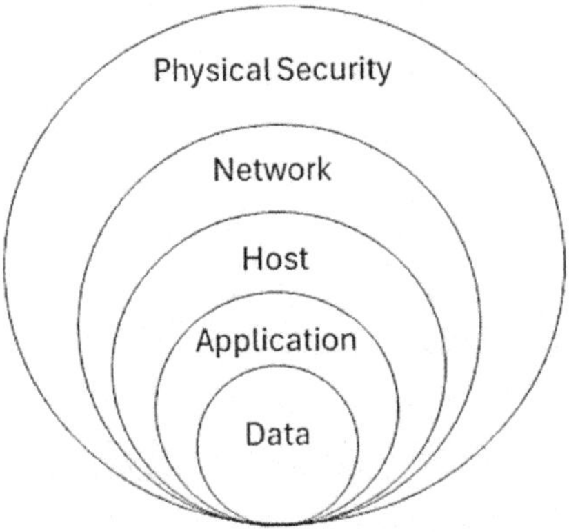

Fundamental Defence-in-Depth Model

Patch Management

Cybersecurity applied to the various layers of your information systems, be it your mobile phone or an extensive corporate system, is not a set-and-forget prospect. Threats evolve, and vulnerabilities that were once unknown float to the surface. To prevent breaches, updates or patches are necessary.

Patch management is the process of applying updates. It involves regularly updating software and hardware to address vulnerabilities and enhance functionality. Unpatched software and even hardware can harbour exploitable weaknesses, making timely updates critical for reducing risk and maintaining a secure environment.

Effective patch management encompasses identifying, testing and applying patches in a timely manner. This can be quite challenging for organisations with many or large corporate systems. The most effective approach is to prioritise patches based on the severity of the vulnerabilities. This entails identifying the organisation's most critical systems and then completing a vulnerability scan on those systems. Patches can then be rolled out to the most critical systems to address the most severe vulnerabilities first. Following this, patches are applied to other less critical systems. Automated patch management tools can help streamline this process by automatically applying patches and updates. Regularly updating software reduces the risk of exploitation and helps maintain the overall security of IT environments.[7]

Now that you have a few basic principles in hand, we will explore the cyber threat environment.

Overview of the Cyber Threat Environment

The cyber threat environment is a dynamic and complex ecosystem where evolving risks continuously challenge individuals, organisations and industries. Cyber threats exploit vulnerabilities in both technology and human behaviour, requiring adaptive and proactive security measures. Understanding this landscape is essential for developing resilient cybersecurity strategies that safeguard digital assets.

Key Characteristics of the Cyber Threat Landscape

Several defining traits shape the modern cyber threat environment:

- **Rapid Evolution**—Cyber threats advance alongside technological innovation, newly discovered vulnerabilities and shifting geopolitical conditions. Attackers continuously refine their methods, rendering static defences inadequate.

- **Expansive Scope**—Cyber threats affect critical infrastructure, businesses and individuals alike. Risks range from system exploits and social engineering tactics to large-scale strategic attacks.

- **Adaptive Nature**—Malicious actors evolve their tactics to bypass security measures, often blending technical and psychological techniques for greater effectiveness.

A well-informed, proactive approach is needed to navigate this ever-changing landscape. Cybersecurity professionals must anticipate emerging threats and implement flexible defences to mitigate risks effectively.

Types of Cyber Threats

Broadly, cyber threats fall into three primary categories: adversarial, non-adversarial and physical/environmental threats. These diverse challenges require security approaches that address both intentional attacks and unintended vulnerabilities (more on this in Chapter 2).

Adversarial Threats: Deliberate attacks from malicious actors, including malware, denial-of-service (DoS) attacks, zero-day exploits and social engineering tactics like phishing. These threats typically come from nation-states, criminal organisations, hacktivists and other threat actors pursuing financial, political or ideological objectives.

Non-Adversarial Threats: Security risks arising from unintentional actions or system issues, such as misconfigurations, software bugs, human error or poor security practices. While not malicious in nature, these vulnerabilities can be equally damaging and often serve as entry points for adversarial attacks.

Physical/Environmental Threats: Risks to digital infrastructure from natural disasters, power outages, hardware failures, physical theft, unauthorised access to facilities and environmental factors like temperature and humidity can disrupt operations and compromise data security through non-cyber means.

Most security incidents involve multiple threat vectors, highlighting the importance of developing security strategies that address technical, human and physical security controls while considering intentional and unintentional risk factors.

Defending Against Cyber Threats

Mitigating cyber threats requires a layered, adaptable security approach. Effective defence strategies include:

- **Threat Intelligence and Monitoring**—Continuously analysing cyber threats to anticipate and mitigate risks.
- **Security Awareness Training**—Educating users on cyber risks, particularly social engineering tactics, to reduce human vulnerabilities.
- **Zero Trust Architecture (ZTA)**—Enforcing strict access controls to limit unauthorised movement within networks.
- **Patch Management and Regular Updates**—Ensuring systems are updated to prevent exploitation of known vulnerabilities.
- **Incident Response Planning**—Developing strategies to detect, contain and recover from security incidents.

The cyber threat environment is a constantly shifting battlefield where attackers refine their tactics to exploit emerging vulnerabilities. Understanding this landscape is the first step toward effective cybersecurity. By recognising key threats, their motivations and countermeasures, individuals and organisations can strengthen their resilience against digital threats. While this section provided a broad overview, later chapters explore specific threats and defence mechanisms in greater detail.

The Evolution of Cybersecurity Practices

In 1988, the Morris Worm infected nearly 10% of the internet,[8] bringing a fledgling digital world to its knees. Today, a single cyber-attack can compromise millions of users or disrupt entire industries. How did we get here? Cybersecurity is not a static discipline but a dynamic field of innovation and adaptation. Its evolution mirrors the rapid technological transformations that have reshaped our digital landscape, reflecting humanity's ongoing struggle to protect critical information assets from increasingly sophisticated threats. From the earliest days of computing to today's complex cyber ecosystems, the journey of cybersecurity reveals a narrative of technological resilience and strategic innovation.

The roots of cybersecurity stretch back to the nascent computing environments of the 1960s and 1970s, where security was primarily a physical concept, involving the control of access to massive mainframe computers and the protection of tangible technological infrastructure. As networks grew and computational capabilities expanded, the threat landscape dramatically transformed. Each technological leap, from time-sharing systems to personal computers, from localised networks to global internet connectivity, introduced new vulnerabilities and challenged existing security paradigms.

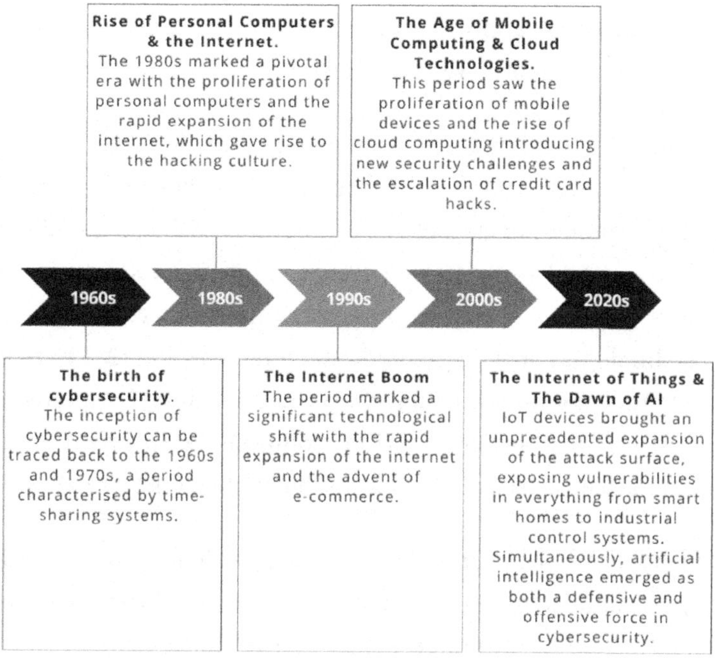

The Evolution of Cybersecurity

The 1990s saw the explosive growth of the internet, bringing a new array of cybersecurity challenges. The interconnected nature of the web made systems more vulnerable to attacks, and the proliferation of email and web applications opened up new vectors for cyber threats. In response, more sophisticated security practices were implemented. Firewalls and IDSs became standard network security components. Their purpose was to monitor and control incoming and outgoing network traffic. Encryption technologies were also adopted to protect sensitive information during transmission.

As the new millennium dawned, the scope and complexity of cyber threats continued to expand. Cybercriminals became more organised and motivated by financial gain, leading to increased phishing attacks, identity theft and financial fraud. This era introduced cybersecurity frameworks and standards, such as *ISO/IEC 27001* and the *Payment Card Industry Data Security*

Standard (PCI DSS). These frameworks standardised security practices and guided organisations to apply security controls and manage risk. Also, the concept of 'defence-in-depth' was developed.

In recent years, cybersecurity practices have evolved in response to emerging technologies and increasingly sophisticated threats. The proliferation of mobile devices, cloud computing, Artificial Intelligence (AI) and the IoT has expanded the attack surface, requiring new security practices. These modern practices focus on proactive threat detection and response. Advanced technologies such as AI and Machine Learning (ML) are standard practice to identify and mitigate threats. At the same time, Endpoint Detection and Response (EDR) systems, SIEM solutions and threat intelligence platforms have arisen as standard tools in the fight against cybercrime.

Looking forward, the future of cybersecurity will likely be defined by even more complex and dynamic challenges. As technologies like quantum computing, advanced AI and increasingly interconnected global systems rise, cybersecurity will continue to be a critical and evolving discipline. The ongoing arms race between defenders and attackers will demand continuous innovation, adaptability and a holistic approach to understanding and mitigating cyber risks. What remains constant is the fundamental human imperative: to protect our digital infrastructure, preserve the integrity of information systems and safeguard the increasingly digital foundations of our global society. Cybersecurity is no longer just a technical challenge but a complex, multidimensional endeavour requiring collaboration across technological, legal, educational and strategic domains.

Summary

Cybersecurity protects systems, networks and data from cyber-attacks that aim to access, alter or destroy sensitive information, extort money or disrupt operations. In our hyperconnected world, cybersecurity is central to our digital lives.

This chapter laid the foundation for understanding cybersecurity in today's digital age. It explored key terminologies and concepts, highlighting cybersecurity's role in protecting data from cyberattacks that can cause financial loss, reputational damage and operational disruption. By providing a top-down overview, this chapter serves as a springboard for further exploration into the field's details.

Mastering cybersecurity requires a solid understanding of its fundamental concepts. This chapter has introduced the building blocks, including the CIA Triad, the distinctions between threats, vulnerabilities and risks, and foundational terminology. These elements have been strategically designed to equip you with the framework necessary to comprehend the practices and strategies discussed throughout this book.

This chapter has set the stage for a deeper dive into the world of cybersecurity. Now let's begin our journey into that world.

Revision Questions

Answers to revision questions can be found in Chapter 20.

1. What is the CIA Triad, and why is it considered fundamental to cybersecurity?

2. Explain the relationship between threats, vulnerabilities and risks in cybersecurity.

3. What are the three types of security controls, and how do they work together?

4. What are some common motivations behind cybercriminal activities?

5. How has the cybersecurity landscape evolved from the 1960s to the present day?

6. What is defence-in-depth, and why is it important in cybersecurity?

7. Why is patch management critical to cybersecurity, and how should it be implemented?

8. What are the four primary categories of cyber threats?

9. What makes Multi-Factor Authentication (MFA) an effective security measure?

10. What are the key characteristics of the modern cyber threat landscape?

Cybersecurity Hacks

Hack 1: Implementing Basic Authentication

The number one cybersecurity hack to protect your information and systems is to use strong, unique passphrases for each account and enable MFA. MFA significantly enhances security by requiring multiple forms of verification, making it much harder for attackers and malware to gain unauthorised access even if they have compromised a password. Start by identifying high-risk accounts and gradually extend MFA to all users.

To assist in creating and managing passphrases and passwords, use the following tips:

1. **Password/Passphrase Strength:** Create a complex passphrase of at least four random words that are long (at least 14 characters in total)—more words would be even better—and include a mix of uppercase and lowercase letters, numbers and preferably special symbols. Avoid using easily guessable information like birthdays, names or common phrases. For example, "Rubberdog21Crystal-CarflyingHome" or "Ant17CarpetSky#NewsGone".

2. **Password/Passphrase Uniqueness:** Never reuse passwords across different accounts. That way, if one is compromised, your other accounts remain secure. A trusted password manager can securely generate and store strong, unique passwords for every login. There are many reliable options available, and using one makes it much easier to avoid password reuse.

Hack 2: Cybersecurity Basics

1. **Keep your software up to date.** This includes your operating system, web browser and other software. Updates often include security patches that can help protect you from attacks.

2. **Be careful what you click on.** Don't open emails or click on links from people or organisations you don't know. Be wary of phishing scams designed to trick you into giving up your personal or financial information. Be extremely cautious of communications requiring an urgent reply.

3. **Back up your data regularly.** This will help you recover your data in case it is lost or stolen.

4. **Be careful what you share online.** Don't share personal information on social media or other public websites.

Chapter 2: The Nature of Threats

– Know Your Enemy –

Objectives

By the end of this chapter, you will be able to:

- Categorise and analyse cyber threats
- Recognise the motivations and capabilities of different threat actors
- Understand methodologies used in various cyber-attack strategies
- Develop a systematic approach to threat identification and mitigation
- Appreciate the interconnected nature of modern cyber threats.

Let's establish a foundational cybersecurity principle: identifying or eliminating all threats, vulnerabilities and risks is impossible. No matter how resilient security measures are, new threats will always emerge, defying full prediction or eradication. Rapid technological advancements and increasingly sophisticated adversarial tactics create a constantly shifting landscape of vulnerabilities. Even the most secure air-gapped systems (computers isolated from all networks) remain susceptible to risks and threats.

So, what defines cybersecurity, and where do you fit in? It's not about constructing an impenetrable defence. Instead, it's about understanding threats, managing risks and protecting critical assets. Your role is to safeguard individuals' and organisations' systems, devices, applications and data while minimising the impact of successful attacks.

Cybersecurity is not just about technology; it hinges on human factors. A strong security culture requires awareness and vigilance across an organisation. Even the most advanced technical controls can be compromised by human error or negligence. Effective cybersecurity requires collaboration among technical specialists, policymakers and end-users to ensure security measures are not only implemented but also understood.

At its core, cybersecurity recognises that absolute security is unattainable. Success lies in continuous threat management, by identifying, prioritising and mitigating risks to an acceptable level.

Traditional approaches to analysing threats often begin with definitions. However, you likely already grasp the concept: threats are ever-present, whether navigating the internet or daily life. Given the rapid evolution of technology and adversarial tactics, rigid definitions can be limiting. A more effective strategy involves categorising threats and applying structured assessment techniques.

By integrating intelligence sources into this practical approach, we can develop a comprehensive, cost-effective and adaptable framework for addressing cyber threats. This strategy helps anticipate dangers, outpace adversaries and implement proactive measures to protect critical infrastructure and assets.

In this chapter, you will examine key threat categories and assessment techniques to strengthen your ability to navigate cybersecurity challenges.

Threat Categories and Common Attack Methods

Over time, various threat taxonomies have been developed to classify and clarify cyber threats. These frameworks are essential for streamlining Cyber Threat Intelligence (CTI) and fostering a shared terminology among cybersecurity professionals. While they offer extensive detail, an in-depth exploration lies beyond the scope of this discussion. However, familiarising yourself with these frameworks can significantly enhance your expertise, and I have included high-value resources at the end of this section to support your learning.

For now, we will focus on broad threat categories to establish a solid foundation. Admittedly, reading through a list of categories can feel dry, but mastering them is vital. This chapter introduces key threat categories, followed by an examination of vulnerabilities in the next chapter. Together, these concepts set the stage for Chapter 4, where we explore the components of cybersecurity risk. Stay with me, and hopefully, I'll see you in Chapter 4.

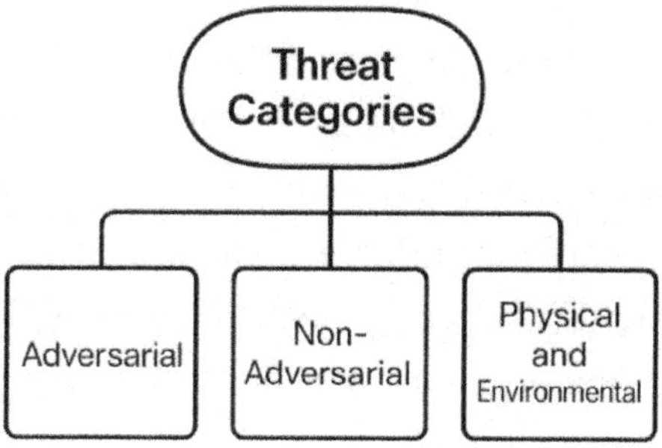

Adversarial Threats

Adversarial threats present a significant challenge for device users worldwide. Attackers employ advanced Tactics, Techniques and Procedures (TTPs) to exploit vulnerabilities, underscoring the ongoing struggle between defenders and malicious actors. These threats rank among the most prevalent in cybersecurity.

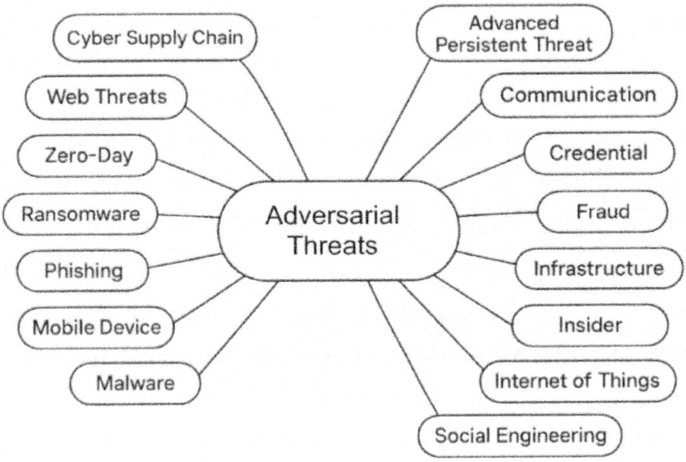

Advanced Persistent Threat: APTs are sophisticated cyberattacks executed by highly skilled, well-funded adversaries, often state sponsored. These attacks aim to covertly infiltrate networks and remain undetected for extended periods, primarily for cyber espionage or data theft. Unlike threats focused on immediate disruption or financial gain, APTs emphasise stealth and persistence, ensuring prolonged access while minimising detection.

Communication Threats: Communication threats target data in transit across networks, including fibre optics, satellites, radio communications and mobile devices. Common methods include Man-In-The-Middle (MITM) attacks, data manipulation and eavesdropping. These threats undermine data confidentiality, integrity and availability, enabling data theft, misinformation and unauthorised access.

Credential Threats: Credential threats arise when attackers steal user credentials, such as usernames and passwords, to gain unauthorised system access. Exploiting the trust associated with authenticated users, these threats make distinguishing malicious activity from legitimate use particularly challenging.

Fraud: Cyber fraud includes deceptive tactics such as business email compromise (BEC) and invoice fraud. Attackers use impersonation and fake documents to steal sensitive information or financial assets. Social engineering techniques, including vendor impersonation, are key in deceiving victims.

Infrastructure Threats: Infrastructure threats target computer systems, networks, data centres and critical infrastructure. Attack methods include DoS attacks, network sniffing and attacks on routers, switches and domain name systems. These threats aim to disrupt, damage or gain unauthorised control over infrastructure components.

Insider Threats: Insider threats originate from individuals with authorised access who compromise an organisation's security, either intentionally (espionage, sabotage) or unintentionally (mishandling sensitive data). Some insiders may also collaborate with external attackers, amplifying the risk.

Internet of Things (IoT) Threats: As IoT adoption grows, so do its security risks. IoT threats include botnet-based large-scale attacks and vulnerabilities like weak encryption or default credentials, which can expose sensitive data and enable unauthorised access.

Malware: As mentioned in Chapter 1, malware refers to malicious software designed to disrupt systems or gain unauthorised network access. It spreads through phishing emails, infected attachments and compromised websites. Common variants include viruses, worms, trojans, spyware, adware and ransomware, each posing distinct risks.

Mobile Device Threats: Mobile and Bring Your Own Devices (BYODs) face threats similar to stationary computing systems, including malicious apps, web-based attacks and network vulnerabilities. Additionally, they are susceptible to physical theft, loss and social engineering attacks like phishing and smishing (SMS phishing).

Phishing: Phishing attacks deceive individuals into revealing sensitive information by impersonating trusted entities in various types of communication, like email. These attacks exploit urgency and psychological manipulation to lure victims into clicking malicious links or downloading harmful attachments. Variants include spear phishing (targeted attacks), whaling (high-profile targets), smishing and vishing (voice phishing).

Ransomware: Ransomware encrypts or blocks access to data until a ransom is paid, making it a growing cybersecurity threat. It spreads through phishing emails and software vulnerabilities. Advanced variants, such as double- and triple-extortion ransomware, involve data theft and further exploitation to pressure victims.

Social Engineering: Social engineering exploits human psychology to gain unauthorised access or extract sensitive information. Attackers often impersonate authority figures or use persuasive tactics such as pretexting, baiting, quid pro quo (exchanging services for information) and tailgating (physical access breaches).

Cyber Supply Chain Threats: Supply chain threats arise from the complexity of ICT and Operational Technology (OT) ecosystems, endangering product security and integrity. Risks include counterfeit components, unauthorised production and malicious software insertion. These vulnerabilities can lead to ransomware attacks, security breaches, operational disruptions and intellectual property theft.

Web Threats: Web threats target websites and online applications, affecting individuals and organisations alike. They include cross-site scripting (XSS), Structured Query Language (SQL) injection, drive-by downloads and watering hole attacks. The OWASP Top 10—published by the Open Web Application Security Project—highlights critical web security risks, including broken authentication, sensitive data exposure and vulnerable third-party components.

Zero-Day Threats: Zero-day threats exploit undisclosed vulnerabilities in software, hardware and supply chains, allowing attackers to breach systems before vendors release fixes. These threats are particularly dangerous due to their unpredictability and lack of immediate countermeasures.

Non-Adversarial Threats

While cybersecurity often emphasises adversarial threats like malware, hacking and cyberattacks, it is also necessary to consider non-adversarial threats. Despite lacking malicious intent, these threats can still disrupt operations and compromise data integrity. Non-adversarial threats include system failures, human errors, workforce shortages, financial disruptions and third-party risks. These threats also include disruptions to essential services such as water, fuel and transportation. Understanding these risks is crucial for maintaining strong cybersecurity defences and mitigating operational disruptions.

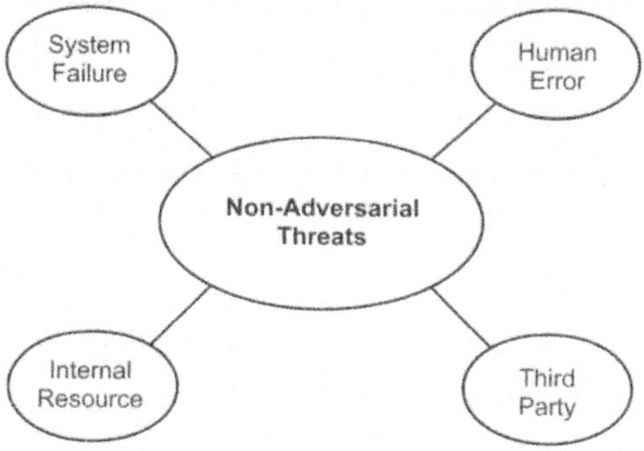

System Failure Threats: Hardware malfunctions, software bugs and communication breakdowns can disrupt functionality and compromise data integrity. These failures may cause service outages, financial losses and reputational damage, with consequences ranging from lost revenue to legal liability, depending on the systems and data affected.

Human Error Threats: Despite being unintentional, human errors introduce vulnerabilities with far-reaching consequences. Common errors include weak passwords, mishandling data, misconfiguring systems and failing to enforce access controls, all of which can lead to unauthorised access, data breaches and compromised security.

Internal Resource Threats: Personnel, expertise or funding shortages pose significant challenges to maintaining operations and meeting objectives. Expanding workloads, employee turnover and financial setbacks can strain systems and weaken organisational resilience.

Third-Party Threats: Disruptions in critical services, such as utilities (water, sewage, fuel, electricity), supply chains, logistics and communication networks, can severely impact operations. Reliance on external providers increases the risk of cascading failures that affect system availability.

Physical and Environmental Threats

Physical and environmental threats extend beyond the cyber domain, threatening systems, infrastructure and personnel. Examples include natural disasters, resource shortages, sabotage, terrorism, theft and civil unrest. Each presents unique challenges requiring proactive mitigation.

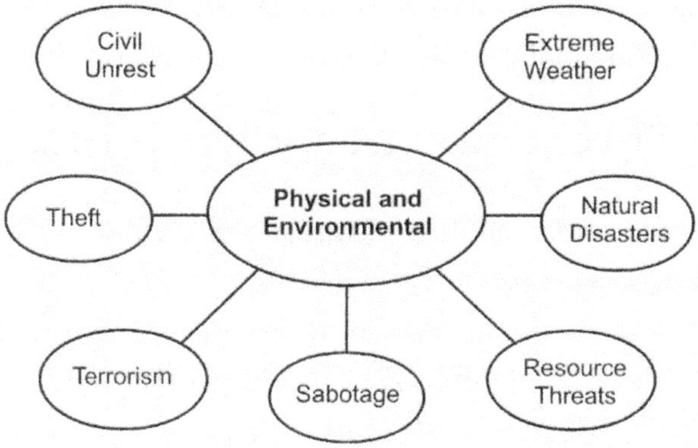

Civil Unrest: Protests and riots can disrupt critical services and expose systems to exploitation by malicious actors, jeopardising data and communications.

Natural Disasters: Earthquakes, tsunamis and volcanic eruptions devastate infrastructure, disrupt operations and complicate recovery efforts.

Extreme Weather: Hurricanes, floods and droughts (intensified by climate change) test infrastructure resilience and require adaptive strategies.

Resource Threats: Energy, water and essential material shortages weaken operational resilience and strain critical infrastructure.

Sabotage: Deliberate system tampering or insider threats can cause extensive harm. Preventive measures like robust access controls, role-based access control and employee vetting mitigate these risks.

Terrorism and Conflict: Terrorism, conflict and geopolitical threats include physical and cyber-attacks driven by political or ideological motives and, at times, the outbreak of war. These threats can endanger public safety and compromise critical infrastructure and systems.

Theft: Theft, including hardware, intellectual property or sensitive data, compromises confidentiality and operational security.

Threat Taxonomy Resources

Enhance your understanding of threat classification with the following useful taxonomic resources:

- European Union Agency for Network and Information Security (ENISA). *ENISA Threat Landscape 2025.* https://www.enisa.europa.eu/publications/enisa-threat-landscape-2025

- Linius, S. (2020). *Cybersecurity & Information Systems Information Analysis Center. Evaluation of Comprehensive Taxonomies for Information Technology Threats.* https://csiac.dtic.mil/articles/evaluation-of-comprehensive-taxonomies-for-information-technology-threats/

- MITRE Corporation. (2019). *CAPEC – Common Attack Pattern Enumeration and Classification (CAPEC).* https://capec.mitre.org/

- National Institute of Standards and Technology. (2012). *Guide for Conducting Risk Assessments. Special Publication (NIST SP) 800-30 Rev 1.* http://dx.doi.org/10.6028/NIST.SP.800-30r1

- Cybersecurity Risk Foundation. (2025). *Threat Taxonomy: A comprehensive threat taxonomy and threat model.* https://crfsecure.org/research/crf-threat-taxonomy/

Strategic Briefing: MOVEit Attack by Clop Ransomware Group

In 2023, the Clop ransomware group exploited a critical vulnerability in Progress Software's MOVEit Transfer tool, a widely used secure file transfer solution, to orchestrate a major cyberattack. Rather than deploying traditional encryption-based ransomware, Clop focused on data theft and extortion.

The incident began in late May 2023 and quickly affected organisations worldwide. The group departed from conventional ransomware tactics of system lockouts, instead threatening to release stolen data unless victims paid ransoms.[9]

This case demonstrates the far-reaching consequences of sophisticated cyber threats and reinforces the importance of proactive cybersecurity measures, timely patching and ongoing vigilance.

Impact and Scope

By mid-2023, the MOVEit attack had compromised 2,667 organisations, affecting approximately 84 million individuals. Notable victims included multinational corporations like Shell and Heidelberger Druckmaschinen, retail entities such as the Boots pharmacy chain, educational institutions like the New York City Department of Education and government agencies like the Louisiana Office of Motor Vehicles and Oregon Department of Transportation. These breaches exposed the sensitive personal data of millions of residents across multiple jurisdictions.[10]

Attack Methodology

Clop exploited a zero-day vulnerability in MOVEit's architecture to gain unauthorised system access across target organisations. After infiltration, the group executed systematic data exfiltration operations, subsequently using the stolen information for extortion. Ransom demands varied from tens of thousands to millions of dollars, depending on the victim's size and the data's sensitivity.

Response and Mitigation

In response, affected organisations were forced to take immediate action, including shutting down their IT systems to prevent further data loss. Incident Response Teams (IRTs) worked tirelessly to identify the extent of the breach, secure compromised systems and restore normal operations. The attack's widespread impact emphasised the critical importance of rapid vulnerability patching and continuous security monitoring across essential software systems.

Lessons Learned

The MOVEit incident demonstrates the dynamic evolution of cyber threats and reinforces key security principles. Organisations must implement rigorous patch management protocols and conduct regular security assessments through vulnerability scanning and penetration testing. Equally important is maintaining comprehensive IRPs to minimise the impact of breaches.

Furthermore, this attack exemplifies how threat intelligence is a crucial preventive tool. Organisations can leverage analytical insights from threat feeds to identify potential vulnerabilities and implement protective measures before threat actors can exploit them. More on this later.

Threat Modelling

Threat modelling, the systematic process of identifying system components and evaluating their vulnerability to potential attacks, is fundamental to modern cybersecurity. If you've used technology, you've likely considered how your systems or devices might be compromised. This intuitive assessment represents the basic principles of threat modelling, though professional practice requires a more structured approach.

As a former system administrator, I initially relied on intuition and reactive measures to address security concerns. This ad hoc approach, while providing quick solutions and valuable hands-on experience in cybersecurity, lacked the systematic rigour necessary for comprehensive security management. Fortunately, the threats I encountered were less sophisticated than today's challenges, and my systems remained secure despite the informal approach. However, this method led to considerable stress and uncertainty about security coverage.

Professional certification, notably the Certified Information Systems Security Professional (CISSP), an internationally recognised credential in information security, transformed my approach to threat management. This formal training provided the framework to shift from reactive problem-solving to methodical risk assessment.

While CISSP requirements are demanding, you might consider beginning with entry-level certifications such as CompTIA Security+ or ISACA's CISM. For aspiring cybersecurity professionals, structured training offers specific methodologies to address complex threats, from vulnerability scanning to incident response protocols. Certifications have become essential in this field, validating your expertise to employers while allowing you to develop a comprehensive understanding of threat modelling and its critical function in risk management.

Threat modelling is the cornerstone of effective risk management, enabling organisations to systematically identify, analyse and address potential vulnerabilities in their systems, networks and data. This methodical approach has become indispensable in today's rapidly changing cyber threat landscape. The following sections examine the key phases of threat modelling, demonstrating how this structured approach strengthens organisational security posture and provides a foundation for a robust cybersecurity strategy.

Threat Modelling Defined

Understanding how potential threats might impact your systems is crucial for effective cybersecurity. This is where threat modelling becomes essential as a systematic approach to securing systems, applications, networks and services.

Microsoft provides a foundational definition: "Threat modelling is an effective way to help secure your systems, applications, networks and services. It is an engineering technique that identifies potential threats and offers recommendations to reduce risk and meet security objectives earlier in the development lifecycle".[11] While this definition captures key elements of threat modelling, its scope requires expansion for contemporary cybersecurity challenges.

The emphasis on the early development lifecycle is too narrow for effective cybersecurity implementation. Threat modelling should not be treated as a one-time activity conducted during the initial stages of the System Development Lifecycle (SDLC). Instead, security professionals must approach it as an ongoing process, continuously revisited and refined throughout each phase to ensure appropriate defences against evolving threats.

This dynamic approach is valuable because cyber threats constantly evolve. Threats identified during initial system development may transform significantly before deployment. Additionally, systems frequently undergo modifications during development, and testing often reveals new vulnerabilities. These changes necessitate ongoing adjustments to both threat models and their corresponding mitigation strategies to maintain system security.

Building on Microsoft's foundation, we can define threat modelling more comprehensively as a proven methodology for enhancing the security of systems, applications, networks, services and data throughout their entire lifecycle. This engineering technique identifies potential threats and provides actionable recommendations to reduce risks and meet security objectives at every stage of the SDLC. In the following chapter, we will examine the SDLC in detail to provide you with a thorough understanding of this useful framework.

The primary goal of threat modelling is to identify and mitigate threats before they become exploitable vulnerabilities. This proactive approach minimises the risks of data breaches, system compromises and other security incidents, ultimately establishing and maintaining a strong security posture for your organisation.

The Threat Modelling Process

Effective threat modelling follows a structured methodology that systematically strengthens your organisation's security posture. The process comprises four distinct phases: design, break, fix and verify. Each phase builds upon the previous one to create a comprehensive security framework. In the following sections, you will learn how to implement each phase and apply them to protect your systems effectively.

The threat modelling process typically involves the following stages[12] and four key questions:[13]

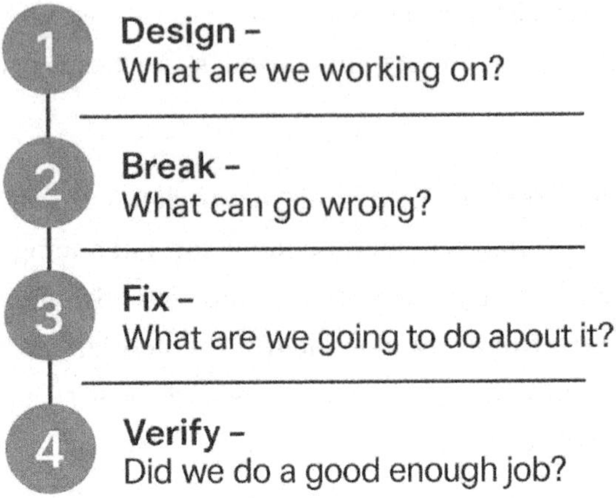

Let's take a closer look at each of these.

Step 1 Design - What Are We Working On?

The first step in threat modelling is developing the system model. This involves diagramming the application or system being reviewed to clearly understand its components, data flows, trust boundaries and potential attack surfaces. While various methods exist for understanding any system, the following key activities form a proven approach to system analysis and are the ones I typically use:

1. **Asset Identification**—Compile a list of system assets, including data, hardware, software components, Application Programming Interfaces (APIs) and processes. This foundational step ensures no critical components are overlooked in your security analysis.

2. **Identify Data Flows**—Create Data Flow Diagrams (DFDs) to visualise information movement within the system, outlining the interactions among different components, processes and external entities. Mastering DFD creation is essential for effective cybersecurity practice, as these visual representations provide practitioners with a comprehensive view

of how information moves through a system. By mapping data flows, trust boundaries and system interactions, DFDs enable security professionals to systematically identify and assess potential threats, making them invaluable for thorough security analysis. While mastering DFD creation takes practice, numerous online resources and tutorials can help you develop this indispensable cybersecurity skill.

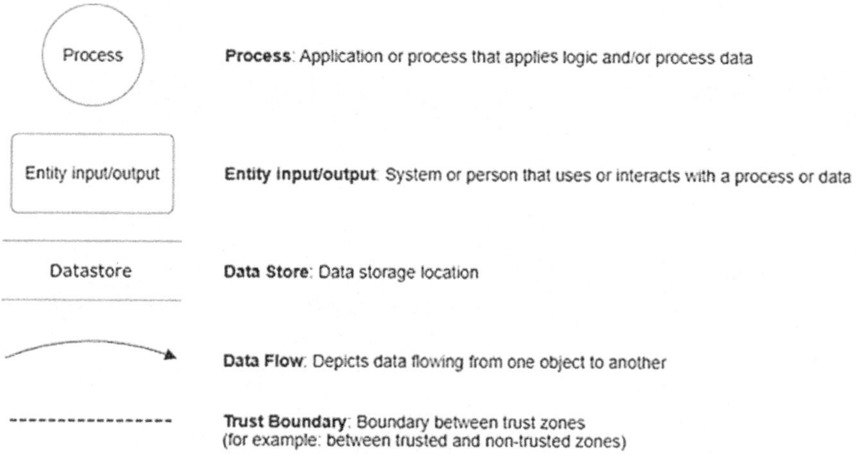

Data Flow Diagram Standard Notation

3. **Trust Boundary Recognition**—Trust boundaries delineate transitions between different security zones within a system, marking points where data crosses between varying levels of trust. Primary examples include boundaries between external networks and internal applications, web applications and database servers, or user-controlled inputs and system processes. These boundaries serve as essential security checkpoints where organisations must implement appropriate controls to protect sensitive assets from untrusted sources.

Example Data Flow Diagram Depicting a Trust Boundary

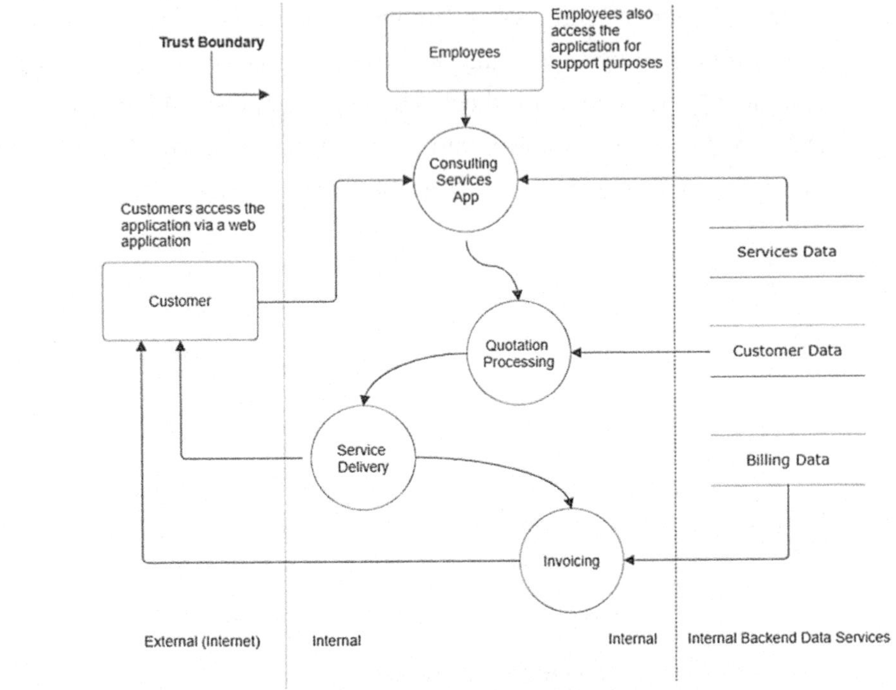

4. **Security Requirements Definition**—Outline the security requirements for your system, including elements such as confidentiality, integrity, availability, authentication and authorisation. While we'll explore these concepts in detail later, understanding them at a high level helps guide your initial threat modelling process. Think of security requirements as your system's protective blueprint; they vary based on its purpose. For example, a platform handling top-secret military data requires security measures that differ significantly from those of an online payment system.

5. **Entry Point Analysis**—Determine all possible entry points into the system, as these are crucial for understanding vulnerabilities. These entry points, which include user interfaces, USB ports, APIs, network interfaces and file inputs, serve as potential avenues for attackers to exploit. Therefore, they require careful examination to identify and

address potential risks. For instance, an unsecured USB port could provide an entry point for malware, while poorly secured APIs might enable unauthorised data access.

Step 2 Break - What Can Go Wrong?

The Break phase of threat modelling systematically identifies potential threats and vulnerabilities that could compromise system security. This step builds on the Design phase by analysing DFDs and system artifacts to assess how information moves across trust boundaries and interacts with various components. The goal is to uncover security gaps and determine where additional safeguards are needed.

A key task in this phase is selecting a threat categorisation framework, such as STRIDE or Attack Trees. STRIDE[14] assesses six specific threat categories: Spoofing, Tampering, Repudiation, Information Disclosure, Denial of Service and Elevation of Privilege. On the other hand, Attack Trees map out potential attack paths, helping organisations understand how threats could materialise. Using multiple frameworks enhances analysis by offering different perspectives on system risks.

Building on the Design phase, this step examines how data moves through the system and across trust boundaries. Analysing the flow of data and its interactions with various components allows us to identify potential vulnerabilities and entry points for threats. The goal is to understand the system's data ecosystem and pinpoint areas requiring additional security measures.

Beyond technical vulnerabilities, this phase also examines threat actor motivations and capabilities. Adversaries range from low-skill opportunists to well-funded, state-sponsored groups. By understanding their objectives, attack methods and available resources, organisations can better anticipate potential threats and implement targeted countermeasures.

Step 3 Fix - Mitigate Identified Threats

In the Fix step, the focus shifts to addressing the threats identified in the Break step. The goal is to reduce risk by implementing security controls and other protective measures to safeguard systems.

Key activities in this step include:

1. **Assess Threat Severity and Impact**—Each identified threat should be evaluated based on likelihood, potential impact and exploitability. Additionally, consider the blast radius—the extent of damage if the vulnerability is exploited. This assessment helps determine the urgency and level of response required.

2. **Prioritise Threats**—Rank threats based on severity, ensuring that the most critical and high-impact risks receive immediate attention. Prioritisation helps allocate limited security resources efficiently, reducing the likelihood of overlooking major threats.

3. **Select and Implement Security Controls**—Apply appropriate technical and procedural security controls to mitigate identified threats:

    - **Technical measures**—Implement encryption, MFA, access controls and system hardening.

    - **Procedural measures**—Establish incident response protocols, enforce least-privilege access and conduct security awareness training.

Several established frameworks and control catalogues guide this process, including:

- ASD. *The Essential Eight*. https://www.cyber.gov.au/resources-business-and-government/essential-cyber-security/essential-eight

- ASD. *The Information Security Manual (ISM).* https://www.cyber.gov.au/resources-business-and-government/essential-cyber-security/ism

- NIST. *SP 800-53 Rev. 5. Security and Privacy Controls for Information Systems and Organizations.* https://csrc.nist.gov/pubs/sp/800/53/r5/upd1/final

- NIST. *Cybersecurity Framework 2.0 (CSF)* https://www.nist.gov/cyberframework

4. **Define or Update Security Requirements**—Translate selected security controls into clear and enforceable security requirements for system design, operations and ongoing management. These requirements provide a structured approach for developers, administrators—and security teams to integrate security from the outset.

5. **Manage Residual Risk:** Not all risks can be eliminated. Organisations should document residual risks, those that remain after implementing security controls, and develop a plan to manage them effectively. Common strategies include:

- Applying compensating controls to reduce residual exposure

- Continuous monitoring for anomalies and suspicious activity

- Considering cyber insurance to mitigate potential financial losses.

Step 4 Verify - Did We Do a Good Enough Job?

The Verify phase serves as the final checkpoint in the threat modelling process, ensuring that security measures are comprehensive, effective and aligned with the identified risks. This phase consists of two primary activities:

1. **Validating the Threat Model**—A systematic review of all components, including DFDs, threat lists, mitigation strategies and security requirements, to confirm that all relevant threats have been properly addressed. This step ensures security controls align with the expected risk landscape and implementation objectives.

2. **Verification Testing**—Security controls must be tested using both manual and automated methods. Penetration testing simulates real-world attacks to uncover exploitable vulnerabilities, while vulnerability scanning employs automated tools to identify potential weaknesses. Additionally, security audits and code reviews further verify that defensive measures are correctly implemented and function as intended.

A verification process helps confirm that the security strategy is both robust and resilient. By carefully reviewing the threat model and rigorously testing defences, we ensure that the four-stage threat modelling process—Design, Break, Fix and Verify—delivers a security posture capable of withstanding real-world threats.

Summary

Cyber threats are an unavoidable reality in today's digital world, constantly evolving to challenge even the most advanced security systems. This chapter explored the primary categories of cyber threats (adversarial, non-adversarial and physical/environmental) and the tailored strategies needed to mitigate them.

Adversarial threats originate from malicious actors such as hackers, cybercriminals and nation-state groups who exploit vulnerabilities using sophisticated Tactics, Techniques and Procedures (TTPs). These threats include Advanced Persistent Threats (APTs), credential theft, malware and communication threats, all of which require proactive defence strategies.

Non-adversarial threats, though not driven by malicious intent, can be equally disruptive. System failures, human errors, insider risks and third-party service disruptions highlight the need for strong operational procedures and risk management frameworks to maintain system integrity and resilience.

Physical and environmental threats extend beyond the digital realm, encompassing civil unrest, extreme weather, natural disasters, resource shortages, sabotage, terrorism and theft. Mitigating these risks requires an integrated security approach that combines cybersecurity controls with physical security measures.

A structured approach is essential for effective threat management. Threat taxonomies, such as those provided by ENISA's Threat Landscape and MITRE's CAPEC, offer systematic ways to classify and analyse threats.

Threat modelling plays a pivotal role in identifying, analysing and mitigating cyber threats to systems, networks and data. This process consists of four key phases:

- **Design**—Understanding the system architecture and security requirements.
- **Break**—Identifying vulnerabilities and potential threats.
- **Fix**—Implementing appropriate security controls and mitigations.
- **Verify**—Testing and validating the effectiveness of security measures.

This chapter has emphasised the importance of a category-based approach to threat assessment. This approach enables cybersecurity professionals to prioritise risks, allocate resources efficiently and develop proactive countermeasures. By applying this structured strategy, organisations can confidently strengthen their security posture and navigate the ever-changing cyber threat landscape.

Revision Questions

1. What is the foundational principle of cybersecurity discussed in the chapter, and why is it important to understand that complete identification and elimination of threats is impossible?

2. How does human nature impact cybersecurity, and what role does fostering a culture of security awareness play?

3. Explain the difference between adversarial and non-adversarial threats with examples from the chapter.

4. What are Advanced Persistent Threats (APTs), and how do they differ from other types of cyber threats?

5. Describe the impact and significance of the MOVEit attack discussed in the chapter.

6. What are the key components of a practical approach to understanding and managing cyber threats as outlined in the chapter?

7. How do insider threats pose a risk to organisations, and what measures can be taken to mitigate these threats?

8. What are some examples of physical and environmental threats, and why is it important to integrate physical security measures with cybersecurity protocols?

9. Explain the process of threat modelling and its significance in the cybersecurity lifecycle.

10. How do threat assessments help in managing cyber threats, and what steps are involved in conducting one?

Cybersecurity Hacks

Hack 1: Create Your Own Cyber Threat Playbook

Develop a playbook that categorises various types of cyber threats, their common indicators and appropriate response actions. This playbook should include sections on adversarial, non-adversarial and physical/environmental threats. Regularly update it with the latest threat intelligence to keep your defence strategies current.

The following is a general example of a Ransomware Cyber Threat Playbook to help you get started. When creating your playbook, focus on a specific threat, such as a particular type of ransomware like CryptoLocker or WannaCry. By tailoring your playbook to address a specific ransomware variant, you can better prepare for its unique tactics and mitigation strategies.

Ransomware Playbook[15]

Objective: To quickly identify, contain and eradicate ransomware infections while minimising data loss and operational disruption.

Detection and Identification

- Monitor for unusual file encryption activities
- Use Endpoint Detection and Response (EDR) tools to identify ransomware signatures.

Containment

- Isolate infected systems from the network
- Disable shared drives and network connections to prevent further spread.

Eradication

- Remove ransomware using anti-malware tools
- Restore affected systems from clean backups.

Recovery

- Verify the integrity of restored data
- Reconnect systems to the network after ensuring they are clean.

Post-Incident Analysis

- Conduct a root cause analysis to understand how the ransomware entered the network
- Update security measures to prevent future incidents.

Hack 2: Get Hands-On

Get involved with online learning platforms to expand your knowledge, sharpen your skills and prepare for real-world threats. These platforms offer opportunities to explore various threat vectors and connect with communities of learners. Joining these communities can help you stay informed about new challenges and learn from others' experiences. Consider exploring:

- **Hack The Box**—A gamified cybersecurity platform with over 2.6 million members, offering virtual machines and challenges that simulate real-world security vulnerabilities
- **TryHackMe**—An interactive platform providing hands-on cybersecurity training through virtual rooms and structured learning paths, enabling users to develop practical skills from beginner to expert levels.

Chapter 3: Vulnerabilities

– The Path to Exploitation –

Objectives

By the end of this chapter, you will be able to:

- Identify and classify different types of vulnerabilities in systems and applications
- Understand how vulnerabilities arise throughout the System Development Lifecycle (SDLC)
- Apply vulnerability assessment methodologies and tools effectively
- Apply strategies for vulnerability management and remediation
- Evaluate the impact of emerging technologies on vulnerability landscapes.

A hairline crack in a dam can lead to catastrophic failure. In the same way, a single security weakness can trigger a devastating data breach. Vulnerabilities in our systems serve as potential entry points that can compromise data, operations and infrastructure.

Cybersecurity rests on three interconnected pillars: threats, vulnerabilities and risks. In Chapter 2, we explored threats, malicious actions aimed at compromising organisational assets and operations. Now, we turn our attention to vulnerabilities, weaknesses in our digital infrastructure that adversaries can exploit to bypass security controls. In the following chapter, we'll examine risks, which represent the potential for loss when threats successfully exploit these vulnerabilities.

Protecting digital systems requires a fundamental understanding of vulnerabilities. Just as an unlocked front door creates an obvious security weakness in your home, vulnerabilities in digital systems represent similar points of exposure. One key to effective cybersecurity is identifying these potential entry points before they can be abused.

This chapter provides a structured approach to recognising vulnerabilities and assessing their impact. It equips you with essential tools and methodologies for identifying weaknesses. Most importantly, it fosters a proactive security mindset because, in cybersecurity, preventing a breach is always better than recovering from one.

Software Vulnerabilities

Software development has grown from basic programs into the intricate applications that power our daily lives. Although this evolution has enabled remarkable capabilities, it has introduced new challenges in maintaining security. Contemporary software resembles a sprawling city; as it expands, securing every entry point becomes increasingly tricky.

Today's software vulnerabilities often stem from this complexity. When developing software, engineers must anticipate countless scenarios and edge cases. They must consider how users might input unexpected data, how network failures could disrupt operations and how simultaneous access to a resource may cause conflicts. Even when employing best practices and security frameworks, the vast number of potential interactions makes it nearly impossible to cover every scenario within typical time and budget constraints. Consequently, vulnerabilities are not only common but inevitable.

Code generated by AI (software created by algorithms rather than humans) adds another dimension to this challenge. While AI-generated code accelerates development, it often lacks the nuanced understanding of the context that human engineers provide. AI models are trained on existing codebases, which may include outdated or insecure practices, inadvertently perpetuating vulnerabilities. Additionally, AI-generated code can struggle with edge cases, failing to anticipate unexpected inputs or complex interactions. Without rigorous human oversight, these automated systems can introduce security flaws that compound existing risks, making it even harder to ensure software integrity.

As software systems become more complex, with multiple integrations and distributed architectures (systems where computing tasks are spread across networked devices), the risk of security vulnerabilities rises significantly. Common issues include:

- Input validation gaps (comparable to leaving your front door unlocked)
- Weak encryption (akin to using a simple lock)
- Outdated security protocols (similar to relying on an obsolete system that attackers can bypass).

Every line of code in your banking app or social media platform represents a potential entry point that could compromise critical systems. These subtle yet significant weaknesses pose real risks to applications relied on by millions daily. By learning to identify and mitigate security gaps throughout the development lifecycle, engineers can turn software from a potential security risk into a resilient and trustworthy system.

Categories of Software Vulnerabilities

Software vulnerabilities represent weaknesses in computer systems that can be exploited by malicious actors. Understanding these categories is essential for security professionals as they form the foundation of effective defence strategies. By mastering vulnerability types, such as injection flaws, authentication weaknesses and insecure data handling, you can prioritise remediation efforts, implement targeted preventative measures and strengthen your security posture. These categorisations offer a systematic approach to identify, assess and mitigate security risks across software applications, which we will explore in the following sections.

Injection Vulnerabilities

These are defined as attacks where malicious code or data is inserted into application inputs, tricking interpreters into executing unintended commands. Common variants include:

- Structured Query Language (SQL) injection
- NoSQL injection
- Lightweight Directory Access Protocol (LDAP) injection.

These vulnerabilities typically stem from insufficient input validation and enable attackers to extract sensitive data, modify database contents or execute arbitrary system commands.

To protect against these vulnerabilities, use parameterised queries (which separate code from data), check all user inputs for validity and clean user-supplied data before your system processes it.

Authentication and Session Management Weaknesses

Authentication and session weaknesses are flaws in systems that manage digital identity verification and user sessions include weak password mechanisms, improper credential storage and exposed session tokens that lead to unauthorised access. Key vulnerabilities involve inadequate session timeouts, the absence of MFA and vulnerable password reset processes that can be exploited for account takeover.

Address login security by requiring multiple verification methods (like passwords plus codes sent to your phone), creating strong password rules, ending inactive sessions automatically and securely storing passwords using modern techniques like Bcrypt or Argon2.

Insecure Data Handling

Insecure data handling is the improper protection of information throughout its lifecycle. This encompasses insufficient data encryption in transit and at rest, inadequate key management, improper implementation of cryptographic algorithms and failure to enforce Transport Layer Security (TLS). Additional risks include inappropriate data retention, unprotected backups and the absence of secure disposal procedures for sensitive information, potentially leading to data breaches even without exploiting other system vulnerabilities.

Protect your data by using strong encryption, managing access keys securely, setting up HTTPS connections correctly and creating clear policies for how long to keep and safely delete data.

Cross-Site Scripting (XSS)

XSS is a prevalent web security flaw that allows attackers to inject malicious scripts into trusted websites. This occurs when applications fail to validate or escape user input properly, allowing unauthorised code execution in users' browsers. XSS attacks may be persistent (stored within the application) or reflected (executed immediately), potentially resulting in session hijacking, keylogging or credential theft through social engineering.

Prevent these attacks by encoding outputs based on where they appear, adding security headers to your website and checking all inputs to stop malicious code from running in users' browsers.

Broken Access Control

Broken access controls are flaws resulting from inadequate enforcement of resource access restrictions. Such vulnerabilities allow attackers to bypass authorisation and access or modify data and functionality beyond their privileges. Common examples include forced browsing, privilege escalation, metadata manipulation and misconfigured Cross-Origin Resource Sharing (a protocol that regulates cross-domain requests), exposing sensitive endpoints.

To fix broken access controls, give users only the permissions they need (see Chapter 12 on the principle of least privilege), create a central system to manage access, verify permissions on the server for all resources and regularly check who has access to what.

Security Misconfiguration

This type of vulnerability arises from improperly configured security controls, such as default credentials, unnecessary features, exposed error messages and unpatched systems. Often resulting from incomplete security hardening, these issues are compounded by inconsistent configurations across environments, unnecessary open ports and excessive permissions that expand the attack surface.

Reduce security misconfigurations by creating and following secure setup guidelines for all systems, checking configurations regularly, automating setup processes and keeping all software updated with security patches.

Cryptographic Failures

Cryptographic failures are deficiencies in protecting sensitive data via encryption, whether at rest or in transit. Such vulnerabilities expose critical information, including passwords, personal data and financial details, to unauthorised access. Common issues include deprecated encryption algorithms, improper certificate validation, weak key generation and inadequate encryption of sensitive data in storage or transmission.

Fix encryption weaknesses by using up-to-date encryption methods, properly managing encryption keys and regularly checking your encryption systems to replace outdated methods.

Server-Side Request Forgery (SSRF)

SSRF allow attackers to induce server-side applications to make requests to unintended locations. These flaws enable attackers to bypass network security controls, access internal services behind firewalls and interact with metadata services in cloud environments. SSRF attacks can lead to data exfiltration, server compromise and lateral movement within internal networks, often by exploiting trust relationships between systems.

To prevent URL-based attacks, strictly validate web addresses, separate different parts of your network, only allow connections to approved destinations and limit automatic redirects in your applications.

Extensible Markup Language (XML) and External Entity (XXE) Processing

This type of vulnerability encompasses security flaws where applications processing XML input are vulnerable to attacks targeting XXE references. These vulnerabilities can lead to the disclosure of confidential data, server-side request forgery, port scanning, denial of service and other system impacts.

XXE vulnerabilities typically arise from outdated XML processors or insufficiently restricted entity resolution in application code.

Protect against XML attacks by setting up XML processors to block external entities and Data Type Definition (DTD) processing, while keeping all XML-related software updated to the latest secure versions.

Cross-Site Request Forgery (CSRF)

CSRF are vulnerabilities that trick authenticated users into performing unintended actions on websites where they're logged in. These attacks exploit the trust websites place in user browsers, forcing them to submit unauthorised requests, including account changes, fund transfers or data manipulation. CSRF vulnerabilities commonly arise from the absence of unique request tokens, failure to verify request origins or reliance solely on cookies for authentication.

Stop CSRF by adding special tokens to forms, verifying where requests come from using cookie settings and requiring users to re-enter credentials for sensitive actions.

Using Components with Known Vulnerabilities

Security risks arising from the inclusion of third-party components (libraries, frameworks, modules) with documented security flaws. These vulnerabilities arise when organisations fail to maintain an inventory of components, neglect regular updates, lack compatibility testing procedures or download components from unofficial sources. Exploitation of these known vulnerabilities can lead to substantial data loss or server takeover with minimal effort from attackers.

Reduce component vulnerabilities by tracking all software dependencies automatically, regularly scanning for security issues and promptly updating outdated components through an organised update process.

While these categories offer a foundational overview of notable software vulnerabilities, they represent only a subset of potential weaknesses. For comprehensive information, consult other authoritative resources such as the Common Weakness Enumeration (CWE) database and the OWASP Top 10. The next section will describe these and other useful resources.

Tools for Identifying and Mitigating Vulnerabilities

The cybersecurity community provides a wide range of resources to help individuals and organisations strengthen security measures and reduce software vulnerabilities. Various industry-standard tools are available to support this effort, some of which are outlined below. Understanding how these tools function will help you navigate cybersecurity standards and make informed choices to protect your organisation.

Common Weakness Enumeration (CWE™) Top 25 Most Dangerous Software Weaknesses

The CWE™ Top 25 Most Dangerous Software Weaknesses provides a comprehensive list of the most critical software vulnerabilities. It is invaluable for developers, offering guidance on prioritising security measures within their code. The CWE™ list helps focus attention on the most severe and widespread issues.[16]

OWASP Top 10

The Open Web Application Security Project (OWASP) Top 10 is a standard awareness document for web application security. It highlights the 10 most critical security risks to web applications, making it a fundamental resource for developers and security professionals. Understanding and addressing these risks is crucial for enhancing application security and preventing common exploits.[17]

OWASP Top 10 for Large Language Model (LLM) Applications

This initiative informs developers, designers, architects, managers and organisations about potential security risks when deploying and managing LLMs—AI systems that process and generate human-like text. It identifies the 10 most critical vulnerabilities in LLM applications, helping you mitigate emerging risks.[18]

Common Vulnerabilities and Exposures (CVE)

The CVE list is a globally recognised database of publicly disclosed cybersecurity vulnerabilities. It provides a standardised reference for known vulnerabilities, facilitating information sharing and improving the coordination of security efforts across the industry. By referencing CVE, organisations can quickly identify and respond to known issues.[19]

Common Vulnerability Scoring System (CVSS)

CVSS is a framework that assigns numerical scores to vulnerabilities based on their severity. These scores help organisations prioritise their responses to vulnerabilities, ensuring that the most critical issues receive prompt attention. CVSS provides a consistent and standardised approach to vulnerability assessment.[20]

Known Exploited Vulnerabilities Catalogue (KEV)

Maintained by the Cybersecurity and Infrastructure Security Agency (CISA), the KEV catalogue lists vulnerabilities actively exploited in the wild. This authoritative source is vital for integrating into your vulnerability management strategy and prioritising fixes for threats that pose immediate risks.[21]

National Vulnerability Database (NVD)

The NVD is the U.S. government's repository of standards-based vulnerability management data. It uses the Security Content Automation Protocol (SCAP), a set of standards for automating vulnerability management, to support security measurement and compliance activities. The NVD is a critical resource for tracking and managing vulnerabilities across diverse systems.[22]

Top Routinely Exploited Vulnerabilities

This advisory, issued by cybersecurity authorities, including the Australian Cyber Security Centre (ACSC), details the CVEs most frequently exploited by malicious actors. The report emphasises the importance of patching older, unpatched vulnerabilities and highlights the continued risk posed by these issues. This publication is generally updated annually.

Developing appropriate secure coding strategies, practices and policies can be daunting. Applying these resources and methodologies is essential for maintaining secure systems and defending against ever-evolving cyber threats.

System Development Lifecycle (SDLC) and Vulnerabilities

The SDLC provides a structured framework for creating and maintaining secure applications and systems. It serves as a roadmap guiding development teams through key phases, from initial planning and requirements analysis, through design and development, to testing, deployment and ongoing maintenance. Although mastering the SDLC requires more detailed study than covered in this text, cybersecurity practitioners must understand its basic principles so they can pinpoint where vulnerabilities emerge during software development. Once you have grasped these fundamentals, explore further resources such as *Software Development Life Cycle (SDLC) Methodologies for*

Information Systems Project Management[23] and *Understanding the SDLC: Software Development Lifecycle Explained.*[24]

Organisations often tailor the SDLC to their specific requirements, ranging from streamlined three-phase approaches (concept, design and implementation) to comprehensive frameworks that integrate with established engineering practices and project management methodologies. For instance, while a small startup might adopt an agile-focused SDLC emphasising rapid iteration, a healthcare organisation might implement a more rigorous framework prioritising security and compliance.

Each phase in the development cycle presents distinct opportunities for security vulnerabilities, making it essential to integrate robust security controls and risk mitigation strategies throughout the process. The following generic SDLC framework serves as a practical model for identifying common vulnerability points across development stages, enabling organisations to implement targeted security measures at critical junctures.

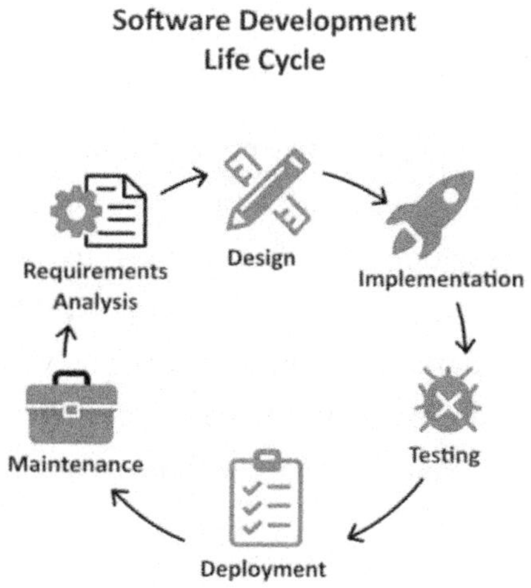

Let's take a closer look at each stage.

Requirements Analysis

The SDLC begins with the Requirements Analysis phase, during which project objectives and functional requirements are defined. At this stage, security requirements are also identified and specified. Failing to do so can create ambiguities that later become vulnerabilities. For example, omitting the requirement for encrypting data transmissions may lead to the use of insecure protocols, thereby exposing sensitive information.

Design

The Design phase outlines the system architecture and software components. Adhering to secure design principles, such as minimising the attack surface (the total points where unauthorised users may attempt entry) and ensuring secure data flow, is essential. During this phase, threat modelling (see Chapter 2) helps you identify potential vulnerabilities and enhance security. A flawed design can introduce inherent weaknesses that are difficult to remediate later. For example, inadequate session management can enable attackers to hijack user sessions, resulting in unauthorised access.[25]

Implementation

During the Implementation phase, developers write the actual code. Secure coding practices are necessary to avoid common vulnerabilities, such as improper input validation, which can lead to SQL injection attacks. Adhering to coding standards and conducting regular code reviews are effective strategies. Additionally, validating the security of APIs (interfaces that enable communication between software components) is vital. Tools like static code analysers, which examine code without executing it, help identify potential vulnerabilities early.

Testing

Security testing is integral to the development process and should include both static and dynamic analysis. Static analysis reviews the code for vulnerabilities without execution, while dynamic analysis tests the application during runtime. Penetration testing (pen testing or ethical hacking) simulates real-world attacks to uncover weaknesses. Integrating these practices into a Continuous Integration/Continuous Deployment (CI/CD) pipeline (which automates the integration of code changes, runs tests and deploys updates continuously) ensures that vulnerabilities are detected and addressed promptly.

It is crucial that all security testing, particularly penetration testing, be conducted with explicit authorisation from system owners. Unauthorised testing, regardless of intent, constitutes illegal activity and may result in prosecution. Even with proper authorisation, situations can become complex, as demonstrated in 2020 when two Coalfire pen testers were arrested despite having appropriate permissions.[26]

Deployment

The Deployment phase involves moving the software to a production environment. Secure deployment practices protect the software by configuring it securely, applying the principle of least privilege (granting access only as necessary) and ensuring all components are up to date. Automation tools help maintain consistent, secure configurations across environments, ensure the deployed configuration adheres to the design and is backed up; a backup enables you to restore the system to a known state if problems arise during deployment.

Maintenance

Maintenance is an ongoing phase that involves monitoring the software for vulnerabilities and applying timely patches. Continuous monitoring is critical for detecting new threats and enabling swift responses. Tools such as Intrusion Detection System (IDS) and Security Information and Event Management (SIEM) solutions are very useful during this phase. Regular updates and effective patch management protect against newly discovered vulnerabilities.[27]

Common Coding Mistakes Leading to Vulnerabilities

Software vulnerabilities typically arise from preventable coding errors that pose immediate threats to modern digital systems. Critical issues include improper input validation and buffer overflows (where applications mishandle data inputs or memory allocation), leading to system crashes, malicious code execution or data breaches. Hard-coded credentials (sensitive information embedded directly in code) create significant security risks by offering attackers direct system access. Poor error handling compounds these vulnerabilities by exposing sensitive system architecture details through verbose error messages.

Building on these weaknesses, sophisticated attacks often exploit XSS and SQL injection vulnerabilities, where malicious actors bypass inadequate input sanitisation to inject harmful code. Weak authentication and authorisation protocols enable unauthorised access, while outdated software components harbour known security flaws. Server-Side Request Forgery (SSRF) allows attackers to manipulate internal server requests to breach network boundaries, while race conditions create security gaps when multiple processes compete for resources without proper synchronisation.

To protect against these threats, development teams must implement comprehensive security measures, starting with the OWASP Secure Coding Guidelines.[28] This includes rigorous input validation, industry-standard encryption methods and systematic security testing throughout the development lifecycle. Early identification and swift remediation transform an organisation's security posture from reactive to proactive.

Static Code Analysis and Dynamic Application Security Testing (DAST)

Software security teams use two key methods to detect vulnerabilities before exploitation. Static code analysis tools like SonarQube and Checkmarx examine source code without running the program, helping identify potential weaknesses during initial development. These tools can detect buffer overflows and SQL injection risks, often reducing remediation costs significantly. DAST complements this approach by evaluating running applications during integration and testing. DAST tools, including Zed Attack Proxy and Burp Suite, simulate real-world attacks to uncover runtime security flaws that static analysis might miss, revealing how attackers might exploit your application in practice.

Implementing both methods creates a comprehensive security testing strategy throughout the development lifecycle. Static analysis excels during code writing and review, while DAST proves most effective during system testing and pre-deployment phases, with each approach compensating for the other's limitations.

Zero-Day Vulnerabilities

A zero-day vulnerability[29] represents a software weakness that developers haven't yet had the opportunity to fix, hence the term 'zero days' to create a patch. When cybercriminals uncover and weaponise these unpatched flaws before security teams can respond, they develop severe security risks. These undisclosed weaknesses give malicious actors a dangerous advantage, often leading to data breaches before organisations can defend themselves.

The 2014 Heartbleed vulnerability demonstrates this threat: attackers exploited a flaw in OpenSSL to steal sensitive data, including passwords and credit card details. More recent examples like Log4j and NotPetya show how zero-day vulnerabilities continue to pose significant risks worldwide, highlighting the need for security monitoring and rapid incident response capabilities.

Hardware Vulnerabilities

Hardware vulnerabilities are a critical yet often overlooked aspect of cybersecurity due to the challenges of addressing them. Unlike software flaws, which can usually be patched quickly, hardware vulnerabilities stem from flaws in physical components, making them costly and difficult to fix. Attackers exploit these weaknesses to gain unauthorised access, manipulate data or disrupt critical services.

A significant yet often underestimated source of hardware vulnerabilities lies in insecure supply chains, particularly in systems handling sensitive data such as financial transactions or medical records. I encountered this issue firsthand as a system manager for a sensitive government network. Ahead of a high-profile event, I needed additional hardware to expand capacity and enhance our defence-in-depth strategy. However, procurement compliance delays led my client to suggest a risky alternative: second-hand hardware from online marketplaces. While this seemed practical under tight budget and time constraints, it posed a severe security threat. Unverified hardware can contain firmware backdoors, rootkits or malicious chips, potentially compromising an entire system. This experience reinforced a crucial lesson: cost and time pressures can lead organisations to overlook security risks, inadvertently introducing malicious components into their infrastructure.

Attackers exploit hardware vulnerabilities using both physical and remote techniques, often targeting weaknesses at the architectural level. Side-channel attacks, such as Spectre and Meltdown[30], exploit flaws in CPU microarchitecture, allowing attackers to access sensitive memory data while bypassing software security controls. Because these attacks compromise fundamental hardware operations, they are particularly dangerous. Similarly, Rowhammer exploited electrical-level memory manipulation, repeatedly accessing specific locations to induce bit flips in adjacent cells.[31] This can lead to data corruption or even privilege escalation, granting attackers unauthorised control.

Challenges in Addressing Hardware Vulnerabilities

Fixing hardware vulnerabilities is considerably more challenging than fixing software flaws. While software patches can be rapidly deployed, hardware fixes often require physical replacement or complex firmware updates that control low-level hardware functions. Deploying firmware updates across large device fleets introduces further risks, as updates may introduce new vulnerabilities or disrupt functionality. The long lifespan of hardware exacerbates these issues, leaving systems vulnerable long after manufacturers end support.

Firmware updates play a vital role in mitigating hardware vulnerabilities and improving security. Manufacturers must rigorously test updates to prevent unintended consequences. Users, in turn, must apply them promptly to minimise risk. Failure to update firmware can render hardware inoperable or expose it to threats. One real-world example of firmware-related risks is the Mirai botnet attack, which exploited IoT devices running outdated firmware.

The Expanding Attack Surface

The expansion of Internet of Things (IoT) devices has significantly increased the attack surface, further complicating security efforts. These devices, from smart home appliances to industrial sensors, often lack robust security measures and run outdated or unpatched firmware. Their interconnected nature means that compromising one device can jeopardise an entire network. The risks are particularly severe in critical infrastructure environments, such as industrial control systems, where a successful attack could have catastrophic consequences.

Hardware vulnerabilities also pose broader security risks. State-sponsored actors have embedded covert vulnerabilities into supply chains, creating persistent threats that organisations may not detect until after a breach. Advanced Persistent Threat (APT) groups specifically target hardware-level exploits, recognising their potential for long-term access to sensitive systems.

Mitigating Hardware Vulnerabilities

Addressing these risks requires a security strategy that spans the entire hardware lifecycle, including:

- Sourcing hardware from verified suppliers to minimise the risk of compromised components
- Applying firmware updates regularly, balancing security improvements with potential operational risks
- Implementing IRPs to address hardware-level threats
- Monitoring emerging attack techniques to stay ahead of adversaries.

As attackers increasingly target hardware-level weaknesses, proactive security measures are essential. The evolving threat landscape requires organisations to prioritise hardware security alongside software defences. Understanding these vulnerabilities and implementing appropriate controls will enable you and your organisation to strengthen defences against new threats while maintaining operational efficiency.

Communication Vulnerabilities

Communication vulnerabilities (critical weaknesses in the digital infrastructure connecting modern systems and users) arise from inadequate security measures in protocols, encryption, messaging systems and networks. These weaknesses create vectors for data interception, manipulation and unauthorised access, threatening organisational security.

Unsecured Communication Protocols

Legacy communication protocols present significant risks to data integrity and confidentiality. These protocols, developed when security was a secondary consideration, typically lack fundamental safeguards such as encryption or authentication mechanisms. Protocols like File Transfer Protocol (FTP), Hypertext Transfer Protocol (HTTP) and telnet (hopefully no one uses telnet anymore!) transmit data in plaintext, exposing communications to Man-In-The-Middle (MITM) attacks and packet sniffing. For example, attackers can easily capture credentials and sensitive data during unencrypted FTP and telnet sessions. Modern organisations should transition to secure alternatives like Secure File Transfer Protocol (SFTP), Hypertext Transfer Protocol Secure (HTTPS) and Secure Shell (SSH) to protect their communications.

Vulnerabilities in Email and SMS Communications

Email and SMS platforms constitute vulnerable attack surfaces due to their ubiquitous use and inherent security limitations. Email systems commonly fall prey to phishing attacks, spoofing and business email compromise (BEC) schemes—sophisticated attacks that can result in credential theft, financial losses and data breaches. SMS messages, lacking end-to-end encryption, remain vulnerable to interception through Signalling System 7 (SS7)[32] network vulnerabilities or International Mobile Subscriber Identity (IMSI) catchers. Organisations should implement effective security measures, including MFA for email systems and secure alternative communication channels.

Unsecured Messaging Platforms

The adoption of consumer-grade messaging solutions for business communications introduces significant security risks. Platforms like WhatsApp, Telegram or improperly configured Slack instances often lack enterprise-grade security features such as encryption, audit capabilities and proper data retention controls. Shadow IT (the unauthorised use of applications) compounds these risks as employees bypass security controls with unapproved communication tools. Organisations should consider implementing and enforcing policies governing secure, enterprise-grade communication tools while actively monitoring for unauthorised applications.

Network Vulnerabilities and Hybrid Environments

Modern hybrid environments, where on-premises and cloud infrastructure intersect, face complex communication security challenges. Misconfigured virtual private networks (VPNs), inadequate Wi-Fi security and improper network segmentation create attack vectors for communication interception and redirection. For instance, VPN misconfigurations may expose internal communications through unsecured tunnels, while outdated Wi-Fi encryption standards like Wired Equivalent Privacy (WEP) facilitate eavesdropping. Zero-Trust Architecture (a security model requiring continuous validation of every communication attempt regardless of origin) provides a framework for securing these hybrid environments. This approach validates each access request thoroughly, even from within the network, protecting increasingly distributed workforces.

The Role of Human Factors

Human behaviour significantly influences communication security. Poor password management, susceptibility to phishing attempts and delayed software updates create security vulnerabilities. Organisations must foster a security-conscious culture through regular training, clear policies and ongoing awareness programs that equip employees to recognise and mitigate common threats.

Addressing Communication Vulnerabilities

Addressing communication vulnerabilities requires a multifaceted strategy integrating strong encryption, secure protocol and heightened user awareness. Organisations should implement current security standards by enforcing Secure Sockets Layer/Transport Layer Security (SSL/TLS) encryption for all network traffic, adopting email authentication protocols like Domain-based Message Authentication, Reporting and Conformance (DMARC) to prevent spoofing and selecting messaging platforms that satisfy enterprise security requirements. Regular security assessments help to detect and respond to evolving threats, while participation in threat intelligence networks provides critical insights into emerging risks. Equally important is strengthening the human element by cultivating a security-conscious workplace culture through targeted training programs and clear communication policies, effectively reducing vulnerability to social engineering and human error.

Configuration Vulnerabilities

Configuration vulnerabilities occur when system settings are improperly established, potentially exposing networks, applications and devices to malicious actors. These security weaknesses commonly stem from unmodified default settings, absent security updates or incorrect configurations implemented during system changes.

Manufacturers typically optimise newly installed hardware and software for user convenience rather than security, creating default configurations that attackers can readily exploit. Key examples include:

- **Default Credentials**—Systems and applications shipped with pre-configured usernames and passwords (e.g., "admin/admin") represent significant security vulnerabilities if left unchanged
- **Open Ports and Weak Firewall Rules**—Unrestricted access to critical services (e.g., SSH, RDP or database ports) increases the risk of unauthorised access
- **Misconfigured Cloud Storage**—Publicly accessible AWS S3 buckets, Azure Blobs or Google Cloud Storage can expose sensitive information
- **Excessive Permissions**—Overly permissive Identity and Access Management (IAM) roles can allow privilege escalation.

System Hardening and Best Practices

System hardening serves as a fundamental strategy for mitigating configuration vulnerabilities by systematically reducing the potential attack surface through:

- Disabling unnecessary services and features
- Modifying insecure default settings
- Applying security patches and updates.

Security frameworks like the Centre for Internet Security (CIS) Benchmarks, the Australian Cyber Security Centre (ACSC) Guidelines for system hardening, NIST 800-53 and ISO 27001 provide structured guidelines for strengthening system configurations. Organisations implementing these best practices ensure a more resilient IT environment.

Configuration Challenges in Cloud Environments

The dynamic nature and shared responsibility model of cloud-native environments introduce unique configuration risks. Frequently observed misconfigurations in these settings include:

- **Unsecured APIs**—API endpoints without proper authentication can expose backend services
- **Overly Broad IAM Roles**—Granting "Administrator Access" to all users increases the risk of privilege abuse
- **Weak Kubernetes Security Configurations**—Misconfigured pod security and network policies can allow lateral movement within a cluster.

To mitigate these risks, organisations should enforce least privilege access and implement Zero Trust Architecture (ZTA) to restrict access based on verification, rather than implicit trust.

Automating Secure Configurations

Automation is critical for maintaining secure configurations at scale. Configuration management tools such as Ansible, Puppet and Chef enable organisations to define security settings as code, ensuring consistent deployment and compliance.

Additional tools that help enforce secure configurations include:

- **Terraform and CloudFormation**—Secure Infrastructure as Code (IaC) management
- **AWS Config, Azure Policy, Google Security Command Center**—Continuous configuration monitoring
- **OpenSCAP and CIS-CAT Pro**—Compliance auditing and security benchmarking.

Despite automation, continuous validation and monitoring remain necessary to detect and remediate misconfigurations before exploitation occurs.

Trade-offs and Practical Considerations

While strict security controls effectively reduce risks, they often create tension with system performance, user productivity and application compatibility. Organisations should thoughtfully balance security imperatives with operational needs. This balance typically involves conducting targeted risk assessments that help prioritise the most critical security configurations while implementing carefully managed exceptions for legacy systems that cannot meet all security requirements. Complementing these approaches, organisations should deploy automated testing regimes to ensure that security measures don't inadvertently disrupt essential application functionality.

Building Resilient Security Postures

Configuration security requires ongoing attention rather than a one-time implementation. A well-rounded, defence-in-depth strategy strengthens security by integrating several key practices:

- Network segmentation establishes clear boundaries, limiting the spread of potential breaches and restricting unauthorised movement between systems and trust boundaries
- Intrusion detection and prevention systems monitor network activity in real-time, identifying and blocking attacks that exploit misconfigurations
- Regular security audits ensure compliance with security standards, helping organisations address vulnerabilities before they can be exploited.

When organisations successfully integrate automation, continuous monitoring and secure-by-design principles, they create an environment that proactively defends against misconfigurations while preserving operational agility.

Vulnerability Management and Patching

Vulnerability and patch management strategies help organisations identify, assess and mitigate security flaws before exploitation. These programs follow structured processes to detect vulnerabilities, evaluate their severity and prioritise remediation. Timely patching is important to ensure known vulnerabilities are swiftly addressed. Through the implementation of a structured approach, organisations can strengthen their cybersecurity resilience, minimise the impact of cyberattacks and protect valuable data and assets.

The following resources can assist in establishing an effective vulnerability and patch management strategy:

- National Institute of Standards and Technology. (2021). *Guide to Enterprise Patch Management Planning*. https://nvlpubs.nist.gov/nistpubs/SpecialPublications/NIST.SP.800-40r4.pdf

- Australian Cyber Security Centre. (2023). *Patching Applications and Operating Systems*. https://www.cyber.gov.au/resources-business-and-government/maintaining-devices-and-systems/system-hardening-and-administration/system-administration/patching-applications-and-operating-systems

Vulnerability Scanning Tools: Essential Components and Functionalities

Once vulnerabilities are identified, organisations need reliable tools to assess and manage them. Vulnerability scanning tools play a critical role in this process by automating the detection of security flaws across systems and applications. Industry-leading scanners like Nessus, OpenVAS and Qualys each offer distinct advantages.

Nessus provides vulnerability assessments, configuration audits and compliance checks through its scanning engine. It offers real-time vulnerability detection across operating systems, network devices, hypervisors and databases. Nessus employs credentialed and non-credentialed scanning methods, allowing security teams to prioritise critical vulnerabilities efficiently. The platform includes pre-configured templates for regulatory compliance frameworks like PCI DSS, *The Health Insurance Portability and Accountability Act* (HIPAA) and CIS benchmarks, enabling organisations to maintain consistent security standards.

OpenVAS, an open-source solution, offers scanning and management features. It provides network vulnerability testing with a regularly updated feed of Network Vulnerability Tests (NVTs), currently exceeding 50,000 tests. OpenVAS supports concurrent scanning across multiple targets and can be configured for various scanning depths and performance requirements. Its architecture allows for customisation to specific needs, while the reporting features include various output formats and risk scoring to help prioritise remediation efforts.

Qualys delivers cloud-based scanning that integrates with security platforms, enhancing visibility and control across hybrid IT environments. Its distributed architecture enables continuous scanning of global networks without hardware management overhead. Qualys offers vulnerability intelligence correlating threat data with asset information for accurate risk prioritisation. The platform provides patch management capabilities, allowing teams to detect and remediate vulnerabilities within a single workflow. Qualys's compliance capabilities include automated assessments against multiple frameworks, with real-time dashboards and documentation to streamline audit processes.

These tools help organisations keep their vulnerability inventory up to date and streamline remediation, strengthening overall cybersecurity resilience.

Vulnerability Disclosure and Responsible Disclosure Programs

Vulnerability disclosure and responsible disclosure programs provide structured methods for identifying and mitigating security risks. These programs enable security researchers, ethical hackers and IT professionals to report vulnerabilities directly to affected organisations or vendors. Establishing clear reporting guidelines and defined response timelines allows organisations to address security flaws before malicious actors exploit them.

The impact of these programs extends to all organisations that develop, purchase or use hardware and software. Responsible reporting gives organisations critical insights into system weaknesses, accelerating patch deployment and improving long-term security. These measures reduce the likelihood of data breaches and enhance system resilience.

Many programs incorporate bug bounty initiatives, offering financial rewards to security researchers who uncover vulnerabilities. Platforms like HackerOne, Bugcrowd and Synack streamline the disclosure process, efficiently managing reports and prioritising vulnerabilities based on risk severity. These platforms bridge the gap between security researchers and operational teams, fostering collaboration that strengthens overall cybersecurity.

Ultimately, vulnerability disclosure programs are necessary to safeguard sensitive data, maintain customer trust and ensure regulatory compliance. Organisations can reinforce their defences by anticipating emerging threats while contributing to a safer digital landscape.

If you're interested in participating, explore these platforms:

- StationX. *Learn Bug Bounty Hunting & Web Security Testing from Scratch.* https://courses.stationx.net/p/learn-bug-bounty-hunting-web-security-testing-from-scratch

- Bugcrowd. *Public Bug Bounty Program List.* https://www.bugcrowd.com/bug-bounty-list/

Vulnerability Patching and Deployment Strategies

Maintaining system security while minimising disruptions requires a structured approach to vulnerability patching. Organisations can achieve this by establishing a consistent patching schedule and applying updates proactively rather than reactively. A well-defined cycle reduces exposure to threats while maintaining system stability.

Before deployment, thorough testing in a controlled environment is essential. Patches can sometimes introduce compatibility issues, degrade performance or disrupt functionality. Testing helps identify and resolve these conflicts before they affect production systems, reducing the risk of downtime.

Prioritisation is key to effective patch management. Not all vulnerabilities pose the same risk, so patches should be applied based on severity, exploitability and potential business impact. Addressing high-risk vulnerabilities first helps mitigate critical threats, while lower-priority patches can follow a structured timeline.

Reliable backups act as a safeguard during the patching process. Keeping up-to-date backups before applying patches ensures quick recovery in case of failures or data corruption. Without a solid backup strategy, an improperly applied patch could cause prolonged disruptions and data loss.

Clear documentation and communication are essential for transparency and accountability. Maintaining detailed records such as patch versions, deployment dates and encountered issues ensures consistency across teams. Keeping stakeholders informed about upcoming updates helps align security efforts with operational needs.

Finally, centralising patch management simplifies deployment across operating systems, applications, drivers and firmware. Automated tools and patch management platforms help verify patch integrity, confirm successful installations and reduce the risk of missed updates. By streamlining this process, organisations can maintain a strong security posture with minimal operational impact.

Deploying Patches on Friday?

Your weekend plans might need a backup!

Summary

Understanding vulnerabilities is fundamental to effective cybersecurity. Like a hairline crack in a dam that can lead to catastrophic failure, even minor weaknesses can open the door to severe breaches. This chapter examined the origins, types and impact of vulnerabilities, emphasising the importance of proactive identification and mitigation.

Vulnerabilities stem from various sources, including software bugs, hardware flaws, misconfigurations and human error. Each can lead to unauthorised access, data breaches or service disruptions. Zero-day vulnerabilities (unknown flaws exploited before public disclosure) pose particularly severe risks, as demonstrated by incidents like the Heartbleed bug that exposed sensitive data across millions of websites.

The rapid evolution of technology has significantly expanded the attack surface. The increasing software complexity, the rise of IoT devices and the adoption of AI and cloud computing have created new attack vectors. With their intricate interdependencies, modern applications are especially prone to injection attacks, authentication weaknesses, XSS and broken access controls. Resources like the *CWE™ Top 25*, the *OWASP Top 10* and the CVE database provide valuable guidance for identifying and prioritising these vulnerabilities.

Hardware vulnerabilities present tough challenges due to their persistent nature. Attacks like Spectre and Meltdown exploit flaws in hardware design, bypassing traditional security measures. Unlike software vulnerabilities, these often require costly firmware updates or physical replacements. The growing number of IoT devices, many running outdated firmware, has further increased these risks.

Configuration vulnerabilities, often caused by default settings, missed updates or poor implementations, are another critical concern. A database left publicly accessible without authentication, for example, can become an easy target. Implementing secure system hardening practices and using configuration management tools helps maintain consistent, secure settings. Adopting a defence-in-depth strategy (layering multiple security controls) further reduces exposure.

This chapter also described the importance of integrating security practices throughout the System Development Lifecycle (SDLC). A proactive, security-first approach ensures vulnerabilities are addressed from design through maintenance. Tools like Nessus, OpenVAS and Qualys automate detection, while platforms like HackerOne and Bugcrowd encourage ethical hackers to report security flaws.

Ultimately, vulnerability management is not just a technical task but a strategic imperative. Organisations implementing comprehensive strategies (encompassing people, processes and technology) are better positioned to protect assets, maintain trust and adapt to the evolving threat landscape.

Revision Questions

1. What are the three interconnected pillars of cybersecurity mentioned in the chapter?

2. What makes zero-day vulnerabilities particularly dangerous?

3. Name and explain three categories of software vulnerabilities discussed in the chapter.

4. Why are hardware vulnerabilities more challenging to address than software vulnerabilities?

5. What are the main phases of the System Development Lifecycle (SDLC), and why is it essential for security?

6. What is the difference between static code analysis and Dynamic Application Security Testing (DAST)?

7. What role do configuration vulnerabilities play in system security?

8. What is system hardening and why is it important?

9. Name three key vulnerability scanning tools mentioned in the chapter and their main features.

10. What is the purpose of responsible disclosure programs?

Cybersecurity Hacks

Remember: Always practice security testing only on systems you own or have explicit permission to test.

Hack 1: Build Your Own Vulnerability Lab

Create a safe, isolated environment to experiment with and understand vulnerabilities:

- Set up a virtual machine using free software like VirtualBox
- Install purposely vulnerable systems like DVWA (Damn Vulnerable Web Application) or WebGoat
- Create a checklist of common vulnerabilities from the chapter (SQL injection, XSS, etc.)
- Challenge yourself to find and exploit these vulnerabilities in your lab
- Document each vulnerability you find and how you'd fix it
- Share your findings with friends or study group.

Hack 2: The Vulnerability Treasure Hunt

Turn vulnerability scanning into an engaging game:

- Download a free tool like OpenVAS
- Create a points system (e.g., 5 points for finding default passwords, 10 for discovering unpatched software)
- Scan your personal devices (phones, laptops, IoT devices)
- Keep a 'vulnerability diary' logging what you find
- Challenge yourself to fix each vulnerability you discover
- Set monthly goals to reduce your 'vulnerability score'
- Celebrate when you successfully patch vulnerabilities.

Chapter 4: Risk Management Essentials Part 1

– Fortifying Your Cyber Defences –

Objectives

By the end of this chapter, you will be able to:

- Understand the fundamental concepts of cybersecurity risk management
- Explain the four phases of the risk management process and how they work together
- Compare major risk management frameworks
- Apply different risk treatment strategies to address identified cybersecurity risks
- Understand how emerging technologies affect cybersecurity risk management strategies.

In 2023, organisations worldwide experienced an estimated $8 trillion in losses due to cybercrime,[33] a figure that is expected to rise in the coming years. As digital systems become more complex and interconnected, businesses of all sizes must adopt a structured approach to identifying, assessing and addressing cybersecurity risks.

Cybersecurity risk management is an ongoing process of identifying, analysing, evaluating and mitigating threats to an organisation's information assets. This structured approach helps security professionals determine which risks require immediate attention, which can be accepted and how limited resources should be allocated to maximise security effectiveness.

Today's cybersecurity landscape presents significant challenges. Attackers continuously refine their methods, using emerging technologies to bypass traditional security measures. Zero-day vulnerabilities, ransomware attacks, supply chain breaches and advanced persistent threats are just some of the risks that organisations face. At the same time, digital transformation initiatives, such as cloud adoption, Operational Technology (OT), IoT integration and remote work, expand attack surfaces, introducing new vulnerabilities that must be systematically managed.

Systematic risk approaches provide structure amid this complexity. Instead of reacting to security incidents as they occur, organisations that implement risk management strategies can anticipate potential threats, prioritise their response based on business impact and align security measures with their strategic goals. This approach shifts cybersecurity from a purely technical concern to a key component of business strategy.

The foundation of effective cybersecurity risk management lies in understanding the relationship between threats, vulnerabilities and impacts. A threat exploiting a vulnerability creates the potential for harm; this potential represents risk. Security professionals must develop analytical skills to assess this relationship across technical environments and business contexts.

Global cybercrime costs are projected to increase to $15.63 trillion by 2029.[34] Beyond financial implications, security breaches can damage organisational reputation, erode customer trust and trigger regulatory penalties. Compliance requirements across industries further emphasise the importance of structured risk management, with frameworks like ASD-ACSC Essential Eight, ISO/IEC 27001 and the NIST CSF providing valuable guidance.

Building on our earlier discussions of threats and vulnerabilities, this chapter introduces key concepts, processes and frameworks for effective cybersecurity risk management. You will learn practical methods for identifying and assessing risks, implementing security controls and maintaining continuous monitoring. Developing these skills will prepare you to help organisations defend against evolving threats while supporting their operational and strategic goals.

Fundamentals of Cybersecurity Risk Management

Key Concepts and Definitions

Effective cybersecurity risk management begins with understanding the core concepts that shape an organisation's security approach. These include risk, threats, vulnerabilities, assets, controls and risk appetite, all of which form the foundation of any security strategy.

At its core, risk represents the potential for loss or harm, typically assessed by evaluating both the likelihood of a threat materialising and the severity of its impact. The widely used formula Risk = Threat × Vulnerability illustrates this relationship. For example, if an attacker (threat) exploits an unpatched software flaw (vulnerability), the organisation may face financial losses, reputational damage and regulatory penalties (the risk).

As described in Chapter 2, a threat is any event, action or condition that could negatively impact an organisation's operations, assets or reputation. Threats can be external, such as cyberattacks, malware and natural disasters, or internal, including human error and insider threats. Complementing this, a vulnerability (see Chapter 3) is a weakness in security measures, system architecture or processes that a threat can exploit.

Assets refer to the organisation's data, devices and infrastructure that require protection. Understanding which assets are most critical allows organisations to prioritise security efforts, ensuring the most valuable resources receive the highest level of defence. These defences are implemented through security controls, which can be:

- Technical controls (firewalls, encryption, intrusion detection systems)
- Administrative controls (policies, training programs, access management)
- Physical controls (surveillance, secure access systems, environmental protections).

Each organisation must define its risk appetite; the level of risk it is willing to accept in pursuit of its objectives. For example, a financial institution handling sensitive customer data may have a low-risk appetite, whereas a tech startup focused on innovation may accept higher risk to prioritise agility. Understanding this balance allows organisations to make informed decisions about risk mitigation strategies.

Organisational Context

Cybersecurity risk management is not just a technical function but a critical component of an organisation's broader strategy. A successful risk management program must align with business goals, prioritise regulatory requirements, address industry-specific challenges and distribute available resources.

Role of Risk Management in Business Strategy

Organisations operate in a risk-versus-reward environment, where cybersecurity decisions impact business growth, innovation and operational stability. Cybersecurity risk management should support business objectives rather than act as a barrier. For example:

- A financial institution must balance security and compliance with customer convenience
- A healthcare provider must secure patient data while ensuring seamless service delivery.

Organisations must proactively address risks by integrating cybersecurity into their business strategy while maintaining operational flexibility. This integration ensures that security measures enhance rather than impede the organisation's core mission and competitive advantages.

Stakeholder Perspectives in Risk Management

Effective cybersecurity risk management requires input from multiple stakeholders, each with distinct priorities and perspectives:

- **Executives and Business Leaders** focus on business continuity, reputation protection, and financial impact assessment.
- **IT and Security Teams** identify technical vulnerabilities, implement mitigation measures and respond to security incidents.
- **Compliance Officers and Legal Teams** ensure regulatory adherence and minimise legal liability.
- **Employees** follow security policies and maintain awareness of evolving cyber threats.

Aligning cybersecurity risk management with diverse stakeholder interests improves decision-making and fosters a security-aware culture across the organisation. This collaborative approach ensures security initiatives receive necessary support at all organisational levels.

Alignment with Organisational Context

Risk management approaches should be tailored to an organisation's unique mission, industry context and operational environment. Several factors influence this alignment:

- **Regulatory Compliance**—Organisations must adhere to industry regulations to maintain legal and operational compliance.

- **Industry-Specific Risks**—Financial services face sophisticated fraud attempts; healthcare organisations must prioritise patient data protection; and manufacturing companies must secure complex supply chains.

- **Cultural and Risk Tolerance Differences**—Some organisations demonstrate high risk aversion, investing substantially in cybersecurity measures, while others prioritise balancing security controls and operational flexibility.

A well-structured risk management program ensures cybersecurity efforts support business resilience, regulatory adherence and long-term strategic goals while reflecting organisational values and priorities.

Resource Availability and Constraints

Organisations must implement security measures within available resources while planning for future investment. Budget constraints require risk management programs to balance costs against security effectiveness, forcing prioritisation of controls that address critical vulnerabilities with demonstrable returns. Personnel expertise considerations involve assessing whether sufficient in-house cybersecurity talent exists or if external support is needed. Additionally, security solutions must integrate seamlessly with the organisation's existing technology infrastructure to avoid compatibility issues. By thoughtfully aligning cybersecurity strategies with these resource realities, businesses can maximise protection without creating unnecessary financial strain or operational disruption.

The Risk Management Process

Risk Identification Techniques and Assessment Approaches

The cybersecurity risk management process requires a structured, iterative approach that enables organisations to systematically identify, assess, treat and monitor potential threats to digital assets. This process empowers security professionals to make data-driven decisions about protective measures while balancing security investments against business objectives.

Risk Identification Techniques

Risk identification serves as the foundation of the entire risk management lifecycle. Organisations cannot properly assess, treat or monitor their cybersecurity risks without identification. Three primary techniques drive risk identification:

Threat Modelling

Threat modelling creates a structured framework for anticipating potential attack vectors before they materialise. This approach systematically analyses systems from an adversary's perspective, examining how malicious actors might compromise valuable assets. Several methodologies guide this process:

- **STRIDE**—This framework evaluates systems across six threat categories: Spoofing, Tampering, Repudiation, Information disclosure, Denial of service and Elevation of privilege. It helps teams identify specific attack vectors rather than vague vulnerabilities.
- **PASTA**—Standing for Process for Attack Simulation and Threat Analysis, this approach takes a risk-centric approach by mapping business

objectives to technical requirements, ensuring threat models address organisational priorities.

- **LINDDUN**—This methodology focuses on privacy concerns through seven categories: Linking, Identifiability, Non-repudiation, Detectability, Disclosure of information, Unawareness and Non-compliance.

When implementing threat modelling, consider beginning with critical systems that manage sensitive data or support fundamental business functions. Document findings that facilitate communication between technical and non-technical stakeholders. Regularly revisit threat models as systems evolve or when new threat intelligence emerges.

Asset Inventories

You cannot protect what you don't know exists. Conducting an inventory of assets creates a catalogue of an organisation's digital resources, including:

- Hardware devices (servers, workstations, networking equipment, IoT devices)
- Software applications (both commercial and custom-developed)
- Data repositories (databases, file shares, cloud storage)
- Network resources (IP ranges, domains, wireless networks)
- Third-party services and dependencies.

Effective asset inventories should document not just the existence of these resources, but also their purpose, ownership, classification level and relationship to business processes. This context proves invaluable when assessing the potential impact of compromised assets.

Modern organisations face challenges with 'shadow IT' (technology deployed without formal IT approval or oversight). Despite these challenges, regular network scanning, integration with procurement systems and cloud monitoring can help maintain accurate inventories.

Vulnerability Assessment Methodologies

Vulnerability assessments systematically identify, classify and prioritise security weaknesses across an organisation's systems as discussed in Chapter 3. These assessments employ various methodologies:

- **Automated scanning**—This method uses tools like Nessus, OpenVAS and Qualys to identify known vulnerabilities across networks, systems and applications. While efficient for covering large environments, automated scans can produce false positives and may miss context-dependent vulnerabilities.

- **Manual testing**—This approach involves security professionals directly examining systems for weaknesses through techniques like code review, configuration analysis and penetration testing. This approach provides deeper analysis but requires specialised expertise and takes more time.

- **Compliance-based assessments**—This method evaluates systems against specific regulatory or industry standards (APPs, ISM, PCI DSS, HIPAA, ISO/IEC 27001, etc.). While valuable for regulatory purposes, these assessments may not address emerging threats outside the compliance framework.

Risk Assessment Approaches

Once risks are identified, organisations must assess their severity to prioritise mitigation efforts. This assessment typically involves evaluating the likelihood of a threat occurring and its potential impact on the organisation.

Qualitative Risk Assessment

Qualitative assessments use descriptive scales (e.g., high/medium/low) rather than numerical values to categorise risks. This approach:

- Relies on expert judgment and stakeholder input
- Uses risk matrices to visualise the relationship between likelihood and impact (see the following risk matrix, also known as a heat map)
- Provides relatively quick insights without extensive data requirements
- Facilitates understanding among non-technical stakeholders.

For example, a qualitative assessment might categorise a ransomware attack as "high likelihood, high impact" based on industry trends and the organisation's current controls. While subjective, this classification provides actionable guidance for prioritisation.

		Impact				
		Negligible	Minor	Moderate	Major	Catastrophic
Likelihood	Almost Certain	Moderate	High	Extreme	Extreme	Extreme
	Likely	Moderate	High	High	Extreme	Extreme
	Possible	Low	Moderate	High	High	Extreme
	Unlikely	Low	Moderate	Moderate	High	High
	Rare	Low	Low	Low	Moderate	Moderate

Quantitative Risk Measurement

Quantitative approaches apply numerical values and statistical analysis to calculate risk levels. Key methodologies include:

- Annual Loss Expectancy (ALE) where ALE (The estimated financial loss an organisation can expect from a specific risk per year) = Single Loss Expectancy (SLE—The projected cost of a single occurrence of a risk event) × Annual Rate of Occurrence (ARO—The estimated frequency at which a risk event is expected to occur in a year)

- Value at Risk (VaR), which estimates the potential financial loss over a specific timeframe at a given confidence level

- Monte Carlo simulations that model various risk scenarios by running multiple probabilistic simulations.

Quantitative assessments require more data but provide concrete financial figures that help justify security investments to executive leadership. For instance, calculating that a potential data breach could cost $2.4 million annually (based on probability and impact) creates a clear threshold for reasonable security spending.

Risk Scoring and Prioritisation

Effective risk management requires prioritisation; organisations cannot address all risks simultaneously. Risk scoring systems help by:

- Standardising how risks are evaluated across different business units

- Incorporating factors beyond likelihood and impact (such as detection capability, exploitability and data sensitivity)

- Creating comparable metrics for diverse risk types

- Establishing thresholds for different response levels.

The Common Vulnerability Scoring System (CVSS) is a widely used framework for assessing and communicating the severity of security vulnerabilities. It provides a standardised approach, but organisations often adapt scoring models to match their specific risk tolerance and operational needs. Customisation allows security teams to prioritise threats based on their unique business context.

When evaluating risks, it's important to differentiate between inherent risk, the risk level before any controls are in place, and residual risk, which remains after existing protections are applied. This distinction helps organisations identify where additional safeguards are necessary and current defences are sufficient.

Using structured risk assessment methods, organisations can better understand their security risks. These approaches help identify and prioritise threats, allowing decision-makers to allocate resources where needed most. This not only strengthens security but also prevents unnecessary spending.

Risk Management Cycle

The cybersecurity risk management cycle represents a structured, iterative approach that enables organisations to improve their security posture continuously in response to evolving threats. Unlike static security implementations, this dynamic process acknowledges that effective risk management requires ongoing attention and adaptation.

Continuous Improvement Principles

At its core, the risk management cycle embraces continuous improvement methodologies derived from established management frameworks like the Plan-Do-Check-Act (PDCA) cycle. This systematic approach ensures that security measures evolve with novel threats and changing business requirements. Key continuous improvement principles in cybersecurity risk management include:

- **Measurement-Driven Decision-Making**—Organisations collect and analyse quantifiable security metrics to assess the effectiveness of existing controls and identify improvement opportunities. These metrics might include vulnerability remediation times, security incident frequency or compliance audit scores.

- **Feedback Integration**—Insights gathered from security incidents, near-misses, audits and stakeholder feedback become catalysts for enhancement. This creates an environment where lessons learned directly inform security strategy refinements.

- **Documentation and Knowledge Management**—Systematic documentation of risk assessments, control implementations and security incidents builds an organisational knowledge base. This repository supports consistent decision-making and preserves institutional knowledge despite personnel changes.

- **Maturity-Based Advancement**—Organisations assess their risk management processes against maturity models to identify strengths, weaknesses and progression opportunities. This structured approach enables targeted improvements that align with organisational goals and industry best practices.

Iterative Risk Management Strategies

Effective cybersecurity risk management operates through cyclical processes rather than linear workflows. This iterative approach enables organisations to adapt to the dynamic threat landscape while maintaining defence mechanisms that remain relevant and effective. Key iterative strategies include:

- **Regular Risk Reassessment**—Organisations establish cadences for risk reviews, typically quarterly for high-priority systems and annually for the broader environment. These reassessments account for new threats, vulnerabilities, business changes and regulatory requirements.

- **Control Effectiveness Validation**—Security teams periodically test implemented controls through vulnerability scanning, penetration testing, tabletop exercises and control audits. These validation activities verify that security measures deliver their intended protections and identify any degradation in effectiveness.

- **Emerging Threat Monitoring**—Dedicated resources monitor threat intelligence sources to identify new attack vectors, vulnerabilities and threat actor techniques. This approach allows organisations to anticipate potential risks before they manifest as security incidents.

- **Incremental Security Enhancement**—Organisations implement targeted security improvements based on risk priorities rather than pursuing comprehensive overhauls. This incremental approach balances security advancement with operational continuity.

Risk Mitigation Strategies

Risk mitigation represents the tactical implementation of security measures designed to reduce identified risks to acceptable levels. These strategies encompass a set of controls that address technical, administrative and physical security dimensions.

Technical Controls

Technical controls leverage technology solutions to prevent, detect and respond to security threats across the organisation's digital environment. These controls serve as the technological foundation of the security program:

- **Network Security Mechanisms**—Implementation of defence-in-depth approaches including firewalls, intrusion detection/prevention systems (IDSs/IPSs), network segmentation and secure gateways. These mechanisms establish security boundaries that limit lateral movement and contain potential breaches.

- **Endpoint Protection Solutions**—Endpoint security deployment, including antimalware tools, host-based firewalls, application control, device encryption and Endpoint Detection and Response (EDR) systems. These measures secure individual devices that might otherwise become entry points for attackers.

- **Identity and Access Management (IAM)**—Implement strong authentication mechanisms (MFA, PAM), authorisation controls (role-based access control, principle of least privilege) and access governance processes. These controls ensure that only authorised individuals can access sensitive systems and data.

- **Secure Development Practices**—Security integration throughout the SDLC, including secure coding standards, vulnerability testing and pre-deployment security validation. These practices build security into applications rather than treating it as an afterthought.

- **Data Protection Controls**—Deployment of encryption (both in transit and at rest), data loss prevention systems, database activity monitoring and data masking technologies. These controls safeguard the confidentiality and integrity of sensitive information throughout its lifecycle.

Administrative Controls

Administrative controls establish the operational framework, processes and human elements of security risk management. These policies and procedures guide organisational behaviour and security governance:

- **Security Policies and Procedures**—Development and maintenance of documentation that establishes security expectations, requirements and processes. These foundational documents provide clarity and consistency in security implementation.

- **Security Awareness and Training**—Implementation of ongoing education initiatives that develop security consciousness among employees, contractors and third parties. These programs transform personnel from potential vulnerabilities into active participants in the security program.

- **Vendor Risk Management**—Establishment of processes to assess, monitor and manage security risks introduced through third-party relationships. These practices extend security considerations beyond organisational boundaries to encompass the entire supply chain.

- **Compliance Management**—Development of frameworks to identify, track and satisfy regulatory, contractual and industry-specific security requirements. These processes ensure alignment between security practices and external obligations.

- **Security Incident Management**—Creation of structured approaches for detecting, investigating, containing, eradicating and recovering from security incidents. These procedures minimise damage when prevention mechanisms fail.

Physical Security Measures

Physical security controls protect tangible assets, including facilities, equipment and supporting infrastructure. These measures bridge the gap between digital and physical security domains:

- **Facility Access Controls**—Implementation of layered physical access mechanisms including badges, biometrics, mantrap systems and visitor management procedures. These controls prevent unauthorised physical access to sensitive areas.

- **Environmental Protection**—Deployment of systems that safeguard equipment from environmental threats, including fire suppression, climate control, water detection and power conditioning. These measures protect against non-human threats that could compromise availability.

- **Media Handling Procedures**—Establishment of processes for securely managing physical media throughout its lifecycle, including secure storage, transportation and destruction. These procedures prevent data breaches through physical means.

- **Equipment Security**—Implementation of controls that protect hardware from theft, tampering and unauthorised access, including cabinet locks, cable locks and asset tracking systems. These measures maintain the integrity of physical technology assets.

A well-structured security framework relies on strategically integrating technical, administrative and physical controls to mitigate risks effectively. This layered approach recognises that no single control is sufficient; instead, security requires a coordinated effort across all domains to reduce vulnerabilities and strengthen overall protection.

Risk Management Frameworks (RMFs) and Standards Overview

Cybersecurity risk management is built on structured methodologies that follow established industry standards. These standards provide organisations with practical guidance for identifying, assessing and mitigating threats while maintaining compliance with regulatory requirements. This section introduces key cybersecurity frameworks that have gained widespread adoption and explains how organisations can select options that align with their specific needs.

If you're already familiar with these concepts, you may wish to skip this introduction, as the next chapter explores them in greater depth. However, if you're new to cybersecurity, this overview provides a good foundation for understanding how systematic strategies enhance security.

Industry-Standard Frameworks

The National Institute of Standards and Technology (NIST) offers two complementary frameworks that serve as cornerstones of modern cybersecurity risk management. The NIST Cybersecurity Framework (CSF) provides an accessible, function-based approach around six core functions: Govern, Identify, Protect, Detect, Respond and Recover. This structure creates an intuitive roadmap that guides organisations through the essential components of a robust cybersecurity program.

In contrast, the NIST RMF takes a more process-oriented approach with its seven-step methodology: Prepare, Categorise, Select, Implement, Assess, Authorise and Monitor. The RMF excels in environments requiring rigorous documentation and formal authorisation processes. While originally designed for U.S. federal agencies, its adaptability has led to widespread adoption across sectors seeking a comprehensive approach to risk management.

The International Organization for Standardization's ISO/IEC 27001 framework establishes requirements for an Information Security Management System (ISMS), providing a systematic approach to managing sensitive information. This globally recognised standard emphasises risk assessment, organisational context, leadership commitment and continuous improvement through its Plan-Do-Check-Act (PDCA) cycle.

Its companion standard, ISO/IEC 27002, provides detailed implementation guidance with specific security controls across domains such as access control, cryptography, physical security and operational security. This prescriptive approach makes ISO particularly valuable for organisations needing clear direction on control implementation while maintaining flexibility to adapt to specific risk profiles.

The Australian Signals Directorate-Australian Cyber Security Centre (ASD-ACSC) have developed frameworks tailored to the Australian context. The Information Security Manual (ISM) provides comprehensive security controls across governance, personnel, physical and information communication technology domains.

The Essential Eight offers a more focused approach, identifying eight critical security controls that effectively mitigate common cyber threats. The Essential Eight provides organisations with clear, actionable priorities that deliver significant security improvements by prioritising application control, patching, macro restrictions, application hardening, administrative privilege management, operating system updates, MFA and regular backups.

Unlike frameworks primarily focused on control implementation, the Factor Analysis of Information Risk (FAIR) model provides a quantitative methodology for measuring and expressing cybersecurity risk in financial terms. FAIR decomposes risk into component factors that can be analysed through probability formulas, enabling organisations to calculate potential loss exposure in monetary values.

FAIR's analytical rigor makes it particularly effective for Cost-Benefit Analysis (CBA) of security investments. It allows organisations to determine whether specific controls deliver appropriate returns. When combined with other frameworks that address control selection and implementation, FAIR provides a powerful tool for prioritising security investments based on quantifiable risk reduction.

Developed by the Information Systems Audit and Control Association (ISACA), Control Objectives for Information and Related Technologies (COBIT) takes a broader governance perspective by integrating IT risk management within larger enterprise governance objectives. Rather than focusing exclusively on security controls, COBIT bridges the gap between technical management and business requirements, ensuring that security considerations influence strategic planning.

COBIT includes a process maturity model that allows organisations to assess their capabilities and establish improvement roadmaps. This makes it effective for demonstrating regulatory compliance and aligning IT security with business goals. The framework's emphasis on accountability structures helps clarify risk ownership across organisational levels.

Comparative Analysis

Framework Strengths and Limitations

Each framework offers distinct advantages while addressing different aspects of cybersecurity risk management. NIST CSF provides an accessible approach that organisations of any size can implement incrementally. However, its flexibility sometimes leaves organisations wanting more prescriptive guidance.

The NIST RMF offers guidance that excels in environments requiring formal authorisation. This rigour can create significant overhead for smaller organisations with limited resources. Similarly, ISO/IEC 27001 provides international recognition but requires substantial documentation and formal audit processes.

The ASD-ACSC Essential Eight stands out for its pragmatic focus on high-impact controls, making it valuable for organisations seeking maximum security improvement with minimal investment. However, its focused scope means it must be supplemented with additional frameworks for comprehensive risk management.

FAIR's quantitative approach transforms security discussions by expressing risk in financial terms but requires specialised expertise to generate accurate calculations. COBIT's governance focus helps align security with business objectives but may seem abstract to technical practitioners.

Selecting Appropriate Frameworks

Organisations should consider several factors when selecting risk management frameworks. Regulatory requirements often narrow the field, as specific industries may mandate particular frameworks. For example, government contractors may need to implement NIST RMF, while multinational corporations might prefer ISO/IEC 27001.

Organisational size and available resources significantly influence framework selection. Large enterprises with dedicated security teams can implement comprehensive frameworks, while small to medium-sized businesses may achieve better results with the focused guidance of the Essential Eight or NIST CSF.

Industry context matters, as certain sectors face unique threats or compliance requirements. The most effective approach often involves combining elements from multiple frameworks to address these distinct needs.

Implementation Considerations

Successful framework implementation requires thoughtful planning and organisational commitment. Rather than attempting wholesale adoption, organisations should consider phased implementation that prioritises high-impact controls addressing their most significant risks.

Cultural factors significantly influence implementation success. Frameworks that align with existing organisational processes often face less resistance than those requiring significant operational changes. Executive sponsorship remains critical, particularly when implementation requires substantial resources.

Many organisations benefit from hybrid approaches that combine elements from multiple frameworks. For example, using the Essential Eight to prioritise immediate control implementation while building toward ISO/IEC 27001 certification provides short-term security improvements and longer-term compliance objectives.

Framework implementation should never become a checkbox exercise focused solely on compliance. The most successful deployments integrate framework principles into organisational culture, creating an environment where security considerations influence decisions at all levels. By treating frameworks as foundational elements rather than compliance hurdles, organisations can build truly resilient security programs that evolve with changing threats and business needs.

Strategic Briefing: The 2023 MGM Resorts Cyberattack

In September 2023, MGM Resorts International, one of the world's largest casino and hospitality companies, fell victim to a sophisticated cyberattack that disrupted operations across its entire portfolio of properties.[35] The incident, attributed to the ALPHV/BlackCat ransomware group, showcased how social engineering tactics could bypass traditional security measures and cause widespread operational disruption.

The attack demonstrated the cascading effects of cybersecurity breaches in the hospitality industry. It highlighted the critical importance of comprehensive security protocols, particularly in organisations managing sensitive customer data and complex operational systems.

Impact and Scope

The cyberattack resulted in a widespread system shutdown affecting MGM's operations across multiple properties. Key impacts included:

- Digital key card systems became inoperable, requiring manual check-ins
- Casino gaming systems were disabled

- Restaurant payment systems were disrupted
- Hotel reservation systems went offline
- Digital displays throughout properties were affected.

The estimated financial impact exceeded $100 million, including direct losses, system recovery costs and lost revenue during the operational disruption. The incident affected thousands of guests across MGM's properties and potentially compromised sensitive customer data.

Attack Methodology

The attackers gained initial access through a social engineering attack targeting an IT service desk employee. By impersonating a legitimate worker who claimed to be locked out of their account, the attackers convinced the help desk to reset credentials, thereby gaining system access. This breach highlighted the critical weakness in the company's human-centric security protocols.

Once inside the network, the attackers moved laterally through MGM's systems, eventually gaining access to critical infrastructure and deploying ransomware that forced the company to shut down its operations to prevent further damage.

Response and Mitigation

MGM executed a comprehensive incident response strategy upon detecting the breach. The organisation's first critical step was to contain the incident by initiating an immediate shutdown of all compromised systems, while engaging specialised cybersecurity firms and law enforcement agencies to investigate the attack.

During the systems outage, MGM maintained basic operations through manual processes, providing regular status updates to stakeholders, including affected customers and investors. The final phase involved a methodical restoration of systems, with each component undergoing rigorous security verification before being returned online. Their response protocol demonstrated MGM's balanced approach to security and operational continuity.

Lessons Learned

The MGM cyberattack revealed critical lessons for hospitality and enterprise organisations about security architecture, human behaviour and operational continuity. The breach demonstrated why organisations must implement Zero Trust Architecture (ZTA) (a security model that requires strict verification of every person and device accessing networks) and mandate MFA for all system access. Effective network segmentation proved essential in containing potential breaches, while systematic security assessments and penetration testing came to prominence as vital tools for detecting vulnerabilities.

Human behaviour stands at the centre of cybersecurity defence. Organisations must deliver targeted security awareness training to all employees and implement rigorous help desk verification protocols. Regular exercises simulating phishing attacks (fraudulent attempts to obtain sensitive information) and social engineering threats, combined with clear security escalation procedures, strengthen an organisation's human security framework.

The attack highlighted how operational resilience determines recovery success. Organisations must maintain tested offline backup systems and regularly updated business continuity plans. This preparation includes documented manual procedures for system outages and comprehensive incident response protocols to guide breach management and stakeholder communication.

Emerging Technologies and Risk Landscapes

As technology evolves at an unprecedented pace, the cybersecurity risk landscape transforms alongside it. New threats emerge in parallel with innovative solutions, requiring security professionals to adapt their risk management strategies continuously. Organisations must not only mitigate advancing threats but also strategically leverage technological advancements to enhance their security posture.

Artificial Intelligence (AI) and Machine Learning (ML)

AI and ML are fundamentally reshaping cybersecurity risk assessment and threat detection capabilities. These technologies excel at processing vast amounts of security data, identifying patterns and anomalies that typically elude human analysts. AI-driven risk assessment tools enable organisations to predict potential security incidents by dynamically categorising risks and adjusting threat levels in real-time.

ML algorithms continuously improve threat detection by analysing historical security incidents and adapting to new attack vectors. However, AI also introduces unique risks, including algorithmic biases that can impact decision-making accuracy, potentially leading to flawed threat assessments. Additionally, adversaries increasingly attempt to manipulate AI models through poisoning attacks designed to evade detection systems. Security teams must implement appropriate safeguards against these emerging challenges.

Cloud Computing Security

Cloud computing has become a core part of modern business operations, necessitating evolved cybersecurity risk management approaches. The shared responsibility model forms the foundation of effective cloud security, clearly delineating which security aspects fall to the provider versus the customer.

Organisations should carefully evaluate risks related to data sovereignty, vendor lock-in and service availability.

Cloud risk management strategies differ significantly from traditional perimeter-based security models, demanding greater focus on identity management, encryption and access control systems. Organisations must implement comprehensive visibility and monitoring strategies to maintain oversight of their cloud assets while ensuring compliance with evolving regulations across different jurisdictions.

Internet of Things (IoT) Security

The proliferation of IoT devices presents distinctive cybersecurity challenges due to their often limited security capabilities. Practical IoT device risk assessment must account for firmware vulnerabilities, insufficient authentication mechanisms and data privacy concerns. Many connected devices become attractive entry points for attackers seeking to breach corporate networks.

Connectivity and vulnerability challenges include addressing the complex interactions between numerous devices operating across multiple protocols. Mitigating IoT security risks requires asset tracking, regular firmware updates and network segmentation to isolate vulnerable devices from critical systems. Organisations should integrate device security into broader risk management frameworks.

Operational Technology (OT) Security

Industrial control systems present unique security challenges as OT environments increasingly connect to IT networks. These systems often run legacy software with extended lifecycles, creating security gaps that require specialised assessment approaches. Risk management strategies must account for the potentially catastrophic physical consequences of OT security failures.

Critical infrastructure protection demands heightened attention as energy, transportation and healthcare sectors face sophisticated threats. Security teams must develop specialised risk assessment methodologies that address the convergence of IT and OT while maintaining operational continuity. This includes implementing appropriate segmentation, monitoring solutions and Incident Response Plans (IRPs) tailored to OT environments.

Zero Trust Architecture (ZTA)

ZTA transforms cybersecurity by replacing traditional perimeter-based defences with a "never trust, always verify" rather than the outdated "trust, but verify" approach. Key principles include continuous verification, least privilege access and assuming breach as the default security posture. Every user and device must validate their identity before accessing organisational resources, regardless of location.

Architectural risk considerations include balancing security requirements with operational needs and user experience. Implementing Zero Trust involves micro-segmentation, strict access controls and continuous user behaviour monitoring. While this strategy significantly reduces risk exposure, organisations must develop careful implementation roadmaps that address legacy systems and existing workflows.

Core Practices for Ongoing Risk Management Success

Effective cybersecurity risk management requires sustained effort through established operational practices:

Continuous Monitoring

Implementing monitoring systems enables organisations to maintain vigilance against new threats. This includes integrating threat intelligence feeds to provide early warning of potential attacks and deploying Security

Information and Event Management (SIEM) solutions to correlate and analyse security data across the enterprise. These systems help security teams identify anomalous patterns that might indicate compromise.

Incident Response Planning

When incidents occur, organisations need predetermined strategies to minimise impact. This involves developing detailed response playbooks that outline specific actions for different threat scenarios and establishing clear communication channels and escalation protocols. Teams should know precisely who to contact, when to escalate issues and how to coordinate response efforts across departments. More on this in Chapter 14.

Risk Culture and Human Factors

Technology solutions alone cannot secure an organisation without addressing human elements. Effective security awareness programs help employees recognise social engineering attempts and understand their role in maintaining security. Complementary training initiatives should provide practical skills development, from identifying phishing attempts to following secure coding practices.

Embedding these core practices into daily operations transforms risk management from an occasional exercise to an integral part of organisational culture. This continuous approach enables organisations to adapt to evolving threats while maintaining appropriate protection for critical assets.

Summary

Cybersecurity risk management is a critical business function, with organisations facing record-breaking financial losses from cyber threats, estimated at $8 trillion in 2023 alone. This chapter explored the key principles, processes and frameworks enabling organisations to effectively identify, assess and mitigate cybersecurity risks.

At its core, risk management involves five essential components: threats (potential harmful events), vulnerabilities (exploitable weaknesses), assets (valuable resources requiring protection), controls (security measures) and risk appetite (the level of risk an organisation is willing to accept). These elements interconnect within the widely used formula Risk = Threats × Vulnerabilities, which is the foundation for quantifying and managing cybersecurity risks.

A structured, four-phase risk management process underpins an effective cybersecurity strategy:

- **Risk Identification** uses threat modelling, vulnerability assessments and industry trend analysis to uncover potential risks.
- **Risk Assessment** evaluates the likelihood and impact of threats using qualitative and quantitative methods, often employing risk matrices to categorise risks from low to extreme.
- **Risk Treatment** involves four primary approaches: avoidance, reduction, transfer and acceptance.
- **Continuous Monitoring** ensures security measures remain effective amid evolving threats.

Organisations leverage established models to implement a risk management framework. The NIST CSF defines six key functions: Govern,

Identify, Protect, Detect, Respond and Recover, while the NIST RMF provides a seven-step process for integrating security throughout an organisation's system lifecycle. International standards such as ISO 31000 and ISO/IEC 27001 also offer structured methodologies for risk management and information security. Region-specific frameworks, like the Australian ASD-ACSC Essential Eight controls, provide tailored cybersecurity guidance.

Emerging technologies introduce both opportunities and risks. AI and ML enhance threat detection but also introduce new attack vectors through algorithmic bias and adversarial manipulation. Cloud computing, OT and the IoT expand the attack surface, necessitating updated security strategies. To counter these risks, the adoption of Zero-Trust Architecture (which replaces traditional perimeter-based security with a "never trust, always verify" model) has become a leading cybersecurity paradigm.

This chapter has emphasised that successful risk management requires ongoing vigilance and adaptation. Organisations are encouraged to establish continuous feedback loops, leverage automation appropriately and maintain regular training programs. This structured approach, combined with awareness of emerging threats and technologies, enables organisations to build resilient cybersecurity postures capable of protecting assets in an increasingly complex digital landscape.

This chapter examines cybersecurity risk management, providing the knowledge and tools to develop, implement and maintain effective risk management strategies that protect organisational assets while supporting business objectives.

Revision Questions

1. What is the relationship between threats, vulnerabilities and risk in cybersecurity?

2. Describe the four phases of the risk management process and their key activities.

3. What are the six core functions of the NIST Cybersecurity Framework (CSF)?

4. Compare ISO/IEC 27001 and the ASD-ACSC Essential Eight frameworks in terms of their approach and focus.

5. What are the Essential Eight controls recommended by the ASD-ACSC?

6. What key lessons were learned from the 2023 MGM Resorts cyberattack?

7. How does AI/ML impact modern cybersecurity risk management?

8. What are the four main risk treatment strategies and when should each be used?

9. How does Zero Trust Architecture differ from traditional security approaches?

10. What core practices should organisations adopt for ongoing risk management success?

Cybersecurity Hacks

Hack 1: Effective Risk Assessment

Develop a matrix to categorise risks based on their likelihood and impact, where likelihood refers to the probability of a risk occurring and impact measures the potential consequences on business operations. This visual tool helps prioritise risks, enabling organisations to allocate resources effectively and focus on mitigating the most critical threats.

Use the matrix to guide decision-making, ensuring that high-likelihood, high-impact risks receive immediate attention while lower-priority risks are managed appropriately. Regularly update the matrix with new data from continuous monitoring, vulnerability assessments and threat intelligence reports. This iterative process ensures that your risk management strategy remains current, adaptive and aligned with shifting cybersecurity challenges.

		Impact				
		Negligible	Minor	Moderate	Major	Catastrophic
Likelihood	Almost Certain	Moderate	High	Extreme	Extreme	Extreme
	Likely	Moderate	High	High	Extreme	Extreme
	Possible	Low	Moderate	High	High	Extreme
	Unlikely	Low	Moderate	Moderate	High	High
	Rare	Low	Low	Low	Moderate	Moderate

Instructions:

1. Categorise Risks—Identify potential risks and place them in the matrix based on their likelihood (Almost Certain, Likely, Possible, Unlikely, Rare) and impact (Negligible, Minor, Moderate, Major, Catastrophic).

2. Prioritise Risks—Focus resources on mitigating risks categorised as High or Extreme.

3. Update—Regularly update the matrix with new data from continuous monitoring and threat intelligence reports.

Hack 2: Quick Threat Modelling with STRIDE

Apply the STRIDE methodology to systematically identify potential attack vectors against a system before threats materialise. This structured approach examines systems from an adversary's perspective across six threat categories: Spoofing, Tampering, Repudiation, Information disclosure, Denial of service and Elevation of privilege.

Select one system or application and evaluate it against each STRIDE category to uncover specific vulnerabilities. This analysis helps identify security gaps that might be overlooked, enabling security improvements before attackers can exploit weaknesses.

STRIDE Analysis Template:

Threat Type	Description	Potential Attack	Current Controls	Risk Level	Action Required
Spoofing	Identity impersonation	Fake login credentials	Multi-factor authentication	Low	Monitor
Tampering	Data modification	Database manipulation	Access controls, logging	Medium	Strengthen
Repudiation	Denying actions	User claims innocence	Audit trails, digital signatures	High	Implement
Information disclosure	Data exposure	Unauthorised data access	Encryption, access controls	High	Review
Denial of service	Service disruption	System overload attacks	Load balancing, monitoring	Medium	Test
Elevation of privilege	Unauthorised access	Admin rights exploitation	Least privilege principle	High	Audit

Instructions:

1. Choose a System—Select a high-value system such as a customer database, email server or financial application.

2. Apply STRIDE Categories—Work through each threat type, identifying specific ways attackers might exploit that system using the template above.

3. Evaluate Existing Controls—Document current security measures and assess their effectiveness against identified threats.

4. Prioritise Actions—Focus on high-risk areas where current controls are insufficient, using this analysis to guide security investment decisions.

Chapter 5: Risk Management Essentials Part 2

– Frameworks and Standards –

Objectives

By the end of this chapter, you will be able to:

- Differentiate between cybersecurity frameworks and standards and explain their roles in risk management
- Compare key cybersecurity frameworks
- Understand the advantages and disadvantages of major international standards
- Identify critical factors when selecting an appropriate cybersecurity framework for an organisation
- Outline the essential steps for the successful implementation of a cybersecurity framework.

Though 'frameworks' and 'standards' are often used interchangeably, they serve distinct yet complementary purposes in cybersecurity. Understanding their differences helps organisations implement effective security measures tailored to their needs.

Cybersecurity frameworks provide guidelines for organisations to identify, protect against, detect, respond to and recover from cyber threats. These principles form the foundation of effective security practice. Frameworks like the U.S. National Institute of Standards and Technology Cybersecurity Framework (NIST CSF) offer flexibility, enabling adaptation across different sectors and scales.

It is essential to distinguish between 'frameworks' and 'standards', as these terms have distinct technical meanings. A standard can function as a framework, but a framework only becomes a standard after formal assessment by an authoritative body such as the International Organization for Standardization (ISO). In the standards section of this chapter, I specifically address frameworks that have received official recognition from standards bodies. I acknowledge this distinction may initially seem nuanced, but this clarification should provide necessary context for the discussion.

Effective risk management requires a structured approach built on established frameworks that evolve alongside emerging threats. This chapter examines three categories: U.S. NIST frameworks, international standards and Australian guidelines from the Australian Signals Directorate-Australian Cyber Security Centre (ASD-ACSC). Each offers unique advantages while sharing fundamental principles of risk assessment and mitigation.

Standards like the *Payment Card Industry Data Security Standard* (PCI DSS) and *The Health Insurance Portability and Accountability Act* define specific technical requirements and codify accepted practices, establishing measurable criteria for security controls and processes.

Selecting an appropriate cybersecurity framework is central to developing a security posture. Methodologies like NIST and ISO/IEC 27001 support risk management while aligning with industry standards. Large enterprises typically implement these to manage complex infrastructures, while small to medium-sized businesses benefit from streamlined options like the Essential Eight, which focus on fundamental security measures.

Resource availability influences framework selection. Organisations with limitations should prioritise frameworks offering clear, actionable controls. ASD–ACSC frameworks provide resource-conscious organisations with targeted strategies suited to their needs.

For Australian organisations, compliance with certain frameworks may constitute a legal requirement. Regulations, including the Australian Privacy Principles (APPs) and Essential Eight, mandate specific security measures to protect data and maintain system integrity.

Adopting internationally recognised frameworks and standards ensures alignment with global best practices. For example, implementing the ISO/IEC 27001 standard, which focuses on information security management systems (ISMS) and risk management, helps businesses streamline security processes and build trust with international partners.

Frameworks and standards create a foundation for organisations to strengthen their security postures, establishing a common language, enabling systematic risk management and aligning security with business objectives. Contemporary frameworks emphasise adaptability, allowing tailored implementation to specific requirements and risk profiles.

The following sections will examine popular cybersecurity frameworks in detail and explore Australian-specific frameworks, highlighting their alignment with global practices and local regulations.

Cybersecurity Frameworks (CSFs)

Cybersecurity presents complex challenges, but established frameworks offer structured guidance to help organisations manage and reduce risks. These frameworks provide a foundation for developing security strategies that align with industry standards, regulatory requirements and emerging threats. By adopting a framework, organisations can enhance their ability to prevent, detect and respond to cyber incidents. In this section, we will examine several widely used cybersecurity frameworks, exploring their key principles, advantages and disadvantages.

U.S. National Institute of Standards and Technology (NIST)

The U.S. NIST establishes comprehensive guidelines to help organisations assess, manage and mitigate cybersecurity risks. These guidelines are structured into several key frameworks, each designed to address specific aspects of cybersecurity and risk management:[36]

- **Integrating Cybersecurity and Enterprise Risk Management (ERM)** aligns cybersecurity risk management with broader enterprise risk strategies to enhance decision-making.

- **NIST CSF** provides a flexible, outcome-driven approach to identifying, protecting against, detecting, responding to and recovering from cybersecurity threats. It is adaptable for organisations at any stage of cybersecurity maturity.

- **Risk Management Framework (RMF)** establishes a systematic, lifecycle-based process for managing security and privacy risks. It helps organisations integrate security practices into their overall risk management activities, with a focus on regulatory compliance.

- **Privacy Framework** supports organisations in managing privacy risks and meeting regulatory requirements for personal data protection.

- **Cybersecurity Supply Chain Risk Management (C-SCRM)** addresses risks associated with suppliers, third-party vendors and other supply chain components to prevent potential vulnerabilities.

- **Workforce Framework for Cybersecurity (NICE Framework)** establishes a common taxonomy for describing cybersecurity work roles, skills and competencies, facilitating more effective workforce development and management.

Among these, the NIST CSF and the RMF are widely recognised as leading tools for building resilient cybersecurity programs. While both frameworks are industry standards, they differ in scope and methodology.

Organisations can also tailor and adapt these frameworks based on their risk profile, regulatory requirements and strategic objectives. Factors such as organisational size, operational complexity and cybersecurity maturity influence the choice of framework.

Understanding the distinct benefits and applications of these frameworks allows organisations to enhance their cybersecurity posture, improve regulatory compliance and better defend against evolving digital threats. Now, let's explore the two most prominent NIST cybersecurity tools, the CSF and RMF, in more detail.

NIST Cybersecurity Framework (CSF)

The NIST CSF offers clear, practical guidance for managing cybersecurity risks across industries, government agencies and organisations of all sizes. It provides a structured, high-level approach that helps organisations effectively understand, assess, prioritise and communicate their cybersecurity efforts.

At its core, the CSF is built around six essential functions: Govern, Identify, Protect, Detect, Respond and Recover. These interconnected components create a resilient, adaptive strategy for managing cybersecurity risks. In the second part of this book, we'll explore each of these functions in detail. For now, here's an overview of how they work within the CSF:

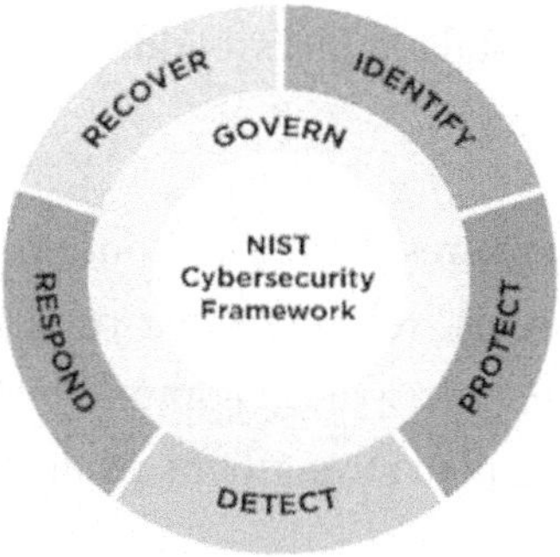

- **Govern**—This function ensures that cybersecurity efforts align with the organisation's goals. It involves defining roles, responsibilities and policies that guide security initiatives. Embedding cybersecurity into broader business strategies allows leadership to make informed decisions that enhance long-term security and resilience.

- **Identify**—Organisations must understand their assets, risks and vulnerabilities. This function involves asset inventories, risk assessments and threat modelling. Clearly identifying what needs protection enables organisations to allocate resources efficiently and mitigate potential threats.

- **Protect**—This function focuses on safeguarding critical systems and sensitive data. It includes access controls, encryption, network security protocols and employee training programs. Effective protection measures reduce the likelihood of cyber incidents.

- **Detect**—The ability to identify cybersecurity threats in real-time is essential. This function involves deploying monitoring tools, intrusion detection systems and conducting regular security audits. Early detection helps minimise potential damage.

- **Respond**—A swift and coordinated response is key when a cybersecurity incident occurs. This function guides organisations in executing IRPs, communicating with stakeholders and implementing containment strategies to reduce harm.

- **Recover:** Recovery ensures business continuity after an incident. It includes restoring systems, refining security measures based on lessons learned and maintaining transparent communication with stakeholders. A well-structured recovery plan strengthens resilience against future threats.

Following these six core functions enables organisations to adopt a structured approach to cybersecurity risk management. The NIST CSF not only provides a framework for protection but also acts as a roadmap for resilience in an evolving digital landscape.

CSF Advantages:

- **Flexibility and Scalability**—The NIST CSF adapts to organisations of all sizes, providing a structured yet non-prescriptive approach that allows customisation based on unique risks, regulations and operations.

- **Alignment with Industry Standards**—It integrates with ISO/IEC 27001, CIS Controls and the Essential Eight, enabling seamless security program enhancement and multi-framework compliance.

CSF Disadvantages:

- **Lack of Implementation Guidance**—The CSF outlines a high-level structure but lacks step-by-step instructions or specific technical controls, requiring interpretation and adaptation.

- **Resource Demands on Small Businesses**—While scalable, limited budgets and personnel may hinder full implementation. Smaller firms often require external support or third-party security services.

Source: NIST. (2024). *The NIST Cybersecurity Framework (CSF) 2.0*. https://nvlpubs.nist.gov/nistpubs/CSWP/NIST.CSWP.29.pdf

NIST Risk Management Framework (RMF)

The *NIST RMF*, as detailed in Special Publication 800-37, presents a seven-step process for integrating cybersecurity, privacy and supply chain risk management into the System Development Lifecycle (SDLC). Initially created for U.S. federal agencies, the RMF has become a widely adopted framework across state, local and private-sector organisations. Its risk-based methodology balances effectiveness, efficiency and legal obligations, ensuring that security practices align with operational goals throughout a system's entire lifecycle.

The RMF's adaptability makes it suitable for diverse technological environments, from Internet of Things (IoT) devices in smart homes to complex industrial control systems in manufacturing plants. It also aligns with the NIST CSF, providing a unified approach to safeguarding critical infrastructure against modern cyber threats like ransomware and supply chain attacks.

The RMF follows seven distinct steps:

- **Prepare**—Establish the foundation for risk management by defining the organisation's risk strategy, determining its risk tolerance and preparing for the effective management of security and privacy risks. This step ensures that all subsequent activities align with business objectives, regulatory standards and operational realities.

- **Categorise**—Based on an impact analysis, categorise the system and the data it processes, stores and transmits. Accurate categorisation ensures that critical information assets receive appropriate protection.
- **Select**—Select baseline security controls from established frameworks and customise them to meet the organisation's specific operational requirements and risk profile. Tailoring controls helps balance security, performance and cost-effectiveness.
- **Implement**—Implement the selected security controls and document their deployment within the system and its operating environment. Detailed documentation supports consistency, accountability and future audits.
- **Assess**—Assess the effectiveness of the implemented security controls to confirm that they are correctly applied, operating as intended and delivering the desired security and privacy outcomes. This evaluation helps identify potential vulnerabilities and areas for improvement.
- **Authorise**—Authorise the system for operation by conducting a comprehensive risk assessment. Before granting approval, decision-makers evaluate the potential impact on organisational operations, assets, individuals, partners and national interests.
- **Monitor**—Continuously monitor the security controls to maintain their effectiveness over time. This process includes tracking changes in the operational environment, responding to emerging threats and adjusting controls to preserve a strong security posture.

These seven steps allow organisations to create a flexible and scalable risk management program. The RMF not only fortifies systems against current and future cyber threats but also supports compliance with legal, regulatory and industry-specific requirements.

RMF Advantages:

- **Comprehensive, Risk-Based Approach**—The RMF provides a structured, repeatable method for managing cybersecurity risks across the SDLC, ensuring security from design to decommissioning
- **Consistency and Compliance**—By following its seven-step process, organisations maintain accountability, enhance regulatory compliance and improve resilience against evolving threats.

RMF Disadvantages:

- **Complex and Resource-Intensive**—The RMF's documentation, assessments and monitoring demand significant resources, posing challenges for organisations with limited expertise or budgets
- **Slower Implementation in Fast-Paced Environments**—Its structured, compliance-driven nature can delay system development, making it less adaptable for fast-paced environments like DevSecOps.

Source: NIST. (2018). *Management Framework for Information Systems and Organizations: A System Life Cycle Approach for Security and Privacy. NIST SP 800-37 Rev. 2.* https://csrc.nist.gov/pubs/sp/800/37/r2/final

Australian Frameworks

Australian Signals Directorate-Australian Cyber Security Centre Information Security Manual (ASD-ACSC ISM)

The ASD-ACSC ISM is the authoritative cybersecurity framework for Australian government entities and organisations handling sensitive information. It establishes security controls to protect critical information assets from changing cyber threats.

The ISM framework is built on seven interconnected components:

- **Risk Management Framework (RMF)**—The ISM provides a structured approach to identifying, assessing and mitigating information security risks. It ensures that risk management strategies align with an organisation's specific operational needs and security requirements.

- **Security Controls**—The ISM covers governance, personnel security, physical security and ICT security and outlines robust controls to safeguard critical data and infrastructure from threats.

- **Compliance and Assurance**—The manual ensures alignment with Australian government policies, standards and legal requirements, reinforcing the consistent protection of sensitive information.

- **Incident Response and Management**—The ISM provides clear guidelines for developing and maintaining comprehensive response plans, helping organisations detect, contain and recover from cybersecurity incidents efficiently.

- **Governance and Accountability**—It emphasises the need for strong oversight mechanisms, ensuring that security responsibilities are well-defined and effectively managed at all levels.

- **Security Awareness Program**—The ISM advocates for ongoing security training and awareness initiatives to cultivate a proactive security culture among employees.

- **Continuous Improvement Process**—The ISM helps organisations strengthen their cybersecurity posture over time by promoting regular assessment and refinement of security controls.

This structured approach ensures that organisations remain resilient against evolving threats while maintaining compliance with national security standards.

ASD-ACSC ISM Advantages:

- **Government-Endorsed and Aligned with Australian Standards**—Developed by the ASD-ACSC, the ISM provides authoritative security guidance for government and critical infrastructure

- **Comprehensive and Regularly Updated**—This manual covers governance, personnel, physical and technical security, and it is frequently updated to reflect emerging threats and best practices.

ASD-ACSC ISM Disadvantages:

- **Highly Prescriptive for Non-Government Organisations**—Designed for Australian government agencies, its strict requirements may be too rigid for private companies.

- **Resource-Intensive Compliance**—Compliance requires extensive security controls, audits and continuous monitoring, making it challenging for smaller organisations.

Source: Australian Cyber Security Centre (ACSC). *ACSC Information Security Manual (ISM)*. https://www.cyber.gov.au/resources-business-and-government/essential-cyber-security/ism

ASD-ACSC Essential Eight

The ASD's Essential Eight framework comprises fundamental cybersecurity strategies that protect Australian organisations against common cyber threats. Based on real-world threat intelligence, these eight measures provide scalable defence mechanisms for organisations of all sizes. The framework includes:

1. **Application Control**—Restricting system execution to approved applications only, preventing malicious software deployment.

2. **Application Patching**—Rapidly deploying security updates to applications to protect against known vulnerabilities.

3. **Restrict Microsoft Office Macros**—Blocking unauthorised macros while permitting vetted, digitally signed macros from trusted sources.

4. **User Application Hardening**—Optimising security settings in web browsers, PDF viewers and common applications to defend against web-based threats.

5. **Restrict Administrative Privileges**—Limiting elevated system access to essential personnel and processes, reducing the impact of potential account compromises.

6. **Patch Operating Systems**—Maintaining current operating system security patches to strengthen system integrity.

7. **Multi-Factor Authentication (MFA)**—Requiring multiple verification methods beyond passwords when accessing sensitive systems.

8. **Regular Backups**—Performing daily backups of critical information stored separately from central systems to enable recovery from ransomware or corruption events.

Essential Eight Advantages:

- **Practical and Effective Cybersecurity Baseline**—The Essential Eight provides prioritised security controls that proactively mitigate common cyber threats.
- **Clear Maturity Model**—The structured maturity model (Levels 0–3) allows organisations to progressively enhance their security posture based on risk appetite and resources.

Essential Eight Disadvantages:

- **Limited Scope Compared to Broader Frameworks**—It focuses on mitigating common threats but lacks incident response, governance and supply chain security coverage.
- **Difficult Implementation for Some Organisations**: Certain controls (application control, privilege restrictions) can be disruptive and complex, especially for SMBs with limited IT resources.

Source: Australian Cyber Security Centre (ACSC). *Essential Eight.* https://www.cyber.gov.au/resources-business-and-government/essential-cyber-security/essential-eight

Other Essential Frameworks

Beyond well-known cybersecurity frameworks like NIST and ISO/IEC 27001, several others play a key role in IT governance and security. These frameworks help organisations manage risk, align IT with business goals and strengthen cyber defences.

Control Objectives for Information and Related Technologies (COBIT), developed by the Information Systems Audit and Control Association (ISACA), focuses on IT governance by ensuring that technology supports business objectives. The CIS Controls, on the other hand, provide a clear, prioritised set of security measures to defend against common cyber threats. While each framework has a specific purpose, together they offer a comprehensive approach to security, compliance and IT management.

Understanding these frameworks equips professionals with the knowledge to choose and implement the best security strategies for their organisation.

COBIT

COBIT, developed by ISACA, is a framework for governing and managing enterprise IT. Designed to bridge the gap between business and IT, COBIT ensures that technology investments align with organisational goals, optimise performance and mitigate risks. Initially created to support IT governance, COBIT has evolved into a widely adopted framework for enhancing compliance, improving decision-making and maximising value from IT resources.

COBIT's adaptability allows organisations to tailor its principles to various industries and regulatory environments, from financial services managing stringent compliance requirements to technology firms optimising IT service delivery. It complements other industry frameworks, such as ITIL and NIST, enabling organisations to establish a cohesive governance strategy that integrates risk management, cybersecurity and operational efficiency.

COBIT is structured around key components that support effective IT governance:

- **Governance and Management Framework** establishes a structured approach to IT governance by defining roles, responsibilities and decision-making processes that align IT with business objectives.

- **Process Model** outlines 40 governance and management processes, providing a standardised method for managing IT operations, risk and performance. These processes ensure consistency in IT practices across the organisation.

- **Business-IT Alignment** aligns IT strategy with business priorities, ensuring that technology investments drive measurable business value while effectively managing risks and compliance obligations.

- **Performance Measurement** implements metrics and maturity models to assess IT performance, enabling organisations to identify gaps, enhance decision-making and continuously improve governance practices.

Following these principles allows organisations to develop a scalable, structured IT governance program that enhances operational efficiency, strengthens risk management and ensures compliance with industry regulations.

COBIT Advantages:

- **Comprehensive Governance Framework**—It provides a structured, standardised approach to IT governance and management, ensuring alignment with business objectives

- **Regulatory Compliance and Risk Management**—It supports organisations in effectively meeting regulatory requirements and managing IT risks.

COBIT Disadvantages:

- **Complex Implementation**—The extensive process model and governance requirements may be resource-intensive for organisations with limited IT maturity.
- **Adaptation Challenges:** Customising COBIT to fit specific organisational needs can be complex, requiring expertise and ongoing refinement.

COBIT empowers organisations to establish a resilient IT governance structure, ensuring that technology investments support long-term business success while maintaining compliance and risk oversight.

Source: ISACA. *COBIT 5: A Business Framework for the Governance and Management of Enterprise IT.* https://www.isaca.org/resources/cobit

CIS Controls

The Center for Internet Security (CIS) Controls, also known as CIS Critical Security Controls, provide a structured approach to securing IT systems and data against cyber threats. Designed as a prioritised set of best practices, these controls help organisations strengthen their security posture by mitigating the most common and impactful attack vectors. Continuously refined based on real-world threat intelligence, CIS Controls offer a dynamic defence strategy adaptable to various operational environments.

CIS Controls are structured into three Implementation Groups (IGs), allowing organisations to adopt security measures progressively based on their size, resources and risk exposure. This tiered approach ensures that organisations can implement effective cybersecurity defences aligned with their operational needs regardless of maturity level.

The CIS Controls framework follows three key principles:

- **Prioritisation of Security Measures**—The controls are ranked based on their potential impact and the likelihood of exploitation. This risk-based prioritisation enables organisations to allocate resources efficiently, focusing first on the most critical security actions.

- **Defence-in-Depth Approach**—CIS Controls collectively create a layered security strategy that mitigates common attack techniques. By addressing fundamental cybersecurity risks, organisations can enhance their resilience against threats like ransomware, phishing and unauthorised access.

- **Continuous Improvement**—The controls are regularly updated to reflect emerging threats and evolving attack patterns. This iterative process ensures that organisations remain equipped to handle modern cybersecurity challenges.

Implementing CIS Controls enables organisations to establish a clear, actionable roadmap for cybersecurity improvement. Their structured yet flexible nature makes them particularly beneficial for organisations seeking to enhance security efficiently while maintaining scalability.

CIS Controls Advantages:

- **Practical and Prioritised Security Guidance**—CIS Controls provide a structured, actionable framework that helps organisations implement effective security measures without unnecessary complexity.

- **Scalability Across Different Organisations**—The tiered IG model ensures that small and large enterprises can tailor security controls to their needs.

CIS Controls Disadvantages:

- **Resource Considerations for Full Implementation**—While the controls offer prioritisation, comprehensive adoption may require significant time, expertise and financial investment, which can be challenging for resource-constrained organisations.
- **Ongoing Maintenance and Updates**—Given the evolving nature of cyber threats, organisations must commit to regularly updating and refining their security practices to maintain effectiveness.

Source: Center for Internet Security (CIS). *CIS Critical Security Controls*. Https://www.cisecurity.org/controls

Cybersecurity Standards

Cybersecurity standards serve as the technical mainstay of information and system protection strategies. These guidelines provide organisations with detailed, actionable measures to implement robust security practices. As we turn the page to cybersecurity standards, we'll unpack their pivotal role in shaping strategies and defence mechanisms and determine how they complement cybersecurity frameworks.

Role of Cybersecurity Standards

Cybersecurity standards provide structured, actionable guidelines organisations can implement to protect their systems, reduce risks and comply with regulations. While broader security frameworks outline high-level principles and best practices, standards specify concrete controls and implementation measures that ensure a consistent, repeatable approach to security.

Cybersecurity standards typically fall into three main categories:

Official Standards—Developed by recognised bodies such as the International Organization for Standardization (ISO), Standards Australia (AS), the National Institute of Standards and Technology (NIST) and the Institute of Electrical and Electronics Engineers (IEEE). These standards establish universally accepted security benchmarks, including ISO/IEC 27001 for information security management and ISO 31000 for risk management.

De facto standards emerge when widespread industry adoption transforms a framework into an expected security practice. The *Payment Card Industry Data Security Standard* (PCI DSS) exemplifies this evolution. Originally introduced as a guideline, it has become a widely accepted requirement for payment security.

Regulatory requirements, though not formal standards, function similarly by requiring compliance with specific cybersecurity controls. *The Australian Privacy Act 1988* and the EU's *General Data Protection Regulation* (GDPR) impose security obligations that influence industry implementations.

Among the most influential standards, ISO 31000 and ISO/IEC 27001 serve as foundational standards. ISO 31000 provides a structured methodology for identifying, assessing and mitigating risks, helping you integrate risk awareness into decision-making. ISO/IEC 27001 establishes a systematic approach for implementing and maintaining an Information Security Management System (ISMS) that protects the confidentiality, integrity and availability of critical information assets.

When you adopt recognised cybersecurity standards, you strengthen your security posture, enhance risk management capabilities and build trust with stakeholders. These standards consolidate collective industry expertise and offer proven methodologies to address evolving cyber threats. Understanding and implementing them is essential for effectively navigating today's complex security landscape.

Now let's take a closer look at these so you can differentiate between them and understand when each should be applied.

ISO 31000: Risk Management

The ISO 31000 establishes principles and guidelines for organisational risk management. This internationally recognised standard helps organisations systematically integrate risk assessment into their decision-making processes. The standard comprises five interconnected components:

1. **Context Establishment**—First, define your organisation's external and internal operating parameters. This initial step determines the scope of risk management activities and establishes specific risk assessment criteria.

2. **Risk Identification**—Teams employ structured techniques, including brainstorming sessions, scenario planning and historical data analysis, to identify potential risks to organisational objectives.

3. **Risk Assessment**—This two-part process involves analysing risks based on their likelihood and potential impact, followed by evaluating their significance to prioritise management responses.

4. **Risk Treatment**—Organisations develop and implement targeted strategies to address identified risks through four main approaches: avoidance, acceptance, transfer or reduction through control measures.

5. **Monitoring and Review**—Risk managers continuously evaluate the changing risk landscape and the effectiveness of existing risk management processes, ensuring the framework maintains its relevance and efficacy.

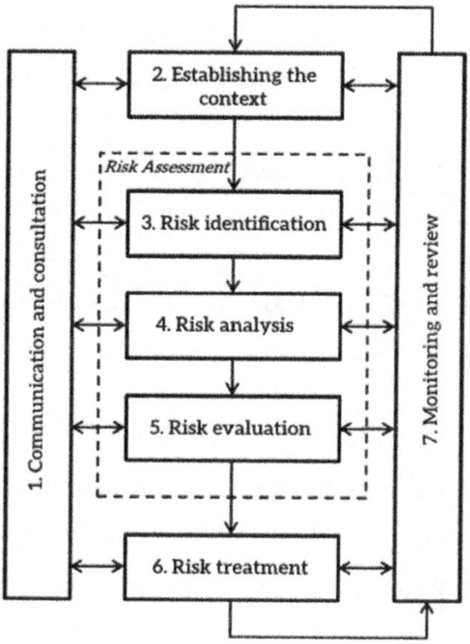

Risk Management Cycle – ISO 31000

ISO 31000 Advantages:

- **Broad Applicability Across Industries**—ISO 31000 applies to organisations of all sizes, from governments to private enterprises. Its principles-based approach is adaptable across different sectors and risk environments.

- **Strategic and Business-Oriented**—Unlike technical frameworks, ISO 31000 integrates risk management with business strategy, helping organisations align risk decisions with objectives and governance.

ISO 31000 Disadvantages:

- **Lacks Specific Implementation Guidance**—ISO 31000 outlines principles but lacks prescriptive steps, technical controls or compliance checklists, requiring organisations to interpret and adapt it.

- **Not Cybersecurity-Focused**—While ISO 31000 covers all risks, it does not specialise in cybersecurity. Organisations may need ISO 27001, NIST CSF or the Essential Eight for cyber-specific guidance.

Source: International Organization for Standardization. *ISO 31000 Risk Management*. https://www.iso.org/standard/65694.html

ISO/IEC 27001: Information Security Management Systems (ISMS)

The ISO/IEC 27001 standard establishes a systematic approach to managing and protecting organisational information through an ISMS. This internationally recognised standard helps organisations safeguard sensitive data while ensuring regulatory compliance. The standard comprises seven core components:

1. **Scope Definition**—First, establish clear ISMS boundaries by considering your organisation's security requirements, objectives and operational processes. This foundation determines which information assets are protected by the system.

2. **Risk Assessment**—Teams conduct systematic evaluations to identify and analyse potential threats to information assets, examining their impact on three critical aspects: confidentiality, integrity and availability of protected data.

3. **Risk Treatment**—Organisations develop and implement specific strategies to address identified risks through four main approaches: mitigation through controls, avoidance of high-risk activities, transfer to third parties or acceptance of residual risks.

4. **Management Commitment**—Leadership must demonstrate active involvement by establishing clear roles, responsibilities and

authorities for information security management, ensuring necessary resources are available for ISMS implementation.

5. **Continuous Improvement**—Security teams regularly monitor, measure and enhance ISMS effectiveness through structured audits and management reviews, adapting to emerging threats and organisational changes.

6. **Documentation Control**—Organisations maintain comprehensive records of ISMS policies, procedures and controls, ensuring all security measures are appropriately documented and accessible to relevant stakeholders.

7. **Legal Compliance**—Teams ensure adherence to relevant legislation, regulations and contractual obligations affecting information security, including specific privacy requirements and industry standards.

ISO/IEC 27001 Advantages:

- **Comprehensive Information Security Management**—ISO/IEC 27001 provides a structured approach to implementing an ISMS, covering risk assessment, security controls and governance.

- **Internationally Recognised and Certifiable**—Certification enhances credibility, regulatory compliance and market competitiveness, demonstrating commitment to security best practices.

ISO/IEC 27001 Disadvantages:

- **Resource-Intensive Implementation**—Certification requires significant investment in time, financial resources and ongoing audits, which can pose challenges for small businesses.

- **Focuses on Governance, Not Technical Controls**—While it establishes a strong governance framework, specific technical controls must be supplemented with NIST SP 800-53 or CIS Controls.

Source: International Organization for Standardization. *ISO/IEC 27001 Information Security Management.* https://www.iso.org/isoiec-27001-information-security.html

Key Standards

Beyond the foundational standards, several other important cybersecurity standards can strengthen your security strategy. While not exhaustive, this selection includes the most significant standards you'll encounter in professional practice:

Official Standards

- **ISO/IEC 27002**—International code of practice that provides detailed guidelines for implementing the information security controls outlined in ISO/IEC 27001
- **FIPS 140-2**—U.S. government standard that specifies security requirements for cryptographic modules protecting sensitive information
- **Common Criteria (ISO/IEC 15408)** —International framework for evaluating security properties of IT products and systems through a standardised certification process
- **NIST SP 800-53**—Comprehensive catalogue of security and privacy controls for federal information systems, widely adopted in non-government settings
- **ETSI TS 102 042**—European standard defining policy requirements for certification authorities issuing digital certificates.

De Facto Standards

- **PCI DSS**—Essential standard for any organisation handling credit card transactions, outlining requirements to secure payment data
- **OWASP ASVS**—Application Security Verification Standard providing a basis for testing web application security controls and guiding secure development
- **SOC 2**—Auditing framework enabling service organisations to demonstrate effective controls for security, availability, processing integrity, confidentiality and privacy.

Regulatory Requirements

- **The GDPR**—European Union regulation with global reach that mandates strict data protection measures and user privacy rights
- ***The Australian Privacy Act 1988***—National legislation establishing how organisations must handle personal information and report data breaches.

As you develop your cybersecurity expertise, understanding when and how to apply these standards will be critical to your professional development.

Selecting and Implementing Frameworks and Standards

Selecting and implementing appropriate cybersecurity frameworks and standards is essential for any organisation seeking to enhance its security posture. This section outlines key considerations when choosing cybersecurity frameworks, specifically focusing on Australian requirements. It examines how to tailor frameworks to an organisation's unique needs and regulatory landscape and provides a step-by-step implementation guide. While this section emphasises Australian contexts, similar approaches apply to frameworks and standards used in other countries.

Factors to Consider When Selecting a Cybersecurity Framework

Industry and Regulatory Requirements

Your framework selection must be guided by compliance with industry-specific and national regulations. Australian organisations adhere to regulations such as the Australian Privacy Principles (APPs), *The Security of Critical Infrastructure (SOCI) Act 2018* and *The Cyber Security Act 2024* (CSA), which, among other things, mandate specific data protection measures. Selecting a framework that aligns with these regulatory requirements ensures legal compliance and minimises penalty risks.

For example, the APPs require organisations to implement comprehensive security measures to protect personal information. A standard like ISO/IEC 27001, which focuses on information security management systems (ISMSs), helps meet these requirements by providing a structured approach to security implementation.

Alignment with ASD-ACSC Guidance

Incorporating ASD-ACSC guidance effectively addresses common Australian cyber threats. The ASD-ACSC's strategies, including the Essential Eight Maturity Framework and the Ransomware Emergency Response Guide, offer practical advice for mitigating risks and strengthening cybersecurity defences. Aligning your chosen framework with ASD-ACSC guidance ensures you adopt approaches tailored to the Australian threat environment.

The Essential Eight includes critical strategies such as application control, patching applications and MFA that protect against prevalent cyber threats. Integrating these strategies into broader frameworks like the CSF enhances overall security resilience.

Organisational Size and Resources

The size of your organisation and the resources available play a crucial role in selecting the right framework. Large organisations with robust IT infrastructure and dedicated cybersecurity teams can benefit from standards such as ISO/IEC 27001 and frameworks such as the ISM or the NIST CSF, which offer detailed guidelines and flexible customisation. In contrast, small and medium-sized enterprises (SMEs) with limited resources may find greater success with frameworks like the Essential Eight or CIS Controls, which focus on key security measures that can be implemented gradually.

Risk Profile

In Chapter 4, we examined risk management and its essential role in assessing an organisation's security posture. From that examination and our exploration of frameworks, you may have concluded that selecting an appropriate framework requires understanding an organisation's specific risks and determining how to allocate limited resources effectively. If you reached this conclusion, congratulations; you are correct.

Understanding an organisation's risk profile forms the foundation for selecting a framework that effectively addresses specific cybersecurity vulnerabilities. When evaluating potential frameworks, consider several critical factors: the fundamental nature of the organisation's operations, the sensitivity level of data being processed and stored, and the current threat landscape relevant to your industry sector. These elements collectively shape which security controls deserve priority, and which framework will provide the most effective protection given your unique circumstances.

The relationship between risk assessment and framework selection creates a strategic feedback loop. Your risk analysis identifies your most significant vulnerabilities, while your chosen framework provides structured approaches to mitigate those weaknesses. This interconnection ensures that security resources target the most consequential threats rather than being distributed across less relevant concerns.

Tailoring the Framework to Organisational Needs

After selecting a framework, you must adapt it to your organisation's requirements and regulatory environment. This customisation involves mapping the framework components to existing processes, identifying security gaps and prioritising actions based on risk assessments and resource availability.

For example, an Australian healthcare organisation in New South Wales might prioritise compliance with *The Health Records and Information Privacy Act* by implementing ISO/IEC 27001 controls focused on data confidentiality and integrity. Similarly, financial institutions might integrate PCI DSS standards within their broader cybersecurity strategy to ensure secure payment card data handling.

Steps for Successful Framework Implementation

Conduct Risk Assessments

Begin implementation by conducting risk assessments. Identify critical assets, evaluate potential threats and vulnerabilities, and assess the potential impact of security incidents. Identifying assets, understanding potential threats and vulnerabilities, and assessing potential impacts are the starting points for establishing an effective cybersecurity program. Risk assessments provide a baseline for developing targeted cybersecurity strategies and prioritising mitigation efforts.

Develop a Cybersecurity Strategy

Develop a cybersecurity strategy aligned with your chosen framework using risk assessment findings. Your strategy should outline objectives, scope and approaches for implementing security measures across the organisation. Effective strategies incorporate specific components, including:

- Cybersecurity implementation plan
- Incident response protocols
- Business continuity procedures
- Staff training and awareness initiatives.

Additionally, clearly define roles and responsibilities for all internal and external personnel and stakeholders involved in the cybersecurity process.

Identify Gaps and Prioritise Actions

Next, identify gaps between your current security posture and the requirements of your chosen framework. Evaluate existing controls, processes and technologies to determine areas needing improvement. Prioritise actions based on risk severity and resource availability to address critical vulnerabilities first.

Implement Controls, Train Staff and Conduct Regular Monitoring

Implementing necessary security controls represents the most critical phase of framework adoption. To maintain a coordinated security posture, deploy appropriate technical solutions, update plans, policies and procedures, and integrate all security measures into organisational operations.

Staff training on cybersecurity practices and their specific security responsibilities is a core aspect of successful implementation. No matter how sophisticated your strategies, plans and procedures are, your organisation risks losing all implementation efforts without properly trained staff who understand their security responsibilities.

Regularly monitoring and evaluating implemented controls helps detect and respond quickly to security incidents. As part of a continuous improvement process, conduct periodic audits and reviews to ensure your cybersecurity framework remains effective and adaptable to emerging threats.

Summary

Cybersecurity frameworks and standards form the essential foundation of effective risk management strategies, providing organisations with structured approaches to identifying, assessing and mitigating security threats. This chapter explored the distinctions between frameworks and standards, examined major implementation options and provided guidance for selecting these tools based on organisational needs and regulatory requirements.

Frameworks and standards serve complementary purposes in cybersecurity. Frameworks like the NIST CSF provide flexible, high-level guidelines that help organisations systematically manage security. In contrast, standards such as PCI DSS and HIPAA define specific technical requirements and measurable criteria for implementing security controls, ensuring consistency and compliance.

The NIST CSF is built around six key functions: Govern, Identify, Protect, Detect, Respond and Recover, offering flexibility for organisations of any size. The NIST RMF complements the CSF with a seven-step process integrating security throughout the System Development Lifecycle (SDLC). While both frameworks excel at different aspects of security management, they share limitations, including resource intensity and implementation complexity.

Australian organisations benefit from nationally specific frameworks. The ASD-ACSC ISM establishes comprehensive controls across seven interconnected components, from risk management to continuous improvement. The Essential Eight framework offers a more focused approach, defining fundamental security strategies, including Application Control, Patching and Multi-Factor Authentication.

Other specialised frameworks serve specific purposes in cybersecurity governance. COBIT bridges business and IT objectives, while the CIS Controls provide prioritised security measures against common threats. These complement international standards like ISO 31000 and ISO/IEC 27001, which establish structured methodologies for risk management and information security systems.

Selecting an appropriate framework requires consideration of multiple factors. Industry and regulatory requirements should guide selection, particularly adherence to APPs and national regulations. Alignment with ASD-ACSC guidance addresses common threats in the Australian context. Organisational size and available resources influence which framework is most appropriate; larger organisations may benefit from comprehensive approaches like ISO/IEC 27001, while smaller entities might find success with the Essential Eight. An organisation's risk profile determines which framework will most effectively address its vulnerabilities.

A good cybersecurity strategy begins with a structured risk assessment to identify key assets, evaluate threats and measure their impact. Organisations can use these insights to create security plans that define goals, implementation steps, incident response procedures and training initiatives. Closing security gaps between current protections and industry standards allows organisations to address urgent vulnerabilities first.

Implementation involves setting up technical protections, updating policies, training employees and continuously monitoring for threats. This chapter emphasised that cybersecurity frameworks are not one-time solutions but ongoing processes that need regular updates and adjustments.

Revision Questions

1. What is the fundamental difference between cybersecurity frameworks and standards, and what roles do they play in risk management?

2. What are the six essential functions of the NIST Cybersecurity Framework (CSF), and how do they work together to create a resilient security strategy?

3. What are the seven steps of the NIST Risk Management Framework (RMF), and how does it differ from the CSF?

4. What are the main components of the Australian ASD-ACSC Information Security Manual (ISM), and what advantages does it offer Australian organisations?

5. What are the eight key strategies of the Australian Essential Eight framework, and why is it particularly beneficial for organisations with limited resources?

6. How does COBIT differ from other cybersecurity frameworks, and what are its key components for effective IT governance?

7. What are the core principles of the CIS Controls framework, and how does its Implementation Group (IG) structure benefit organisations of varying sizes?

8. What are the five interconnected components of ISO 31000, and how does it differ from ISO/IEC 27001?

9. What key factors should organisations consider when selecting a cybersecurity framework, particularly in the Australian context?

10. What are the essential steps for successfully implementing a cybersecurity framework, and why is each step important?

Cybersecurity Hacks

Hack 1: Essential Eight Quick Start Implementation

Instead of implementing all cybersecurity controls simultaneously, adopt the Australian Essential Eight framework as your foundation. Create a tracking spreadsheet listing the eight fundamental controls: Application Control, Application Patching, Restrict Microsoft Office Macros, User Application Hardening, Restrict Administrative Privileges, Patch Operating Systems, MFA and Regular Backups.

Begin with the three most critical controls: Multi-Factor Authentication, Regular Backups and Patch Operating Systems. These deliver immediate security benefits while being relatively straightforward to implement. For each control, establish three maturity levels: "Not Implemented", "Partially Implemented" and "Fully Implemented".

Monitor your progress weekly, acknowledging achievements as controls advance from "Not Implemented" to higher levels. This systematic approach prevents being overwhelmed and provides swift security improvements without demanding extensive resources or technical expertise. As your confidence increases, progressively implement the remaining five controls, prioritising those that address your organisation's most significant vulnerabilities first.

Hack 2: Framework Mapping for Clarity

Create a visual mapping document that connects similar controls across cybersecurity frameworks to clarify their relationships and prevent duplication of effort.

Begin with a spreadsheet listing NIST CSF's core functions (Identify, Protect, Detect, Respond, Recover) in the first column. For each function, list the key controls recommended by the framework. Then add columns for additional frameworks you're working with (such as ISO/IEC 27001, Essential Eight or CIS Controls) and map their corresponding controls.

For example, under the "Protect" function, you might map NIST CSF's access control requirements to ISO/IEC 27001's A.9 Access Control section and the Essential Eight's "Restrict Administrative Privileges" control. This visual mapping enables you to:

- Understand conceptual connections between frameworks
- Identify the most frequently recommended controls across frameworks
- Recognise which controls deliver the greatest cross-framework compliance benefits.

This approach transforms abstract framework knowledge into practical understanding.

Chapter 6: Confidentiality

– Stealth Mode –

Objectives

By the end of this chapter, you will be able to:

- Explain the fundamental role of confidentiality within the CIA Triad
- Identify and analyse various threats to data confidentiality
- Understand the legal and regulatory frameworks governing data confidentiality
- Apply effective techniques for securing communication channels and protecting sensitive information
- Define confidentiality programs and Incident Response Plans (IRPs).

Chapter 1 introduced the CIA Triad—Confidentiality, Integrity and Availability—as the foundation of information security. These three principles shape security policies, strategies, architectures and controls worldwide. Each element is so central to cybersecurity that I have dedicated a separate chapter to explore them in depth. Over the next three chapters, we will examine each component, beginning with confidentiality, to provide you with a strong foundation in cybersecurity.

Confidentiality is central to cybersecurity, ensuring sensitive information remains safe from unauthorised access and disclosure. A single breach can compromise personal privacy, expose proprietary business data and erode trust. In this chapter, we will build on the concepts introduced in Chapter 1 and examine how confidentiality protects personal data, intellectual property and proprietary business information.

You'll learn to identify and counter critical threats, from advanced cyber-attacks and unauthorised access to deceptive social engineering tactics. Our discussion will also cover practical approaches to securing information.

By mastering these fundamental principles, cybersecurity practitioners can defend sensitive information and foster a culture of trust and security that upholds information protection.

The Data Privacy Imperative

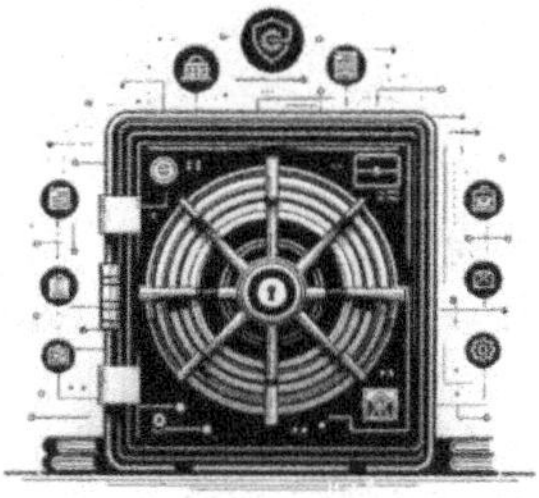

Privacy protection forms the foundation of trust, individual autonomy and the integrity of our digital society. When this protection fails, the consequences for individuals can be severe and long-lasting. Personal identity theft, financial fraud, emotional distress and the violation of fundamental privacy rights often leave victims struggling for years to recover their security and peace of mind.

Major incidents demonstrate these devastating impacts at scale. The 2017 Equifax breach exposed the personal information of 147 million individuals, while Yahoo breaches affected over three billion user accounts.[37] In Australia, the recent Medibank Private and Optus breaches further illustrate these widespread risks. For affected individuals, such breaches typically necessitate years of credit report monitoring, replacement of compromised identification documents and management of the psychological impact of personal information exposure.

These cases have prompted more stringent regulatory frameworks worldwide. The GDPR in Europe, the Australian Privacy Principles (APPs) and *The Health Insurance Portability and Accountability Act* (HIPAA) in the U.S. now enforce serious penalties for privacy failures. Organisations face fines of up to €20 million or 4% of annual global turnover under the GDPR,[38] while Equifax agreed to a $700 million settlement following their breach.

Beyond direct financial penalties, organisations suffer damaged reputations, lost customer trust and potential credit-rating downgrades that impair their ability to raise capital. As your digital footprint expands, protecting personal data becomes increasingly essential, not just for corporate compliance but for preserving individual dignity and freedom in our interconnected world. Understanding these impacts will be critical for your future professional decision-making regarding data management.

Importance of Data Confidentiality

In an era of frequent and highly publicised data breaches, from the exposure of millions of social media profiles to major financial institution hacks, understanding and protecting data confidentiality is more important than ever. For individuals and organisations, safeguarding sensitive information is vital for preventing identity theft, financial loss, reputational damage and regulatory penalties.

Why Confidentiality is Vital for Individuals and Organisations

For Individuals: Safeguarding Privacy and Security

Imagine waking up to find your bank account drained or your medical records leaked online. These scenarios, once rare, are becoming increasingly common as cyber threats evolve. Each time you enter credit card details on a website, send an email or log into social media, you create a digital footprint that, if inadequately secured, becomes vulnerable to exploitation. Various aspects of your life may be exploited including:

Financial data—A single exposed credit card number can trigger fraudulent transactions or identity theft. Cybercriminals use stolen financial data to make unauthorised purchases, apply for loans or sell your information on the dark web.

Medical Information—Leaked medical histories can affect employment opportunities and insurance coverage. In one widely reported case, a health database breach exposed the mental health records of thousands of patients, leading to workplace discrimination and social stigma.

Personal Communications—Private conversations through emails, direct messages or video calls can be intercepted and misused. Recent high-profile hacking incidents demonstrate how leaked messages can damage reputations, end careers and even influence political outcomes.

Understanding these risks enables you to take proactive steps, including using strong passwords, enabling two-factor authentication and carefully managing the information you share online.

For Organisations: Protecting Business Interests and Reputation

For companies, confidentiality breaches can result in financial losses, legal consequences and lasting reputational damage. Consider a recent case in Australia including Optus, Medibank, Medisecure and Ticketmaster. These breaches exposed customer data and financial reports, triggering lawsuits, regulatory fines and a significant decline in consumer trust.

Key areas where confidentiality is vital for businesses include:

Intellectual Property and Trade Secrets—Companies invest years in research and development. When confidential product designs, software code or marketing strategies leak, competitors gain unfair advantages, undermining innovation and profitability.

Customer Trust and Data Protection—Consumers expect responsible data handling. Security lapses that expose customer details, such as credit card numbers or purchase histories, can lead to severe penalties under data protection legislation like *The Australian Privacy Act*.

Employee Privacy—Businesses must safeguard employee records, including payroll data and performance reviews. Failures in this area can result in identity theft, legal disputes and diminished staff morale.

Financial Security and Compliance—Publicly traded companies must protect their financial reports to prevent insider trading and market manipulation. Mishandled financial data can trigger regulatory action, erode investor confidence and create business instability.

Organisations can mitigate threats and cultivate long-term trust by integrating confidentiality practices into daily operations, such as encrypting sensitive files, implementing access controls and conducting cybersecurity training.

A Shared Responsibility

Data confidentiality extends beyond technical concerns; it represents a shared responsibility. Whether you manage your personal digital footprint or handle sensitive information for a business, the risks of exposure remain real and far-reaching. The increasing frequency of cyberattacks and data leaks demonstrates the necessity for preventative measures. By staying informed and implementing security best practices, you can protect yourself and your organisation from the costly and often irreversible consequences of a breach.

Impact of Confidentiality on the CIA Triad

The CIA Triad is the foundational model in information security. While each component is essential, confidentiality often serves as the first barrier, protecting sensitive information and supporting integrity and availability.

Confidentiality and Integrity

Confidentiality and integrity are deeply interlinked, as measures to safeguard data often ensure its accuracy and reliability. For example, when a bank uses end-to-end encryption for online transactions, it not only keeps the data confidential but also preserves its integrity, ensuring that transaction details remain unaltered from sender to recipient.

The importance of this relationship was demonstrated by the exploitation of vulnerabilities in Ivanti Connect Secure and Ivanti Policy Secure devices. These vulnerabilities, actively exploited by threat actors in late 2023 and into 2024, allowed for unauthorised remote access, leading to data breaches and potential manipulation of affected systems. This incident is an example of how a failure in maintaining confidentiality, through unpatched vulnerabilities, directly leads to a compromise of data integrity. Attackers gaining access allowed them to potentially alter configurations, inject malicious code and exfiltrate sensitive data, thus violating confidentiality and integrity.

Moreover, cryptographic protocols, such as Public Key Infrastructure (PKI), exemplify how confidentiality safeguards verification processes. By protecting cryptographic keys from unauthorised access, these protocols maintain the integrity of digital signatures and authentication mechanisms. If the confidentiality of private keys were compromised, the integrity of digital signatures would be invalidated, rendering authentication mechanisms unreliable.

Confidentiality and Availability

While confidentiality and availability may seem contradictory, they are complementary when balanced correctly. Effective confidentiality measures ensure data is accessible to authorised users while remaining protected from malicious actors.

During the 2022 Colonial Pipeline ransomware attack, hackers exploited weak confidentiality controls to access and encrypt critical systems, resulting in fuel shortages across the U.S. Southeast. This incident demonstrates how a breach of confidentiality can directly compromise availability and disrupt essential services.

On the other hand, technologies such as Role-Based Access Control (RBAC) and Multi-Factor Authentication (MFA) exemplify how confidentiality and availability can work in harmony. These systems ensure that only authorised personnel can access sensitive data, which maintains system security while preventing unauthorised disruptions.

Furthermore, maintaining system stability is a direct outcome of strong confidentiality controls. In 2020, Zoom faced "Zoom-bombing" incidents where meetings were disrupted due to inadequate privacy settings. By implementing stronger confidentiality controls like passcodes and waiting rooms, Zoom protected users' privacy and ensured stable and secure meeting environments.

Tactical Briefing: Password Spray Attack

A password spray attack is a common method attackers use to breach confidentiality and gain unauthorised access. Unlike traditional brute-force attacks that target a single account with numerous password attempts, password spraying involves trying a small number of common passwords against many different accounts. This method helps avoid account lockouts and detection mechanisms. Understanding the approach used in a password spray attack can help you prepare and implement defensive measures to protect against such attacks. So, what is involved in this type of attack?

1. Reconnaissance

   - Objective: Use social media and company websites to gather information about the target system and user accounts. Scanning tools like Nmap can be used to scan the target network for active services and endpoints.

2. Username Enumeration

   - Objective: Compile a list of valid usernames.

3. Password List Preparation

   - Objective: Create a list of commonly used passwords.

4. Executing the Attack

   - Objective: Attempt to log in using the compiled usernames and password list.

- Process: To avoid detection, try a small number of common passwords ("Password123", "Welcome1") against many usernames.

5. Post-Exploitation

- Objective: Maintain access, expand control and install backdoors or create new accounts to ensure continued access.

6. Covering Tracks

- Objective: Remove attack traces to avoid detection by deleting or altering logs.

Note: The password spray attack technique is for authorised penetration testing only and should not be attempted without explicit permission, as unauthorised use is likely illegal.

Threats to Confidentiality

Understanding threats to data confidentiality is paramount for maintaining strong cybersecurity. Cyber threats arise from various sources, each employing different techniques to compromise sensitive information. Expanding on the cyber threat discussion in Chapter 2, this section explores critical threats specific to data confidentiality, including unauthorised access, data interception, social engineering, physical breaches and human error, an often-underestimated risk. This analysis will equip you with the knowledge to enhance your defences and safeguard sensitive data.

Unauthorised Access

Unauthorised access remains a dominant threat to data confidentiality. Malicious actors, including hackers, malware and insiders, exploit vulnerabilities to gain entry. Hackers deploy tactics such as brute-force attacks, credential theft and exploiting software flaws. Malware (spyware, ransomware and keyloggers) steals sensitive information. Insider threats from malicious employees or negligent contractors pose a severe risk due to their access privileges. For instance, a 2023 IBM study found that insider-driven attacks were the costliest, averaging USD 4.90 million,[39] underscoring the importance of stringent internal security protocols.

Interception

Network interception occurs when malicious actors capture data transmitted across networks by exploiting vulnerabilities in communication channels. These attackers can position themselves between senders and receivers using Man-In-The-Middle (MITM) attacks or employ packet sniffing to capture data packets traversing networks. Without proper encryption, attackers can access sensitive information, including login credentials, personal details and financial data. Security professionals use tools such as Wireshark, tcpdump, BetterCAP and NetworkMiner to detect and analyse these threats.

TEMPEST - Eavesdropping

TEMPEST stands for Telecommunications Electronics Materials Protected from Emanating Spurious Transmissions. Electronic devices emit unintended electromagnetic signals during routine operation, creating a vulnerability attackers can exploit using TEMPEST surveillance techniques. With specialised signal processing equipment, these attackers can intercept and decode emissions from computer monitors, keyboards and other peripherals, potentially capturing sensitive data from considerable distances. This surveillance method presents a serious risk to organisational security, particularly in environments dealing with classified or sensitive information, where conventional physical security measures often fail to address the threat of electromagnetic leakage.

In one notable demonstration, researchers used Software-Defined Radios (SDRs) to capture electromagnetic emissions from HDMI cables, successfully reconstructing images of the content displayed on a digital monitor.[40]

Social Engineering

Social engineering attacks exploit human psychology to manipulate individuals into compromising security through information disclosure or unsafe actions. In phishing attacks, cybercriminals send deceptive but legitimate-looking emails to trick recipients into revealing sensitive data or activating malicious links. A more targeted variant, spear phishing, uses personalised messages to attack specific individuals or organisations. Pretexting represents another technique where attackers create convincing scenarios to extract confidential information. These human-focused attacks bypass traditional technical security measures, making security awareness training essential.

According to the Verizon 2020 Data Breach Investigations Report, social engineering attacks such as Phishing and Pretexting accounted for 73% of cybersecurity incidents.[41]

Physical Security Breaches

Physical security breaches involve unauthorised access to devices or locations containing sensitive information, including theft of laptops, documents, mobile devices or intrusion into secure areas. Organisations must implement layered physical security controls, including access management systems, surveillance monitoring and secure storage protocols. Device theft presents a dual threat through hardware loss and potential data exposure, emphasising the need for encryption and remote data wiping capabilities.

According to the World Security Report, physical security incidents cost companies $1 trillion in 2022.[42]

Human Error

Human error significantly impacts data confidentiality through unintentional actions like misdirected communications, incorrect security configurations and weak password practices. Organisations must implement security awareness programs teaching data handling protocols, phishing recognition and password management. Regular training reinforces these practices while building a security-conscious culture that reduces breach risks.

Today, organisations face diverse confidentiality challenges, from technical vulnerabilities to human factors. These include unauthorised system access, network interception, human error and physical breaches requiring specific security controls. Protecting sensitive information requires an integrated security approach combining technical safeguards, physical access controls and staff training programs.

Strategic Briefing: PharMerica Data Breach

In March 2023, PharMerica Corporation, a leading provider of pharmacy services in the United States, suffered a major data breach that exposed the personal information of 5.8 million individuals. This incident underscored critical vulnerabilities in healthcare cybersecurity, revealing the far-reaching consequences of unauthorised access to sensitive medical records.

Beyond the immediate exposure of protected health information (PHI), the breach also raised concerns about compliance with HIPAA and the broader implications for patient trust and regulatory penalties.

Impact and Scope

The breach compromised a vast amount of Personally Identifiable Information (PII) and medical records, including:

- Full names and addresses
- Dates of birth and Social Security numbers
- Detailed prescription history
- Health insurance information
- Other protected health data covered under HIPAA.

With 5.8 million individuals affected, the breach heightened risks of identity theft, medical fraud and insurance exploitation. Additionally, the incident disrupted trust in PharMerica's ability to safeguard patient data, potentially leading to long-term reputational and financial repercussions.[43]

Attack Methodology

The unauthorised access occurred between March 12 and March 13, 2023, when an unidentified cybercriminal group infiltrated PharMerica's systems. The breach went undetected until March 21, indicating a critical gap between the initial compromise and its discovery.

While ongoing investigations have attributed the attack to the Money Message ransomware group, the precise entry method remains unclear. However, cybersecurity analysts suggest that phishing attacks, credential theft or unpatched system vulnerabilities may have played a role.

Response and Mitigation

Upon discovering the breach, PharMerica moved quickly to contain the damage and prevent further unauthorised access. Their priority was securing compromised systems, ensuring that any vulnerabilities exploited during the attack could no longer be used. Cybersecurity specialists were brought in to conduct a forensic investigation to determine the full extent of the breach and identify potential entry points.

At the same time, PharMerica established a structured communication plan to notify affected individuals, providing clear guidance on protective measures. To help mitigate the risks of identity theft and fraud, the company offered credit monitoring services and encouraged individuals to activate fraud alerts.

Beyond direct customer response, PharMerica also collaborated with law enforcement agencies and regulatory bodies to ensure compliance with reporting requirements. The company remained in close contact with these entities, assisting in investigations and demonstrating a commitment to transparency.

While these actions mitigated immediate risks, the breach demonstrates areas where stronger preventive measures could have reduced exposure.

Lessons Learned

The PharMerica breach offers critical insights for healthcare organisations and entities handling sensitive personal data. Key takeaways include:

- Proactive Threat Monitoring—Implement advanced threat detection systems to identify intrusions earlier.

- Robust Data Encryption—Strengthen encryption protocols for stored and transmitted sensitive information.
- Regular Security Audits—Conduct routine vulnerability assessments to identify weaknesses.
- Incident Response Optimisation—Improve detection and response timelines with streamlined protocols.
- Access Control Reinforcement—Enforce stricter authentication measures to prevent unauthorised access.

Additionally, healthcare organisations must prioritise staff training to recognise cyber threats and develop comprehensive disaster recovery plans that ensure operational resilience in the face of cyberattacks.

Confidentiality in Communication

Without adequate safeguards, sensitive information transmitted over networks remains exposed to interception, unauthorised access and other cyber threats. This section examines how to secure communication channels and outlines essential protective measures to keep data confidential, such as encryption, secure messaging platforms and network protocols. It also presents real-world examples of secure communication practices in corporate settings and personal interactions.

Importance of Securing Communication Channels

Today, we are constantly connected through multiple channels, from mobile phones and laptops to smart TVs and desktop computers. This hyperconnectivity allows us to exchange personal messages, financial details and even medical records, often without considering security risks. Social platforms such as TikTok, Reddit, Discord and many others facilitate real-time interactions, but they also present vulnerabilities that cybercriminals can exploit.

Unsecured communications, whether through emails, instant messaging or data transfers, create risks, including data breaches, identity theft and corporate espionage. The consequences can be severe, leading to financial losses, reputational damage and legal repercussions.

The rise of remote work and mobile connectivity has further amplified these risks. Employees frequently access corporate networks from various locations and devices, expanding the potential attack surface for cyber threats. Organisations should implement sturdy security measures, such as end-to-end encryption, virtual private networks (VPNs) and MFA, to safeguard sensitive communications. Securing communication channels is not only necessary for protecting data but also for maintaining trust between businesses, employees and customers.

Techniques for Securing Communication Channels

To protect data confidentiality, several proven techniques can be implemented to secure communication channels. The five most effective methods include:

Encryption

Encryption protocols, such as end-to-end encryption for emails, instant messaging and file transfers, ensure that only authorised recipients can access sensitive content. This prevents unauthorised interception and protects information integrity throughout transmission.

Multi-Factor Authentication (MFA)

MFA strengthens security by requiring multiple forms of verification before granting access. Even if a password is compromised, an attacker still needs the secondary authentication factor, such as a time-sensitive code sent to a mobile device or a physical security key. This significantly reduces the risk of unauthorised access.

Virtual Private Networks (VPNs)

VPNs establish encrypted tunnels for data transmission, shielding communications from interception when using public networks. These tools are relied on by remote workers accessing corporate resources, as they mask IP addresses and encrypt all network traffic, preventing eavesdropping and MITM attacks.

Secure Collaboration Platforms

Platforms explicitly designed for secure collaboration enhance teamwork while maintaining data confidentiality. Features such as encrypted file sharing, granular access controls and comprehensive audit logs help organisations protect sensitive discussions and documents. These platforms often include version control and data loss prevention capabilities to further safeguard information.

Secure Messaging Applications

Applications like Signal, WhatsApp and Telegram offer end-to-end encryption for text messages and voice calls, ensuring that only the intended recipients (be sure you know who they are!) can access the content. This protection is particularly critical for sensitive communications, such as in healthcare settings where patient confidentiality is legally required.

Examples of Secure Communication Implementation

Different scenarios require different approaches to secure communication. Here are a few examples:

Corporate Environments

Securing confidential communications is essential in business settings for protecting intellectual property, trade secrets and sensitive data. Organisations implement security measures like VPNs, secure email gateways and encrypted communication platforms. For example, a technology company may use encrypted email services and VPNs to prevent unauthorised access to internal communications and data transfers.

Personal Use

For individuals, securing personal communications is equally important in safeguarding privacy and sensitive information. Using secure messaging apps with end-to-end encryption provides an added layer of protection. Additionally, VPNs help prevent data interception when connecting to public Wi-Fi networks. For example, travellers using airport or hotel Wi-Fi can use a VPN to encrypt their connection, reducing the risk of cyber threats from unsecured networks.

In summary, securing communication channels is a fundamental aspect of data protection. Techniques such as encryption, VPNs and secure messaging ensure that transmitted data remains confidential and protected from interception. Implementing these security measures is vital in today's digital landscape, whether for corporate or personal use.

Legal and Regulatory Aspects of Confidentiality

The legal and regulatory frameworks surrounding data privacy and confidentiality are critical in determining how organisations manage sensitive information. These frameworks establish obligations that require entities to protect personal data, uphold individuals' privacy rights and ensure timely reporting of breaches. For cybersecurity practitioners, a good understanding of these regulations is essential for maintaining compliance and safeguarding confidential data.

Prominent Regulations: The Privacy Act 1988, GDPR and HIPAA

Three of the most significant regulations governing data confidentiality are *The Privacy Act 1988* in Australia, *The General Data Protection Regulation* (GDPR) in the European Union and *The Health Insurance Portability and Accountability Act* (HIPAA) in the United States. Each framework enforces strict data protection measures, ensuring personal information is handled securely.

The Australian Privacy Act 1988 and Australian Privacy Principles (APPs)

The *Privacy Act 1988* serves as Australia's primary data protection law, designed to safeguard individuals' privacy while regulating how organisations collect, store and use personal information. It applies to federal government agencies, larger businesses with annual turnover exceeding $3 million and certain smaller entities handling sensitive data. The Act established 13 APPs that provide comprehensive guidelines:

- **APP1, Transparency**—Organisations must maintain clear privacy policies that disclose how they collect, use and manage personal data in compliance with APPs.

- **APP2, Anonymity and Pseudonymity**—Individuals should be able to interact anonymously or use pseudonyms where practicable.
- **APP3, Data Collection**—Only necessary personal information should be collected through lawful and fair means.
- **APP4, Unsolicited Information**—Unsolicited personal information must be destroyed or de-identified if not required for legitimate purposes.
- **APP5, Notification**—Individuals must be informed when their data is collected, including the purpose and potential recipients.
- **APP6, Use and Disclosure**—Personal data should only be used for the primary purpose for which it was collected unless an exception applies under law.
- **APP7, Direct Marketing**—Strict limitations exist on using personal data for direct marketing, with opt-out mechanisms required.
- **APP8, Cross-Border Disclosure**—Organisations must ensure that personal information sent overseas receives protection equivalent to the APPs.
- **APP9, Government Identifiers**—Restrictions apply to adopting, using and disclosing government identifiers.
- **APP10, Data Quality**—Collected data must be accurate, complete, up to date and relevant for its intended use.
- **APP11, Data Security**—Personal data must be protected from misuse, interference, loss and unauthorised access or disclosure.
- **APP12 and APP13, Access and Correction**—Individuals have the right to access and correct their personal data held by organisations.

Violations of *The Privacy Act* may result in enforcement actions by the Office of the Australian Information Commissioner (OAIC), including financial penalties up to $2.1 million for serious breaches and compliance directives.[44]

The General Data Protection Regulation (GDPR)

Implemented in May 2018, the GDPR is one of the most comprehensive data protection laws worldwide. It governs how organisations collect, process and store the personal data of EU citizens, with an emphasis on data security, transparency and individual rights. Key aspects of the GDPR include:

- **Data Minimisation and Retention**—Organisations must limit data collection to what is strictly necessary and retain it only for as long as required for specified legal purposes.
- **Explicit Consent and Data Subject Rights**—Individuals must provide clear, informed consent before their data is processed. They also have the right to access, rectify, erase, restrict processing and transfer their personal data.
- **Breach Notification**—If a data breach risks individuals' rights and freedoms, organisations must report it to supervisory authorities within 72 hours and notify affected individuals without undue delay.
- **Data Protection Officers**—Many organisations must appoint dedicated personnel to oversee GDPR compliance and serve as contact points for data subjects and authorities.

Violations of the GDPR can result in substantial fines, up to €20 million or 4% of the organisation's annual global turnover, whichever is higher.[45]

The Health Insurance Portability and Accountability Act (HIPAA)

Enacted in 1996, HIPAA establishes stringent privacy and security standards for Protected Health Information (PHI) in the U.S. healthcare sector. The Act is enforced by the Department of Health and Human Services (HHS) and includes several requirements:

- **Privacy Rule**—Establishes nationwide standards for safeguarding PHI, balancing patient privacy with appropriate information disclosure
- **Security Rule**—Mandates administrative, physical and technical safeguards for electronic PHI, requiring risk analyses and management strategies
- **Breach Notification Rule**—Requires covered entities to notify affected individuals, HHS and, in certain cases, the media if a breach of unsecured PHI occurs affecting 500 or more individuals
- **Minimum Necessary Standard**—Requires that access to PHI be limited to the minimum information needed to accomplish the intended purposes.

Penalties for HIPAA violations range from financial fines of $100 to $50,000 per violation (with annual caps) to criminal charges for severe breaches, making compliance essential for healthcare providers and their business associates.[46]

Role of Data Breach Notification Laws

Data breach notification laws ensure that individuals are promptly informed when their data is compromised. These laws vary by jurisdiction but generally require organisations to notify affected individuals and regulatory authorities within a specified timeframe. The aim is to mitigate harm by allowing individuals to take protective measures, such as changing passwords or monitoring credit reports.

In Australia, data breach notification laws operate under the Notifiable Data Breaches (NDB) scheme, which was incorporated into *The Privacy Act 1988* in February 2018. This scheme requires organisations and agencies covered by *The Privacy Act* to notify affected individuals and the OAIC when an eligible data breach is likely to result in serious harm. Organisations must conduct a reasonable and expeditious assessment within 30 days upon discovering a potential breach.

In the U.S, data breach notification laws exist primarily at the state level, with all 50 states having enacted their own requirements. California leads with legislation through *The California Consumer Privacy Act* (CCPA) and its enhancement, *The California Privacy Rights Act* (CPRA). These laws require businesses to notify affected California residents when their personal information is compromised, specifying notification timing, content requirements and potential penalties for non-compliance. Other notable state laws include the New York *SHIELD Act* and Virginia's *Consumer Data Protection Act*, creating a complex compliance landscape for organisations operating across multiple states. Unlike Australia's unified federal approach, the U.S. lacks comprehensive national data breach notification legislation, though sector-specific federal laws like HIPAA govern specific industries.

Challenges in Keeping Up with Regulations

Ensuring compliance with data protection regulations presents significant challenges for organisations, particularly multinational corporations that must navigate multiple regulatory frameworks simultaneously. Keeping pace with evolving legal requirements is further complicated by the rapid advancement of technology, which often outstrips the development of regulatory standards. For example, *The Australian Privacy Act 1988* was enacted before widespread internet use and decades before emerging issues such as using personal images to train AI systems became prevalent.

Organisations must implement compliance strategies to address these challenges effectively. These strategies include conducting regular risk assessments and compliance audits, delivering targeted employee training programs and deploying advanced data protection measures tailored to specific regulatory requirements. Furthermore, maintaining current knowledge about regulatory updates, cross-border data transfer restrictions and industry best practices is essential for mitigating legal risks and avoiding significant penalties, which can reach millions of dollars and cause lasting reputational damage.

Implementing a Confidentiality Program

Imagine your most sensitive data falling into the wrong hands—now imagine having the power to prevent it. A comprehensive confidentiality program is indispensable for any organisation committed to safeguarding sensitive information. A well-designed program not only protects data but also fosters trust, mitigates risks and ensures regulatory compliance. Implementing such a program isn't just a precaution; it's a critical defence against data breaches and insider threats.

An effective confidentiality program comprises several key components, each vital in securing information assets.

Security Policy

A clear and comprehensive security policy forms the foundation of any confidentiality program. This policy should outline data handling procedures, access controls and employee responsibilities in protecting sensitive information. It must specify:

- Who has access to specific data
- Under what conditions access is granted
- Protocols for requesting and approving access.

For instance, a financial institution may implement a policy requiring dual authorisation before granting access to customer financial data. This approach, known as dual-control security, ensures that no single individual can compromise sensitive information.

To remain effective, security policies should undergo regular reviews and updates to address emerging threats and evolving regulatory requirements.

Employee Training

Regular security awareness training is essential in mitigating human error, one of the most significant risks to data confidentiality. Employees must be educated on:

- Best practices for handling sensitive data
- Recognising phishing attacks and social engineering tactics
- Creating and maintaining strong passwords.

According to the 2023 Data Breach Investigations Report, 74% of security incidents involve human error through phishing, mishandling data or weak credentials. Interactive training sessions, real-world simulations and continuous education programs can significantly enhance employees' ability to recognise and respond to security threats.

Incident Response Plan (IRP)

Even with strong preventive measures, security incidents can still occur. An effective IRP ensures a swift and organised approach to identifying, containing and recovering from data breaches. This plan should include:

- Procedures for detecting security incidents
- Communication protocols with stakeholders and regulatory bodies
- Mitigation strategies to reduce damage and restore operations.

For example, healthcare organisations following HIPAA regulations must report breaches to the appropriate authorities within specified timeframes. Regular testing and refinement of the response plan help organisations stay prepared for evolving threats.

Data Loss Prevention (DLP)

DLP solutions are critical for monitoring, identifying and blocking unauthorised data transfers. These tools help enforce security policies by:

- Scanning network traffic, email communications and endpoints for policy violations
- Preventing sensitive data from being sent through unsecured channels
- Detecting and responding to potential data exfiltration attempts.

Organisations handling intellectual property, trade secrets or financial data should consider DLP a fundamental security measure. For example, a company might use DLP software to block employees from sending confidential documents via personal email accounts, reducing the risk of accidental or intentional data loss.

Auditing and Monitoring

Continuous auditing and monitoring play a vital role in ensuring data confidentiality. Regular audits help:

- Identify new security risks and vulnerabilities
- Verify compliance with internal policies and external regulations
- Detect unauthorised access or suspicious activities.

Advanced monitoring solutions, such as Security Information and Event Management (SIEM) systems, provide real-time visibility into security events, allowing organisations to respond proactively. By integrating ML-driven threat detection, companies can further enhance their ability to identify and mitigate risks.

Continuous Improvement

Security is not a one-time implementation but an ongoing process. As cybersecurity expert Bruce Schneier states, "*Security is a process, not a product*". Organisations must continually assess and enhance their confidentiality programs by:

- Staying informed about emerging cybersecurity threats
- Investing in new security technologies
- Fostering a culture of security awareness among employees.

Prioritising continuous improvement allows organisations to protect sensitive data better and maintain the trust of their customers, employees and stakeholders.

Summary

Confidentiality is central to cybersecurity, ensuring that sensitive information is protected from unauthorised access and disclosure. As digital transformation accelerates, the need to safeguard confidentiality has grown, spanning legal, ethical and operational dimensions. This chapter examined the essential role of confidentiality in cybersecurity, the threats it faces and the strategies for its protection.

Confidentiality is a fundamental pillar of the CIA Triad—Confidentiality, Integrity and Availability. Restricting access to authorised individuals not only safeguards sensitive data but also reinforces integrity and availability, preventing unauthorised modifications and disruptions. High-profile breaches, such as those affecting Equifax and Yahoo, illustrate the severe financial, reputational and legal consequences of compromised confidentiality.

Numerous threats challenge confidentiality, including unauthorised access, interception, social engineering, physical security breaches and human error. Hackers, malware and insider threats pose persistent risks. Interception techniques, such as Man-In-The-Middle (MITM) attacks and packet sniffing, exploit unprotected data transmissions. Social engineering manipulates human psychology to bypass security measures, while physical breaches, such as stolen devices, highlight the importance of physical security. Human error, a leading cause of data breaches, underscores the necessity of comprehensive training programs.

This chapter has mentioned various protective measures used to counter these threats. Access control mechanisms, including RBAC and MFA, ensure that only authorised users can access sensitive data. Encryption secures data by converting it into an unreadable format, protecting it during storage and transmission. Numerous other methods, ranging from network segmentation to secure coding practices, further enhance confidentiality. In reality, the breadth of confidentiality measures is vast enough to warrant an entire book.

Understanding and implementing these techniques is critical for an effective cybersecurity strategy.

Legal and regulatory frameworks also play an important role in upholding confidentiality. Regulations such as *The General Data Protection Regulation* (GDPR), *The Privacy Act 1988* and *The Health Insurance Portability and Accountability Act* (HIPAA) establish stringent data protection and breach notification requirements. Compliance with these laws is essential for organisations, not only to avoid legal and financial penalties but also to maintain public trust and protect individuals' privacy.

In an era where cyber threats continuously evolve, maintaining data confidentiality requires a proactive and adaptive approach. Organisations must balance security with operational efficiency, investing in continuous education, advanced security measures and regulatory compliance. This chapter has provided a foundational understanding of confidentiality within the CIA Triad, setting the stage for a deeper exploration of cybersecurity principles. The next chapter will focus on integrity, the second pillar of the CIA Triad, further expanding your understanding of information security.

Revision Questions

1. What is the primary role of confidentiality in cybersecurity?

2. Explain the legal implications of data breaches under *The General Data Protection Regulation* (GDPR) and *The Health Insurance Portability and Accountability Act* (HIPAA).

3. Describe the CIA Triad and the interdependence between its three components: Confidentiality, Integrity and Availability.

4. Provide real-world examples of data breaches and discuss their impacts on organisations and individuals.

5. Why is data confidentiality critical for both individuals and organisations?

6. Identify and describe the various types of threats to data confidentiality, including unauthorised access, interception, social engineering and physical security breaches.

7. What role does human error play in compromising data confidentiality, and how can it be mitigated?

8. Discuss the different methods for protecting data confidentiality, including access control, data classification and encryption.

9. What are the key legal and regulatory frameworks governing data confidentiality, and what are their main requirements?

10. What are the essential components of a comprehensive confidentiality program and why is continuous improvement necessary?

Cybersecurity Hacks

Hack 1: Implementing Role-Based Access Control (RBAC)

RBAC is a method for restricting system access to authorised users based on their role within the organisation. It minimises the risk of unauthorised access by ensuring users can only access the information necessary for their job functions.

Steps:

1. Identify Roles—Define all the roles within your organisation and specify the access level required for each role.

2. Assign Permissions—Map each role to specific permissions that grant access to relevant data and systems.

3. Implement RBAC—Use an RBAC system or software to enforce these permissions across your network. For hands-on experience, check out Azure RBAC (https://learn.microsoft.com/en-us/azure/role-based-access-control/).

4. Review Regularly—Periodically review and update roles and permissions to adapt to changes in job functions or organisational structure.

Hack 2: Encrypt Sensitive Data

Encryption transforms data into an unreadable format using algorithms, ensuring that only authorised parties with the decryption key can access the original information.

Steps:

1. Identify Sensitive Data—Determine which data needs encryption based on its sensitivity and regulatory requirements.

2. Choose Encryption Tools—Select appropriate encryption tools and algorithms (e.g., AES-256) that provide strong protection.

3. Encrypt Data at Rest—Apply encryption to stored data, including databases, file systems and backups. Take a look at AxCrypt (https://axcrypt.net/) to help with this step.

4. Encrypt Data in Transit—Use encryption protocols to protect data transmitted over networks. While SSL/TLS (commonly used in HTTPS) is widespread for securing web traffic, consider implementing more advanced encrypted connections for data transit. Explore the SSH (Secure Shell) protocol in Linux for a more robust and versatile secure communication solution. Example: To establish an encrypted SSH connection in Linux (ssh username@remote_host). See "How SSH establishes secure communication", https://www.redhat.com/sysadmin/ssh-secure-communication.

5. Manage Encryption Keys—Implement a secure key management system to regularly store, distribute and rotate encryption keys.

Chapter 7: Integrity

– The Source of Truth –

Objectives

By the end of this chapter, you will be able to:

- Define data integrity and explain its importance as a component of the CIA Triad
- Identify and analyse various threats to data integrity
- Examine techniques for maintaining data integrity
- Implement appropriate integrity controls for different systems
- Understand the legal and regulatory requirements related to data integrity.

In the previous chapter, we examined confidentiality, the first component of the CIA Triad. This chapter focuses on integrity, an equally vital pillar. Integrity ensures that data remains accurate, consistent and unaltered throughout its lifecycle, safeguarding it from unauthorised modifications that could lead to severe consequences. This protection is essential not only for system functionality but also for maintaining trust in digital infrastructures. In the next chapter, we will explore availability, the third and final component of the CIA Triad.

Compromised data integrity can devastate organisations through financial losses, operational disruptions and reputational damage. Consider the 2016 Bangladesh Bank cyberattack, where hackers infiltrated the bank's systems and manipulated transaction records to steal $81 million from its Federal Reserve Bank of New York account.[47] In healthcare, compromised records can disrupt patient care systems, create operational chaos and endanger lives through incorrect treatment protocols or medication.

Data integrity plays a fundamental role across cybersecurity domains, ensuring the effectiveness of security controls and system reliability. Take intrusion detection systems, for example. These tools rely on accurate, unaltered data to identify threats. If an Intrusion Detection System (IDS) database is compromised, it may fail to detect security breaches, leaving systems vulnerable. Additionally, regulations discussed in the previous chapter, such as the GDPR and HIPAA, impose strict requirements for data accuracy, reinforcing the legal imperative of maintaining integrity.

The 2001 Enron scandal exemplifies how compromised data integrity can trigger widespread economic turmoil. Although primarily a financial fraud case, the scandal involved extensive manipulation of accounting records, ultimately leading to Enron's collapse. This crisis prompted the introduction of *The U.S. Sarbanes-Oxley Act (SOX)* in 2002, which established stricter corporate transparency and accountability standards; principles that now extend beyond financial reporting into cybersecurity practices.

This chapter examines various threats to data integrity, ranging from sophisticated cyberattacks to inadvertent human errors. Understanding these risks allows organisations to implement robust security measures. We will explore key integrity verification techniques, including hashing algorithms, digital signatures and data logging. These methods help detect unauthorised changes while ensuring data accuracy and reliability across systems and applications.

Importance of Data Integrity

Ensuring data integrity is just as critical as maintaining confidentiality and availability. Data integrity refers to the accuracy, consistency and trustworthiness of information throughout its entire lifecycle. With rising cyberattacks, system errors and human mistakes, protecting data integrity has become paramount for accurate decision-making, operational efficiency and maintaining stakeholder trust.

Without proper integrity safeguards, organisations risk basing critical decisions on corrupted or manipulated information. Data integrity violations can lead to significant financial losses, damaged reputations, regulatory penalties and potentially life-threatening consequences in sectors like healthcare or aviation. Implementing strong input validation, access controls, audit trails, data validation mechanisms and secure backup systems provides a multi-layered defence against integrity threats.

Why Data Integrity is Vital for Individuals and Organisations

Data integrity forms the foundation of reliable digital systems. This concept affects both personal activities and organisational operations in profound ways.

For Individuals

People rely on data integrity in nearly every aspect of daily life, from financial management to healthcare and personal communications.

Personal Finance: Accurate financial data is essential for budgeting, tax filing and investment decisions. If bank records are altered due to cyberattacks or system errors, incorrect account balances or fraudulent transactions could lead to serious financial consequences.

Health Records: Reliable medical records ensure that diagnoses and treatments are based on accurate information. Inaccuracies in patient data can lead to inappropriate treatments or misdiagnoses, highlighting the critical nature of integrity in healthcare systems.

Personal Communications: Integrity in digital communications ensures that messages remain unaltered from sender to recipient. When the content of an email or message is tampered with, it can lead to misunderstandings or security breaches, affecting personal relationships and trust.

For Organisations

For businesses and institutions, data integrity serves as a pillar of operational reliability, compliance and reputation.

Operational Data: Accurate and consistent operational data enables effective decision-making. For example, a manufacturing company relies on precise inventory and production figures to manage its supply chain efficiently. Lapses in data integrity can lead to production delays or financial losses.

Customer Trust: Maintaining accurate customer records is necessary for building and sustaining trust. Inaccurate customer data can lead to billing errors, misdirected communications or even fraudulent activities, eroding consumer confidence.

Financial Reporting: Organisations depend on integrity in financial records to ensure accurate reporting and regulatory compliance. Discrepancies in financial data may result in legal consequences, regulatory penalties and damaged reputation among investors and stakeholders.

Intellectual Property: The integrity of proprietary research data, product designs and strategic plans is critical to preserve competitive advantage. Unauthorised changes or corrupt data can compromise innovations and weaken market position.

The Role of Integrity in the CIA Triad

As we've discussed, the CIA Triad is the foundational cybersecurity model. While all three components work together, integrity ensures that data remains accurate and unaltered throughout its use.

Integrity and Confidentiality

Data confidentiality measures, such as encryption, access controls and authentication, not only prevent unauthorised access but also protect data integrity. When encrypted data is tampered with, decryption fails, revealing unauthorised modifications. Similarly, digital signatures confirm that data remains unchanged, reinforcing trust in online transactions.

Integrity and Availability

While availability ensures that data is accessible to authorised users, integrity guarantees that accessible data remains accurate and uncorrupted. Techniques such as checksums, hash functions and digital signatures verify whether data has been altered. In financial systems, these verification measures prevent fraud and ensure the reliability of transactions.

Overall, data integrity is the foundation of trust and reliability in information systems, affecting everything from personal finance and healthcare to corporate operations and cybersecurity. By prioritising data accuracy and consistency, you can reduce risks, prevent fraud and maintain operational efficiency in your personal and professional activities. As cyber threats evolve and data volumes grow, maintaining integrity becomes essential for security, trust and long-term success.

Threats to Data Integrity

What happens when your critical or personal data is altered without your knowledge? Data integrity threats create serious vulnerabilities in information systems. This section explores the primary risks to data integrity, from sophisticated cyberattacks to common human errors, and examines methods used to compromise information accuracy. Understanding these challenges will help you develop effective protection strategies and strengthen your data security measures.

Cyberattacks

Cyberattacks represent a significant threat to data integrity, involving malicious activities designed to alter, delete or manipulate information. Ransomware is particularly devastating, encrypting data and making it inaccessible until a ransom is paid. Even when decryption tools exist, they often prove unreliable, leaving many victims unable to recover their original data.

The NotPetya attack of 2017 demonstrates the scale of this threat, targeting organisations globally and causing widespread data corruption and operational disruption. Equally concerning are data manipulation attacks, where adversaries modify records to commit fraud, espionage or sabotage. These attacks pose particular risks to financial institutions and government agencies where data accuracy is central to their work.

Hardware Failures

Hardware failures constitute another significant risk to data integrity. Disk errors, power outages and mechanical malfunctions can corrupt files, sometimes irreversibly. A sudden power loss during data writing may leave files incomplete or damaged. This highlights the importance of proper shutdown procedures, such as using "shutdown /s" on Windows or "sudo shutdown now" on Linux, to prevent corruption.

As hardware components age and degrade, the risk of irretrievable data loss increases. Organisations must implement proactive hardware maintenance, redundancy strategies and reliable backup procedures to ensure data can be restored to its original state after a failure occurs.

Software Bugs

Software bugs can introduce inconsistencies and inaccuracies that compromise data integrity. Errors in code often lead to improper data processing, incorrect storage or unintended overwrites that damage information quality.

For example, a bug in a Database Management System (DBMS) might cause faulty indexing, resulting in lost or corrupted records. Software development professionals consistently find that all complex systems contain bugs; the challenge lies in identifying and addressing these issues before they affect critical data.

To mitigate these risks, organisations should implement rigorous testing protocols, version control systems and regular software updates. Proactive debugging and comprehensive quality assurance processes help detect and resolve issues before they impact data integrity.

Human Error

Human error remains one of the most common yet underestimated threats to data integrity. Accidental data entry mistakes, unauthorised modifications and improper data handling can cause severe operational disruptions.

In 2018, a major airline experienced widespread operational chaos due to a single incorrect data entry. This error led to numerous flight cancellations and delays, demonstrating how even minor human mistakes can have significant consequences across an organisation.

You can reduce these risks by implementing training programs, automation tools for routine tasks and stringent access controls that limit the potential impact of human error on data integrity.

Insider Threats

Insider threats (malicious actions by employees or trusted individuals) pose a severe risk to data integrity. Insiders can deliberately manipulate or delete critical data with privileged access for financial gain, revenge or competitive advantage.

In 2024, a former employee of a Sydney-based company was arrested for intentionally altering sensitive operational data, causing disruptions that took weeks to rectify. This case highlights the importance of implementing strict access controls, continuous monitoring systems and thorough employee vetting processes to prevent insider threats.

In summary, protecting data integrity requires a comprehensive understanding of potential threats and vulnerabilities. Each risk requires specific defensive measures, to protect against cyberattacks, hardware failures, software bugs, human error or insider threats. By implementing multi-layered security protocols, redundancy systems and continuous monitoring, you can safeguard your data against corruption and manipulation. These protective measures ensure the long-term reliability of your information systems and maintain trust in your data assets.

Increasing Sophistication of Attackers

The cyber landscape is evolving at an unprecedented pace, with attackers continuously refining their methods and leveraging emerging technologies to compromise data integrity. This rapid escalation presents a formidable challenge for cybersecurity professionals, who must adapt their defences to counter increasingly sophisticated threats.

Advanced Attack Vectors

Cybercriminals harness AI and ML to automate vulnerability discovery and craft highly convincing phishing campaigns. Generative AI has revolutionised social engineering, enabling attackers to produce context-aware messages that mimic an organisation's communication style, allowing them to bypass traditional security filters. In 2023, AI-generated spear phishing campaigns achieved opening rates nearly three times higher than conventional attempts.

Advanced Persistent Threats (APTs), often backed by state-sponsored or well-funded actors, employ multi-stage methodologies, custom malware and Living-off-the-Land (LotL) techniques (exploiting legitimate system tools to evade detection). The supply chain attack against SolarWinds (see Chapter 9 Strategic Briefing) exemplifies this level of sophistication, as attackers compromised the software build system and maintained persistent access for nine months before detection.

Evasive Techniques and Supply Chain Attacks

Traditional security measures struggle against the increasing prevalence of polymorphic malware, which continuously alters its code signatures while maintaining functionality, and fileless attacks, which operate entirely in memory, leaving no trace on disk. Research indicates that between 2020 and 2023, fileless attacks surged by 900%, disproportionately targeting financial institutions and healthcare records.[48]

Cybercriminals are also shifting their focus toward software supply chains, embedding malicious code into legitimate applications via development environments and update mechanisms. Open-source components present a particularly vulnerable attack surface, as demonstrated in 2023 when multiple widely used JavaScript libraries were compromised, affecting thousands of applications.

Effective Countermeasures

Organisations must implement a multi-layered security approach to protect against threats to data integrity. Key strategies include Zero-Trust Architecture, which verifies every user and device attempting to access resources; behavioural analytics to detect subtle anomalies in system and user activities; rigorous supply chain verification to ensure third-party components meet security standards; AI-powered security automation for rapid threat detection; and continuous employee education programs on recognising and responding to new threats.

As data integrity risks become more sophisticated, your security measures must evolve accordingly. This requires vigilance through regular system monitoring, adaptability in updating security protocols and a proactive defence-in-depth strategy. By implementing these measures, you can better safeguard your data against current threats while preparing for future challenges in the increasingly complex digital landscape.

Strategic Briefing: The Equifax Data Breach

In 2017, a single security flaw led to one of the most significant data breaches in history, exposing the personal information of 147 million individuals and shaking consumer trust in financial institutions.[49] The breach, caused by an unpatched vulnerability in the Apache Struts web application framework, revealed how fundamental security oversights can result in catastrophic data exposure. Its long-term consequences impacted both Equifax and millions of affected individuals, demonstrating the urgent need for proactive cybersecurity measures.[50]

Impact and Scope

The breach's repercussions extended across the financial services industry, affecting millions worldwide. Hackers accessed an unprecedented amount of sensitive information, including names, birth dates and addresses. Additionally, approximately 209,000 credit card numbers and sensitive documents for 182,000 consumers were exposed.[51]

Financially, the breach cost Equifax over $1.7 billion by 2021.[52] However, the company suffered severe reputational damage beyond monetary losses, eroding customer trust and diminishing its market value.[53] The breach's legacy continues today, as affected individuals remain at risk of identity theft and financial fraud.

Attack Methodology

Understanding the impact requires examining how attackers infiltrated Equifax's systems. Exploiting the publicly disclosed vulnerability (CVE-2017-5638) in Apache Struts, attackers gained unauthorised access due to Equifax's failure to apply available patches for several months.[54] An expired SSL certificate further compounded the issue, preventing malicious traffic detection.

Once inside, attackers maintained access for 76 days, moving laterally through Equifax's systems while systematically exfiltrating vast amounts of sensitive data.[55] The prolonged intrusion highlighted major weaknesses in Equifax's security monitoring, as the breach remained undetected until July 29, 2017, long after the initial compromise.[56]

Response and Mitigation

Equifax's initial response faced widespread criticism for its delayed disclosure and lack of transparency. The company waited six weeks before informing the public, intensifying scrutiny over its crisis management.

In response, Equifax launched a dedicated website to help consumers check their exposure status and provided free credit monitoring services. A large-scale call centre was also established to handle inquiries. Internally, the company implemented urgent security enhancements, including patching vulnerabilities and strengthening monitoring protocols.

The severity of the breach led to executive resignations, including the CEO, CIO and CSO. These leadership changes marked the beginning of a major overhaul in Equifax's security and risk management approach.

Lessons Learned

The Equifax breach provided critical insights into security governance, technical controls and incident response. Key takeaways include:

- **Vulnerability Management**—Organisations must enforce structured patching schedules, maintain accurate asset inventories and conduct routine security assessments.

- **Threat Detection and Monitoring**—Continuous network monitoring, robust encryption and proactive certificate management are essential for preventing unauthorised access.

- **Governance and Policies**—Effective security strategies require clear policies, regular reporting to leadership and strong third-party risk management.

- **Incident Response Preparedness**—Organisations must have well-documented response plans, conduct regular crisis simulations and establish communication channels with law enforcement and cybersecurity firms.

The Equifax breach is a watershed moment in cybersecurity history, reinforcing the devastating consequences of security negligence. It remains a powerful reminder that in today's evolving threat landscape, strong cybersecurity measures are not optional but imperative.

Techniques for Maintaining Data Integrity

Maintaining data integrity is essential, ensuring that information remains accurate, consistent and trustworthy. A variety of techniques safeguard data, each addressing different aspects of security. These methods, from advanced encryption to efficient checksums, protect against corruption and tampering. While each technique has unique strengths and limitations, when combined, they provide comprehensive protection.

This section explores the primary methods for maintaining data integrity, outlining their benefits, limitations and practical applications.

Hashing

Hashing generates a unique digital fingerprint, or hash, for a dataset. These hash functions, also called one-way functions, convert data of any size into a fixed-length string of characters. Even the slightest change in data results in a significantly different hash value, making it easy to detect unauthorised modifications.

Because hash values cannot be reverse engineered to reveal the original data, hashing is widely used to secure passwords, verify file integrity and protect blockchain transactions. For example, the Secure Hash Algorithm (SHA-256) is a common standard in cybersecurity.

While hashing effectively detects changes, it does not prevent data alteration; it only alerts users to modifications. For this reason, you'll often need to combine hashing with other security measures for complete protection.

Digital Signatures

Digital signatures ensure both the authenticity and origin of data. Using cryptographic algorithms, they bind a signer to a piece of data, confirming that it has not been altered since the signature was applied.

This technique combines hashing with asymmetric encryption (a private key signs the hash of the data), while a corresponding public key verifies the signature. You'll encounter digital signatures in secure email communications, legal contracts and software distributions.

A key advantage of digital signatures is non-repudiation; the signer cannot deny their involvement. However, their effectiveness relies on the security of the private key. If this key is compromised, data integrity is at risk.

Checksums

Checksums are a lightweight method for detecting accidental data corruption. They generate a value based on the contents of a dataset, allowing for quick integrity verification.

Unlike hashing, checksums are primarily used for error detection rather than security. They play a crucial role in network communications and file transfers, ensuring data remains intact during transmission. For instance, the Transmission Control Protocol (TCP) employs checksums to verify the integrity of packets sent over a network.

While efficient, checksums offer limited protection against intentional tampering, making them best suited for detecting unintentional errors rather than securing sensitive information.

Data Logging and Auditing

To detect unauthorised modifications, organisations implement data logging and auditing. Logging records all data interactions, capturing details, including access times, user identities and alterations. Auditing analyses these records to identify anomalies or suspicious activities.

For example, DBMSs use transaction logs to ensure that any changes can be traced and, if necessary, rolled back to a previous state. This enhances accountability and enables prompt detection of integrity breaches.

When implementing logging and auditing in your systems, consider that these processes require significant storage and processing resources, which affects scalability.

Access Controls

Access control determines who can access and modify data. By implementing well-defined policies, you can prevent unauthorised modifications and reduce security risks.

Techniques such as role-based access control (RBAC) and discretionary access control (DAC) help enforce data security. RBAC restricts access based on an individual's role within an organisation, ensuring employees only interact with data relevant to their duties.

Strong access controls minimise the risk of insider threats and unauthorised changes, but you must carefully manage them to balance security with usability.

Data Validation

Data validation is an essential but often underestimated security measure. By enforcing predefined rules, it ensures that only accurate and safe data enters a system.

For example, validation can prevent malicious code injections by restricting the types of inputs a system accepts. In financial applications, you might implement validation rules requiring that all transactions include a valid account number and a positive transaction amount.

By preventing errors and security vulnerabilities at the point of entry, data validation significantly enhances overall integrity and reliability.

Selecting the Right Approach

When designing your data integrity strategy, consider that each technique offers distinct advantages for different situations. Hashing and digital signatures provide security guarantees for sensitive information. Checksums offer efficient error detection for data transmission. Logging and auditing create accountability and traceability. Access controls and data validation prevent unauthorised or invalid data entry.

The most effective approach combines multiple techniques based on your specific data sensitivity, operational requirements and risk environment. By implementing complementary controls, you can ensure data integrity remains intact while maintaining system performance and usability.

Tactical Briefing: Hashing and Digital Signatures

Understanding Hashing

Overview:

As described in the previous section, hashing is a fundamental tool for ensuring data integrity. Even the slightest modification to a file produces an entirely different hash, making tampering immediately detectable.

Exercise 1: File Hash Verification

Objective:

Demonstrate how hashing can verify file integrity by detecting modifications.

Task Steps:

1. Choose a hashing tool—Use a free hash calculator or built-in OS commands:

    a. Windows: certutil -hashfile <filename> SHA-256

    b. Linux/macOS: sha256sum <filename>

2. Compute an initial hash—Select a document and calculate its SHA-256 hash.

3. Modify the file—Introduce a small change, such as adding a space.

4. Recompute the hash—Run the hash function on the modified file.

5. Compare results—Observe how even a minor change results in a completely different hash value.

Key Takeaways:

- Hashing detects unauthorised modifications
- Even small changes produce entirely new hash values
- While hashing verifies integrity, it does not prevent alterations.

Understanding Digital Signatures

Overview:

Digital signatures confirm both the authenticity and integrity of data. By using encryption, they ensure that signed data remains unchanged since the time of signing.

Exercise 2: Digital Signature Experience

Objective:

Explore how digital signatures validate document authenticity and detect tampering.

Task Steps:

1. Select a document—Choose a PDF file you created or received.
2. Digitally sign the document—Use a free PDF tool, such as Adobe Reader, to apply a digital signature.
3. Modify the signed document—Introduce a minor change, such as adding a space.

4. Verify the signature—Check how the signature is invalidated after modification.

5. Analyse verification details—Review tampering alerts and security warnings.

Key Takeaways:

- Digital signatures verify authenticity and integrity
- Any modification to a signed document invalidates the signature
- Private key security is essential for maintaining trust.

Ensuring Data Integrity in Different Systems

Data integrity is a non-negotiable when it comes to safeguarding accurate, consistent and reliable information across systems. Whether managing databases, file systems or communication networks, organisations must implement strong integrity controls to prevent corruption, unauthorised modifications and data loss. This section outlines key techniques and best practices to safeguard data integrity within these environments.

Databases

Databases underpin decision-making and business operations, storing vast amounts of critical information. Without proper integrity controls, databases become vulnerable to corruption, inconsistencies and cyber threats. The following mechanisms help ensure data reliability:

Integrity Constraints

These enforce rules that preserve data accuracy and consistency:

- Primary Keys ensure each record in a table is unique, preventing duplicate entries
- Foreign Keys maintain relationships between tables, ensuring referential integrity
- Unique Constraints prevent duplicate values within specific fields, reinforcing data consistency.

For example, in an employee database, a foreign key linking a department ID to an employee record ensures all employees belong to valid departments, preventing orphaned records.

Transaction Logging

Transaction logs record all database operations, allowing systems to restore data after failure or corruption. Techniques like Write-Ahead Logging (WAL) ensure changes are logged before being committed, providing a reliable rollback mechanism during unexpected failures.[57]

Data Backups

Regular backups are a fundamental safeguard against data loss from accidental deletion, corruption or cyberattacks. A multi-layered backup strategy includes:

- Full backups, creating a complete database copy
- Incremental backups, storing only changes made since the last backup
- Differential backups, recording changes since the last full backup.

By combining these backup methods, organisations can quickly recover lost or compromised data.

File Systems

Understanding where and how files are stored, on local servers, cloud infrastructure or personal devices, is central to ensuring file system integrity. File corruption, unauthorised modifications and accidental deletions represent common risks. The following techniques help mitigate such issues:

File System Permissions

Permissions determine who can read, write or execute files. Organisations can prevent unauthorised access and file modifications by setting appropriate permissions, thereby protecting data integrity. Linux-based systems use a combination of owner permissions, groups and other permissions to control access. For example, permission on a Linux file may appear as: -rw-r--r-- 1 user group 4096 Jul 10 09:30 example.txt. This indicates that the owner can read and write the file (-rw), while others have read-only access.

Access Control Lists (ACLs)

ACLs provide granular access control, allowing administrators to define detailed permissions for specific users or groups. Unlike standard file permissions, ACLs enable complex access policies, making them valuable for organisations managing sensitive data.

Journaling File Systems

Journaling file systems, such as ext4 (Linux) and NTFS (Windows), maintain a transaction log that records changes before they are applied. If a system crashes or loses power unexpectedly, the journal allows the file system to recover without data corruption.[58]

For example, ext4 journals metadata changes, reducing filesystem inconsistency risks. Meanwhile, NTFS logs both metadata and file content changes, offering higher protection levels.

Communication Systems

Communication systems facilitate data exchange between users, applications and services. Ensuring the integrity of transmitted data is essential to prevent unauthorised tampering, interception or corruption. The following tools assist to safeguard communication systems:

Data Encryption

Encryption transforms data into an unreadable format, ensuring only authorised parties can decrypt it. Transport Layer Security (TLS) encrypts internet communications, preventing eavesdropping and data manipulation.

Message Authentication Code (MAC)

A MAC is a cryptographic checksum that verifies a message's authenticity and integrity. It is generated using a secret key and message content, allowing recipients to detect unauthorised modifications.[59]

For example, HMAC (Hash-Based Message Authentication Code) is widely used in secure communications, ensuring messages remain unaltered during transmission.

Secure Communication Protocols

Secure protocols protect data integrity by encrypting transmissions and authenticating sources. Common examples include:

- HTTPS (Hypertext Transfer Protocol Secure)—Encrypts web traffic to prevent MITM attacks

- SSH (Secure Shell)—Ensures encrypted remote access to servers and devices
- VPNs (Virtual Private Networks)—Create encrypted tunnels for secure data transmission over the internet.

By implementing these measures, organisations can safeguard sensitive information from cyber threats and unauthorised alterations.

Considering Specific Requirements and Vulnerabilities

When integrating integrity controls, they must be tailored to each system's unique needs and vulnerabilities. Different systems face distinct threats, requiring customised security strategies.

For example, databases handling high transaction volumes and complex relationships require strong integrity constraints, transaction logging and access control mechanisms to maintain consistency and security. In contrast, file systems focus on preventing unauthorised access and protecting sensitive files, often relying on strict permission settings and encryption.

Moreover, different systems are vulnerable to specific attacks. Databases frequently face Structured Query Language (SQL) injection, where malicious actors manipulate queries to gain unauthorised access or alter data. Implementing parameterised queries and enforcing input validation ensures user-supplied inputs cannot be exploited for harmful queries.

File systems, however, face ransomware threats, where attackers encrypt critical data and demand payment for decryption. Regular, encrypted backups (preferably stored offline) along with ransomware detection tools serve as essential safeguards against these attacks.

Ultimately, selecting and implementing security measures based on a system's specific risks and operational requirements is critical to ensuring data integrity and resilience.

Legal and Regulatory Aspects of Data Integrity

Data integrity extends beyond technical requirements to serve as a legal and regulatory obligation. Various laws mandate stringent controls to maintain data accuracy, consistency and reliability. Organisations must achieve compliance to avoid legal penalties and preserve stakeholder trust. This section examines the key aspects of this critical relationship.

Compliance with Industry Regulations

Compliance with industry regulations protects data integrity, ensuring that sensitive information remains accurate, complete and unaltered. These regulatory frameworks establish strict requirements for organisations handling sensitive data, reinforcing the need for strong security measures and reliable governance practices.

For example, in the United States, HIPAA requires healthcare providers to implement safeguards that protect patient data integrity, ensuring medical records remain trustworthy and unmodified. In Australia, the Therapeutic Goods Administration (TGA) enforces *Data Management and Data Integrity* (DMDI) guidelines for therapeutic goods manufacturers.[60] Additionally, *The Privacy Act 1988*, through the Australian Privacy Principles (APPs), outlines data integrity obligations for businesses across multiple industries.

On a global scale, the *Payment Card Industry Data Security Standard* (PCI DSS) mandates comprehensive security controls, such as encryption and access restrictions, to protect credit card data from tampering.[61] Non-compliance can lead to severe penalties. For instance, HIPAA violations can result in fines of up to $1.5 million per category per year. Compliance is therefore not just a regulatory necessity but a fundamental aspect of organisational security.

Beyond avoiding financial and legal repercussions, regulatory compliance strengthens overall data protection and builds stakeholder trust.

Data Retention and Audit Trails

Think avoiding records is a loophole? Not quite. Data retention policies and audit trails are essential for ensuring legal compliance and data integrity. Regulations require organisations to retain data for specified periods and maintain detailed logs of access and modifications.

For example:

- Australia's *Telecommunications (Interception and Access) Act 1979* mandates telecom companies retain certain data for two years
- *The Corporations Act 2001* and Australian Taxation Office (ATO) require companies to keep financial and tax records for at least five years
- The U.S. *Sarbanes-Oxley Act (SOX)* requires public companies to retain financial records and audit trails for seven years to ensure transparency.

Audit trails document who accessed data, when and what changes were made. These logs are useful for detecting unauthorised modifications, investigating breaches and proving regulatory compliance during audits.

Electronic Records and Signatures

The legal validity of electronic records and digital signatures is governed by laws such as *The Electronic Signatures in Global and National Commerce (E-SIGN) Act* in the U.S. and Australia's *The Electronic Transactions Act 1999*. These laws confirm that electronic records and signatures are legally binding when they meet authenticity and integrity requirements.

Digital signatures, secured through cryptographic techniques, ensure that electronic documents remain unaltered after signing. This legal framework supports secure, reliable electronic transactions and strengthens data integrity in digital communications.

Aligning Data Integrity Practices with Legal and Regulatory Frameworks

Ensuring data integrity requires aligning security measures with legal frameworks. This includes implementing encryption, access controls and audit trails to prevent unauthorised modifications and maintain accuracy.

Regulatory landscapes evolve, and organisations must stay informed to remain compliant. For instance, the EU's GDPR introduced stricter data integrity measures. Non-compliance can result in multi-million-dollar fines, highlighting the importance of proactive regulatory adaptation.

Integrating legal requirements into data integrity practices allows organisations to mitigate risks, ensure compliance and safeguard their most valuable asset—trustworthy data.

Summary

Integrity is the cornerstone of data reliability, ensuring information remains accurate, consistent and unaltered throughout its lifecycle. It safeguards data from unauthorised modifications that could lead to severe consequences, including financial losses and compromised decision-making. Data integrity directly impacts operational efficiency and strategic outcomes, as corruption can result in misleading conclusions with potentially catastrophic repercussions. This chapter explored the techniques, mechanisms and challenges involved in maintaining and verifying data integrity.

Cryptographic hashes are fundamental to ensuring data integrity. These mathematical algorithms generate a unique digital fingerprint for each dataset, enabling verification of data consistency. Even slight alterations produce entirely different hash values, immediately indicating tampering or corruption. Hash functions like SHA-256 play a central role in secure communications, digital forensics and blockchain technology.

Digital signatures preserve data integrity by combining hashing with public key cryptography. They authenticate data origin while verifying that content remains unaltered, securing sensitive information and strengthening its credibility. In financial transactions and legal documents, digital signatures ensure record authenticity, reducing fraud risk.

Checksums detect errors in data transmission and storage by verifying that transmitted data matches the original input. Though less secure than cryptographic hashes, they offer efficient integrity verification in non-critical applications such as file transfers and embedded systems.

Maintaining data integrity requires mitigating threats, including cyberattacks, hardware failures and human errors. The Equifax breach demonstrates the consequences of insufficient controls; a single vulnerability exposed sensitive information for millions. To avoid facing a similar crisis, organisations must implement proactive measures like continuous monitoring, access restrictions and effective backup protocols.

Access controls and data validation mechanisms protect integrity by restricting access to authorised personnel and verifying data at multiple checkpoints. System-specific approaches enhance security: databases employ constraints and transaction logging, while file systems use permissions and journaling. Secure communication channels provide additional protection against unauthorised manipulation.

Legal and regulatory frameworks enforce data integrity standards, including HIPAA, PCI DSS and GDPR. These regulations establish compliance requirements for stringent data protection. Failure to meet these standards can result in legal penalties, financial losses and erosion of stakeholder trust.

Maintaining data integrity combines technological solutions, policy enforcement and best practices. As cyber threats grow increasingly sophisticated, data integrity principles remain vital to resilient information systems, safeguarding digital assets for organisations and individuals alike.

Revision Questions

1. What is data integrity, and why is it a critical component of the CIA Triad?

2. What are the primary consequences of compromised data integrity?

3. Can you provide a real-world example where data integrity was compromised and describe the impact?

4. What are some common threats to data integrity, and how do they differ?

5. How does hashing help in maintaining data integrity?

6. What role do digital signatures play in ensuring data integrity?

7. Why are checksums necessary for data integrity, and what are their limitations?

8. What is the significance of data logging and auditing in maintaining data integrity?

9. How do access controls contribute to data integrity?

10. What are the legal and regulatory requirements related to data integrity, and why are they important?

Cybersecurity Hacks

Hack 1: Create a Personal "Integrity Checklist" System

Develop a simple, reusable checklist that you can apply to any system or project to assess data integrity controls quickly. This consolidates this chapter's concepts into a practical tool:

The 5-Minute Integrity Assessment:

- **H**ashing—Are files/data using hash verification?
- **A**ccess—Who can modify this data, and is it restricted?
- **L**ogging—Are changes being tracked and auditable?
- **B**ackups—Is there a recent, tested backup available?
- **V**alidation—Are inputs being checked before entry?

Instead of conducting lengthy security reviews, use this "HALVB" acronym to rapidly identify integrity gaps in any system. This checklist approach helps you systematically apply complex concepts without missing critical elements, making data integrity assessment faster and more thorough.

Hack 2: Use the "Digital Fingerprint" Method for Personal Files

Implement a simple file integrity monitoring system using built-in OS tools to automatically detect tampering in your most critical documents.

Quick Implementation:

- Create a batch script (Windows) or shell script (Linux/Mac) that generates SHA-256 hashes for your important files
- Schedule it to run weekly and compare against previous hashes
- Set up email alerts for any changes to critical files like contracts, certificates, or financial records

Example Windows batch script:

```
for %%f in (C:\Important\*.pdf) do (
    certutil -hashfile "%%f" SHA256 >> integrity_log.txt
)
```

This gives you enterprise-level file integrity monitoring without expensive software. It's perfect for freelancers, small businesses, or personal use, automatically alerting you to ransomware, accidental modifications, or unauthorised access to your most valuable documents.

Chapter 8: Availability

– Always On, Always Ready –

Objectives

By the end of this chapter, you will be able to:

- Define availability as a fundamental component of the CIA Triad
- Identify various threats to availability
- Understand key High Availability (HA) principles
- Explain the importance of Disaster Recovery (DR) planning and business continuity measures in ensuring availability
- Appreciate network and cloud availability considerations.

All three components of the CIA Triad are equally critical in establishing a strong foundation for information security. This chapter builds upon our earlier discussions of Confidentiality and Integrity, completing our exploration of cybersecurity's core principles with the final pillar: Availability. Availability ensures that systems and information remain accessible when needed, enabling users to perform their tasks without interruption. In today's hyperconnected world, where digital access underpins organisational operations, maintaining consistent availability plays a big role in business continuity and user confidence.

The consequences of downtime can be severe. Unexpected system outages lead to financial losses, reputational damage and operational inefficiencies. Financially, disruptions result in lost revenue, increased operational costs and penalties for failing to meet Service Level Agreements (SLAs). Reputationally, customers may lose trust in an organisation's ability to provide reliable services.

High-profile incidents further underscore the critical importance of availability. Cyber-attacks such as Distributed Denial-of-Service (DDoS) and ransomware attacks have repeatedly demonstrated their capacity to cripple businesses. For instance, the 2017 WannaCry ransomware attack infected over 230,000 computers across 150 countries, disrupting major sectors, particularly healthcare. In the UK, hospitals had to cancel appointments and divert emergency patients, illustrating how availability breaches can directly impact essential services and potentially endanger lives.

Organisations implement HA and DR strategies to counteract these risks and ensure resilience. HA designs systems that continue operating even amid hardware or software failures. Core techniques include redundancy, load balancing and failover mechanisms, which prevent a single point of failure from disrupting services. For example, redundant servers and network paths enable seamless transitions to backup components when failures occur, maintaining uninterrupted operations.

DR equips organisations to swiftly restore services after major disruptions. Effective DR strategies incorporate frequent backups, geographically dispersed recovery sites and routine disaster simulations to test preparedness. The 2011 earthquake and tsunami in Japan demonstrated the value of comprehensive disaster recovery plans. Many data centres remained operational due to resilient infrastructure, strategic backup systems and pre-established contingency plans. This event highlighted the importance of DR frameworks that address not only IT infrastructure but also power, communication and resource distribution systems.

In this chapter, you will examine various threats to availability, from sophisticated cyberattacks to human errors, and explore the techniques organisations use to safeguard their systems. Striking a balance between availability and stringent security measures is a key part of creating a resilient cyber environment. By studying past incidents and implementing proactive strategies, you can help organisations mitigate risks, maintain operational stability and ensure continuous access to critical services and data.

Importance of Data Availability

System downtime and data inaccessibility can severely affect individuals and organisations. Cloud outages disrupt business operations, while cyberattacks can cripple critical infrastructure. As a fundamental cybersecurity principle, availability ensures uninterrupted access to digital resources, supporting personal and enterprise needs.

Why Availability is Vital for Individuals and Organisations

For Individuals

From online banking to cloud storage, people expect uninterrupted access to essential digital services. Ensuring availability minimises disruptions and reduces risks.

Financial Services: Online banking and digital payments must function reliably to prevent transaction failures and financial disruptions. A single outage can block access to funds, delay critical payments and damage customer trust.

Healthcare Access: The availability of medical records and telehealth services is critical for patient care. System failures can delay treatment decisions and compromise health outcomes, particularly in emergency situations.

Cloud Storage and Personal Data: Individuals store vital documents, photos and backups on cloud services. Availability issues can prevent access to important files during critical moments, highlighting the need for local backups and alternative access methods.

For Organisations

For businesses and institutions, availability is fundamental for productivity, customer satisfaction and regulatory compliance. Downtime leads to financial losses, reputational damage and operational setbacks.

E-commerce and Online Services: Digital businesses must remain operational 24/7. Even brief outages can cost millions in lost sales and damage brand reputation, as customers quickly seek alternatives when services are unavailable.

Manufacturing and Supply Chains: Just-in-time (JIT) manufacturing, where components arrive precisely when needed in production, depends on continuous uptime to prevent costly delays. System failures can halt entire production lines and create ripple effects throughout supply chains.

Government and Critical Infrastructure: National security and public services require reliable system access. Availability failures in emergency response systems, utility networks or communication infrastructure can threaten public safety and essential services.

The Role of Availability in the CIA Triad

Within the CIA Triad, Confidentiality and Integrity protect data, and Availability ensures users can access it when needed.

Availability and Integrity

Availability and integrity are closely linked; unreliable access can compromise data consistency. System failures due to cyberattacks or corruption can jeopardise both availability and integrity.

For instance, in 2017, the NotPetya malware attack crippled systems worldwide, rendering data inaccessible and corrupting critical files beyond recovery. Comprehensive backups and disaster recovery plans are vital to maintaining availability and data integrity.

To safeguard integrity, organisations implement error-checking mechanisms, data replication and multi-region redundancy. Cloud providers distribute data across multiple centres, ensuring both availability and protection against data loss.

Availability and Confidentiality

While confidentiality restricts data access to authorised users, availability ensures those users can access it without disruption. Striking the right balance is the challenge.

In 2023, a DDoS attack targeted Australian government websites, overwhelming systems and disrupting public services. While security measures protected sensitive data, the attack highlighted the importance of load balancing, rate limiting and traffic filtering in preserving availability.

Conversely, strict security protocols, such as MFA and access control policies, enhance confidentiality while ensuring seamless access for legitimate users.

Organisations also rely on Business Continuity Planning (BCP) and DR strategies to maintain availability. Geographically distributed data centres, automated failover mechanisms and real-time monitoring help mitigate disruptions caused by cyber threats or natural disasters.

In summary, availability is indispensable for modern digital systems. Whether ensuring financial transactions process smoothly, healthcare services remain functional or critical infrastructure operates without disruption, availability serves as a foundational cybersecurity pillar. By implementing resilient network architectures, redundancy strategies and proactive risk management, you can help organisations safeguard systems, prevent costly downtimes and ensure essential services remain accessible when they are needed most.

Threats to Availability

Imagine the frustration of a streaming service crashing just before a season finale. In today's digital environment, uninterrupted access to systems and data isn't just a convenience; it's perceived by many consumers as a necessity. Yet, various threats can disrupt operations, leading to significant downtime and economic loss. Understanding these risks is required for building resilient defences and ensuring the continuous availability of critical resources.

Types of Threats

Cyberattacks

Cyberattacks pose one of the most persistent and evolving threats to availability. They target systems with the intent to disrupt, compromise or destroy critical infrastructure. The three most common and dangerous forms of cyberattacks are ransomware, denial-of-service (DoS) and Distributed Denial-of-Service (DDoS) attacks.

Denial-of-Service (DoS): A DoS attack floods a targeted system, service or network with excessive traffic from a single source, rendering it inaccessible to legitimate users. These attacks can lead to downtime for businesses, financial losses and frustration for users.

Distributed Denial-of-Service (DDoS): A more sophisticated version of a DoS attack, DDoS uses multiple compromised computers (botnets) to attack a single system, making mitigation significantly more challenging. One of the most prominent real-world examples occurred in 2016, when a DDoS attack on the Australian Bureau of Statistics (ABS) website disrupted the national census, preventing millions of Australians from participating.[62] The attack demonstrated how critical infrastructure could be severely impacted by targeted cyber threats.

Ransomware: A form of malware that encrypts essential data, preventing access until a ransom is paid. Ransomware attacks have surged globally, with the U.S. being the most targeted country, accounting for 50.8% of ransomware cases in Q1 2024.[63] However, Australia is also heavily impacted, with ransomware attacks costing the economy $2.59 billion annually.[64]

Emerging Threats: Ransomware is always developing, with new attack methods such as triple extortion (where attackers demand ransom from multiple stakeholders) and Ransomware-as-a-Service (RaaS) (where cybercriminals sell ransomware kits to other hackers). These advancements make ransomware a continuously growing threat to your organisation's data availability.

Hardware Failures

Hardware failure remains a significant challenge for ensuring uninterrupted availability. Infrastructure components, such as disks, network switches, servers (both physical and virtual) and power supplies, are all prone to degradation and failure over time.

Since predicting which components might fail and when is difficult, organisations employ several protective measures:

Redundancy—Critical systems often use failover mechanisms, where a secondary system automatically takes over when the primary system fails.

Automated Monitoring—Many enterprises rely on real-time monitoring tools to detect hardware degradation before failures occur.

Backups and Disaster Recovery (DR) Plans—Routine data backups ensure that organisations can quickly restore operations in case of a failure.

One critical metric used to anticipate failures is Mean Time Between Failures (MTBF)—a measure of the average time between inherent failures during normal operation. MTBF helps businesses predict hardware lifespan and schedule timely replacements and upgrades.

In addition to standard hardware malfunctions, natural disasters such as hurricanes, earthquakes, floods and fires pose a severe risk to physical infrastructure. When critical hardware is damaged due to environmental factors, service disruptions can last for days or even weeks, highlighting the necessity for disaster recovery planning.

Software Bugs

Even the most meticulously designed software systems are not immune to coding errors, logic flaws and hidden vulnerabilities that can trigger anything from minor glitches to catastrophic failures. In an era where software underpins critical services, from transportation networks to financial transactions, a single bug can send ripples through an entire industry.

A notable example occurred in 2019, when a bug in Sabre's airline reservation system caused widespread flight cancellations and delays across multiple airlines.[65] Similarly, in 2021, a configuration error in Facebook's backbone network caused a six-hour outage affecting billions of users worldwide.[66] Such incidents demonstrate the need for:

- Thorough testing and debugging before deployment
- Automated error detection and correction systems
- Regular software updates and security patches to fix vulnerabilities
- Comprehensive change management processes to prevent configuration errors.

As you develop and maintain software systems, remember that thorough testing, code reviews and gradual deployment strategies are essential practices for ensuring availability. Given the increasing complexity of modern software, ensuring code reliability and system resilience is necessary for maintaining continuous availability of your critical services.

Human Error

Despite technological advancements in automation and AI, human error remains one of the most persistent threats to availability. A single misconfiguration, an overlooked detail or a lapse in protocol can result in major service outages, data loss or security breaches. Unlike cyberattacks, which require external threats to bypass defences, human error often originates from within, making it harder to predict and prevent.

Perhaps no incident illustrates this better than the 2021 Facebook outage, which disrupted Facebook, WhatsApp and Instagram for over 5.5 hours, cutting off communication for 3.5 billion users worldwide. The cause? A misconfigured router update—a seemingly small error with immense consequences. The event demonstrated how even the most sophisticated global platforms remain vulnerable to the simplest of mistakes.

To mitigate the risks associated with human error, organisations implement:

- Comprehensive employee training programs
- Strict procedural guidelines and system access controls
- Automation of repetitive tasks to minimise human intervention
- Rigorous change management processes with multiple approval checkpoints.

Advanced Persistent Threats (APTs)

APTs rank among the most elusive cyber threats, infiltrating networks and remaining undetected for months or years. Unlike conventional cyberattacks seeking immediate disruption, attackers orchestrate APTs with long-term objectives, including extracting sensitive information, sabotaging critical infrastructure or manipulating digital systems.

APTs employ multiple attack vectors designed to circumvent standard security measures. Phishing campaigns are common entry points, deceiving employees into revealing login credentials that enable deeper network penetration. Zero-day exploits allow threat actors to operate undetected within compromised systems. Once established, these cybercriminals methodically expand their access, silently gathering intelligence or altering operations for strategic advantage.

A notable example occurred in 2020 when the SolarWinds attack compromised numerous government agencies and private companies worldwide. The attackers remained undetected for months, gaining access to sensitive systems by compromising a trusted software update mechanism. This APT demonstrated how even well-protected organisations can fall victim to carefully orchestrated, patient attacks.

The interconnectivity of global systems means a single vulnerability can trigger cascading failures. APTs have demonstrated the capacity to disable power grids, disrupt financial markets and compromise government agencies. This potential for widespread damage makes continuous monitoring and layered security approaches essential for protecting vital digital infrastructure.

Insider Threats

While external cyberattacks dominate headlines, some of the most damaging security breaches originate from within an organisation itself. Whether intentional or accidental, insider threats pose a significant risk to system availability, as they exploit privileged access to compromise security, disrupt operations or leak sensitive data. Unlike external threats, insiders do not need to bypass firewalls or intrusion detection systems; they are already inside the gates.

A Ponemon Institute study revealed that insider threats cost U.S. businesses an average of $15.4 million annually,[67] with incidents steadily increasing. While some cases involve disgruntled employees deliberately sabotaging systems, others stem from careless mistakes, such as an IT administrator accidentally misconfiguring a database, leading to widespread downtime. In industries where data confidentiality and system uptime are critical, even minor lapses in judgment can have far-reaching consequences.

To combat insider threats, organisations implement strict access control policies, limiting employees' ability to modify critical systems without proper authorisation. Behavioural monitoring tools help detect unusual activity patterns, allowing security teams to identify potential threats before they escalate. Cybersecurity awareness programs further reinforce the importance of caution and compliance, ensuring that employees remain a line of defence rather than a point of vulnerability.

As you prepare for your career in cybersecurity, understanding such internal threats is just as important as recognising external attacks. The principles of least privilege access and separation of duties will be valuable tools in your professional toolkit for maintaining system availability.

Strategic Briefing: 2023 Record-Breaking DDoS Attack

In late 2023, a groundbreaking DDoS attack set new records and revealed the growing sophistication of cyber threats targeting system availability.[68] This attack exploited the RST_STREAM frame feature in the HTTP/2 protocol, allowing attackers to rapidly open and close streams, overwhelming servers by exhausting their computational resources. The method effectively circumvented traditional DDoS defences, including rate limiting and standard blocklists, highlighting the necessity for more advanced protection strategies.

The attack demonstrated the devastating potential of protocol-level exploits. It underscored the key role of adaptive defence mechanisms, particularly for organisations providing critical cloud infrastructure and services supporting global internet operations.

Impact and Scope

The DDoS attack triggered widespread service disruptions across major cloud service providers with significant consequences:

- Service outages affecting AWS, Google and Cloudflare platforms
- Unprecedented traffic volumes that overwhelmed existing mitigation systems
- Cascading failures impacting dependent services and applications
- Extensive website and application unavailability for end users

- Disruption of critical business operations across multiple industries.

Financial implications were substantial, with downtime costs averaging $8,850 per minute for affected businesses,[69] while larger organisations potentially faced losses reaching hundreds of thousands of dollars per hour. Beyond immediate financial damage, the incident affected countless users globally and eroded customer confidence in cloud service reliability.

Attack Methodology

The attackers exploited a vulnerability in the HTTP/2 protocol by manipulating its RST_STREAM frame feature. Their technique involved rapidly opening and closing connection streams, creating an efficient attack vector that required minimal resources to generate maximum impact. This methodology effectively bypassed conventional DDoS protection measures not designed to detect such protocol-level manipulation.

The attack achieved unprecedented scale, exceeding previous records. While a 2021 attack on an Azure customer (mitigated by Microsoft) had reached 3.45 terabits per second with 340 million packets per second, the HTTP/2 Rapid Reset attack employed more sophisticated methods focused on exhausting server resources rather than simply overwhelming bandwidth capacity.

Response and Mitigation

Upon detecting the attacks, major cloud providers activated emergency response protocols. Initial measures included implementing temporary protocol modifications and deploying specialised filtering rules to identify and block malicious traffic patterns while maintaining legitimate user connections.

Service providers maintained partial operations throughout the attack through traffic prioritisation and resource isolation strategies. They provided consistent status updates via service health dashboards and direct communication channels. The response culminated in coordinated industry action, with affected providers sharing threat intelligence and collaboratively developing protocol-level fixes to prevent similar attacks in future.

Lessons Learned

This record-breaking DDoS attack revealed critical insights for cloud providers and enterprise organisations regarding protocol security, defence architecture and operational resilience. The incident demonstrated the necessity for adaptive defence systems capable of identifying and responding to novel attack patterns in real-time. Protocol-aware traffic analysis proved vital for recognising malicious patterns, while cross-provider collaboration emerged as crucial for developing appropriate mitigations.

Technical architecture forms the foundation of effective DDoS defence. Organisations must implement multi-layered protection strategies that combine traditional volumetric filtering with protocol-level inspection. Regular security assessments focusing on protocol implementation vulnerabilities, paired with stress testing defensive systems under extreme conditions, significantly strengthen an organisation's technical security posture.

The attack illustrated how operational preparedness directly influences recovery success. Organisations must maintain incident response playbooks addressing availability threats and regularly update service continuity plans. This preparation should include documented degradation procedures for partial service operation and comprehensive communication protocols to guide stakeholder updates during extended outages.

Techniques for Ensuring Availability

When systems fail, businesses face significant consequences, from financial losses to reputational damage. This is why forward-thinking organisations implement strategic availability-enhancing measures rather than leaving resilience to chance. Redundancy, clustering and disaster recovery plans all serve as essential components in maintaining system availability. This section explores the key mechanisms that keep businesses operational even during disruptions. While we address immediate response strategies here, Chapters 14 and 15 provide a comprehensive discussion of recovery and business continuity planning.

High Availability (HA) Strategies

Redundancy

Implementing redundancy effectively across critical system components is fundamental to achieving HA. Redundant systems typically follow an active-passive configuration, where standby components remain idle until failure activates them. This approach eliminates single points of failure, ensuring continuous operation without service disruptions.

Key redundancy measures include additional hardware, software and network components that activate automatically when needed. Data centres routinely implement redundant power supplies, multiple network connections and backup storage systems to maintain uninterrupted service. Extending redundancy beyond local infrastructure, through geographically distributed data centres, for example, significantly enhances overall resilience and strengthens operational stability.

Clustering

Clustering integrates multiple servers into a unified system, typically functioning in an active-active configuration. Unlike passive redundancy, clustering enables all nodes to process workloads simultaneously, optimising computational efficiency. When a node fails, another immediately assumes its workload, maintaining seamless service availability without noticeable interruption.

Organisations commonly deploy clustering for mission-critical applications such as database management, high-demand file storage and web services, where continuous uptime is essential. Beyond enhancing availability, clustering improves system performance by distributing workloads intelligently, reducing latency and preventing server overload.

Load Balancing

Load balancing distributes incoming network traffic across multiple servers, stopping any single server from becoming overwhelmed. This strategy is particularly valuable during peak usage periods and is critical in mitigating DoS attacks that target a single entry point.

Implemented through hardware appliances or software solutions, load balancers continuously monitor server health and direct traffic to the most available resource. Major technology companies like Amazon and Google employ sophisticated load balancing technologies to manage their extensive global infrastructure. By eliminating processing bottlenecks and optimising resource allocation, load balancing enhances overall system responsiveness and user experience.

Disaster Recovery (DR) Strategies

Backups and Recovery

System failures and cyber threats frequently cause data loss, making backup and recovery strategies a core consideration. Regular, systematically structured backups ensure rapid data restoration following corruption, deletion or security breaches.

Organisations should implement a multi-layered backup approach, incorporating full system backups, incremental updates and differential snapshots to maintain data integrity. Storing these backups in multiple secure locations, including off-site data centres, cloud storage and physically secure vaults, protects against data loss from cyberattacks, fires or natural disasters. A well-designed backup and recovery strategy minimises downtime and safeguards operational continuity.

Disaster Recovery (DR) Planning

A DR plan provides a structured framework for restoring critical systems and infrastructure following unexpected disruptions. This document should be accessible in both digital and physical formats to ensure key personnel can execute recovery procedures without delay.

Regular testing and frequent updates are necessary to maintain the plan's relevance in response to emerging threats and evolving IT environments. Beyond technical recovery measures, a DR strategy builds organisational confidence in resilience and preparedness, supporting overall business continuity.

Business Continuity Planning (BCP)

BCP extends beyond IT infrastructure to ensure that all essential operations continue during major disruptions. This process involves identifying key services, personnel and dependencies, then developing sustainable strategies to maintain them under adverse conditions.

Effective continuity plans address a wide range of potential disruptions, from cyberattacks to environmental disasters and large-scale infrastructure failures. These plans should include clear communication protocols for stakeholders, regulatory bodies and the public, particularly when operational challenges might affect an organisation's reputation. Integrating business continuity planning with disaster recovery strategies enables organisations to develop an approach to resilience that minimises both financial and operational risks.

Testing and Validation

Regular testing provides the most reliable evaluation of HA and DR strategies. Systematic assessment under realistic conditions confirms that redundancy mechanisms, failover processes and recovery plans function as intended.

Effective testing methods include controlled failover drills, complete backup restorations, simulated cyberattacks and collaborative tabletop exercises. These assessments help organisations identify vulnerabilities and refine response strategies before actual disruptions occur. For instance, routine failover testing can reveal weaknesses in backup configurations, allowing teams to implement necessary improvements. A structured testing schedule ensures that HA and DR strategies remain current, effective and aligned with operational requirements.

Cost-Benefit Analysis (CBA)

Organisations often perceive HA and disaster recovery investments as expensive until they fully assess the financial impact of downtime. Unplanned outages result in lost revenue, reputational damage and regulatory penalties, making proactive investment in HA and DR strategies financially prudent.

A detailed CBA helps balance expenses associated with infrastructure, software and skilled personnel against the potential financial consequences of service disruptions. According to the Ponemon Institute's 2021 industry report, unplanned downtime can cost an estimated $9,000 per minute on average. By comparing these costs with the expenses of implementing HA and DR solutions, organisations can make informed investment decisions. Prioritising critical systems and services for HA optimises resource allocation and ensures maximum Return On Investment (ROI).

Network Availability and Security

Imagine your network as the circulatory system of your organisation; when it functions optimally and securely, information flows freely, powering every aspect of your operations. Just as a body needs protection from pathogens, your network requires defences against an ever-evolving array of threats. This section examines the essentials of network resilience, exploring key elements to maintain both availability and security. From preventing cyber-attacks to ensuring system reliability, we'll demonstrate how to create a network that balances functionality with security.

Network Security Controls

Network security controls form the foundation of overall network security and availability. Primary defences include firewalls, intrusion detection systems (IDSs) and Intrusion Prevention Systems (IPS). Firewalls function as gatekeepers, filtering incoming and outgoing traffic based on predetermined security rules. Stateful firewalls examine the state of active connections alongside packet filtering, maintaining security by blocking malicious traffic while allowing legitimate communications.

IDSs monitor network traffic for suspicious activity and alert administrators to potential threats, while IPSs detect and actively prevent malicious traffic. These systems play a valuable role in identifying and mitigating threats, including malware, DoS attacks and unauthorised access attempts.

Additional network security controls include VPNs, which encrypt connections between remote users and the network; Network Access Control (NAC), which verifies devices before permitting network access; and Security Information and Event Management (SIEM) systems, which consolidate and analyse security data. Implementing these layered defences ensures networks remain continuously monitored and protected against advancing threats.

Network Monitoring

Cyber threats operate continuously, making round-the-clock network monitoring essential. Real-time analysis of network activity helps you detect abnormal patterns, identify vulnerabilities and maintain operational stability. When you examine traffic flows, monitor system performance and flag irregular behaviour, you can quickly pinpoint issues such as bandwidth constraints, unauthorised access attempts and hardware failures. There are many tools available that provide the visibility needed to implement this approach, allowing you to address potential problems before they develop into significant disruptions or security breaches.

Network Segmentation

Consider a security breach as a digital incursion. How far will its effects spread through your network? That's your blast radius. By strategically segmenting your network, you create containment boundaries that limit how far an attack can propagate.

Network segmentation involves dividing a network into smaller, isolated segments to contain potential attacks. By separating critical systems and sensitive data from less secure areas, you prevent attackers from moving laterally within the network. This containment approach strengthens security by reducing the attack surface and limiting the potential damage of a breach.

A practical implementation example is the Demilitarised Zone (DMZ), an isolated network segment that functions as a buffer between the public internet and your private internal network. It typically hosts public-facing services, such as web and email servers, while shielding internal systems from direct external access. This structured approach to segmentation provides layered protection for your most valuable digital assets.

Balancing Security and Performance

Consider your network as a city. Too many security checkpoints, and traffic stalls; too few, and you're inviting trouble. This balance is critical. Emphasise security too heavily, and you'll frustrate users with delays and restrictions; prioritise convenience excessively, and you risk exposing your organisation to cyber threats.

Achieving the optimal balance requires a thoughtful approach that addresses both security requirements and operational needs. Excessively strict controls can impede efficiency and user experience, while inadequate security leaves networks vulnerable to attack.

Cloud-based security solutions offer advantages in maintaining this balance. Services like DDoS protection can dynamically scale to absorb large-scale attacks without degrading network performance. These services deliver effective protection while preserving the flexibility to handle varying traffic volumes, enabling you to adjust security measures according to current threat levels and specific business requirements.

Cloud Availability and Considerations

Just as your physical network requires protection, your cloud environment demands strategic planning to ensure continuous operation. Ensuring availability in cloud environments is essential for maintaining uninterrupted services and business continuity. To achieve effective cloud availability, you need to understand shared responsibilities, evaluate Service Level Agreements (SLAs), implement cloud-based disaster recovery solutions and select dependable providers that align with your organisation's specific requirements. This section examines how each of these elements contributes to a resilient cloud strategy.

Cloud Impact on Availability

Cloud computing enhances availability through distributed resources and scalable infrastructure, while introducing a shared responsibility model between providers and your organisation. Providers typically manage physical infrastructure, including data centres and hardware, while your organisation retains responsibility for data, applications and configurations. This division of duties requires a clear assignment of responsibilities to prevent security gaps that might compromise service continuity. Understanding this relationship is your first step toward maintaining consistent cloud availability.

Shared Responsibility Model

Understanding the shared responsibility model is fundamental to cloud security and availability. This model clearly defines which security tasks belong to the provider versus you, as the customer. Providers typically secure the underlying cloud infrastructure, hardware, networking and data centres, while your organisation must secure data, applications and access controls within the cloud environment.

The division of responsibilities varies by provider and service model. When your workload is hosted on Software as a Service (SaaS), Platform as a Service (PaaS) or Infrastructure as a Service (IaaS), each allocates different levels of control and responsibility. For instance, Amazon Web Services (AWS) secures the underlying infrastructure, but you must manage your data security and access permissions.[70] Clear documentation of these responsibilities ensures both parties maintain their roles in preserving availability.

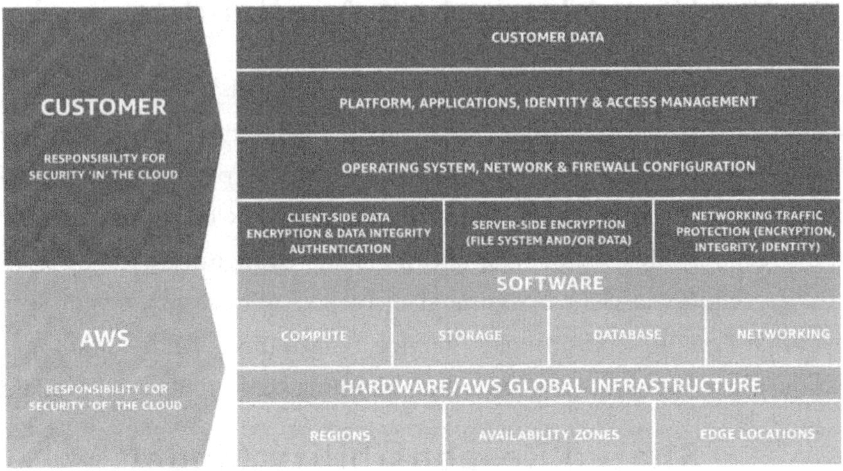

AWS Shared Responsibility Model - https://aws.amazon.com/compliance/shared-responsibility-model/

Service Level Agreements (SLAs)

SLAs define the critical terms of service between cloud providers and your organisation. They establish clear expectations and engagement rules and specify commitments from each party, including uptime guarantees, recovery timeframes and penalties for non-compliance.

When evaluating providers, prioritise those offering high uptime guarantees and clearly defined recovery times. For example, Microsoft Azure provides an SLA of 99.99% uptime for virtual machines, translating to minimal potential downtime.[71] You should thoroughly review SLAs to understand service levels and the implications of potential disruptions before making commitments. These agreements serve as your formal assurance of service quality and availability.

Cloud-Based DR Solutions

Cloud-based disaster recovery solutions offer powerful, cost-effective options for ensuring availability. These solutions enable you to replicate and back up data and applications in the cloud, allowing quick restoration during disruptions.

Services such as AWS Elastic Disaster Recovery[72] and Azure Site Recovery[73] provide automated failover and failback capabilities, reducing downtime and minimising data loss. The scalability and flexibility of cloud DR solutions make them particularly valuable for your organisation when seeking efficient, reliable recovery options that can adapt to changing needs. By implementing these solutions, you gain peace of mind and practical protection against service interruptions.

Choosing Reliable Providers

Many cloud providers now offer specialised services designed for HA. When selecting a provider, you should evaluate their security measures, DR plans and reliability track record. Key assessment factors include data centre redundancy, geographic diversity and incident response capabilities. Google Cloud Platform (GCP), for example, operates a global network of data centres with multiple redundancies, ensuring HA during regional outages.[74]

Additionally, assess each provider's compliance with industry standards and regulations relevant to your field, along with their customer support capabilities. Conduct thorough evaluations to select providers that align with your specific availability and security requirements. This careful selection process helps ensure your services remain resilient and accessible, even during unexpected disruption.

Summary

As we conclude our exploration of Availability, you can now appreciate its fundamental role as a principle of the CIA Triad. This often overlooked component ensures operational continuity, keeping systems and data accessible precisely when needed. Throughout this chapter, you've discovered the complexity of availability, from its far-reaching impact on business operations to the sophisticated strategies and technologies designed to maintain it. These concepts equip you with valuable insights for creating resilient systems that sustain business operations, even during challenging circumstances.

Availability directly influences business continuity and organisational reputation. When you experience downtime, whether from cyberattacks, hardware failures, software bugs or human error, your organisation faces significant financial losses and reduced productivity. For large enterprises, a single hour of downtime can cost hundreds of thousands of dollars, affecting both immediate revenue and long-term customer trust. Understanding these implications helps you justify and prioritise availability investments within your organisation.

In this chapter, you have learned that maintaining continuous availability requires an approach spanning prevention, mitigation and recovery strategies. Two critical approaches stand out: HA and DR. HA focuses on designing interruption-resistant systems through redundancy, clustering and load balancing to eliminate single points of failure. DR enables your organisation to recover quickly from major disruptions through systematic backups, recovery planning and business continuity procedures. By implementing both strategies together, you create effective protection against diverse availability threats, from minor outages to major disasters.

A key challenge you'll face is balancing availability with security measures. Excessively strict security controls can hinder system performance and accessibility, while inadequate security leaves systems vulnerable to availability-compromising attacks. Achieving this balance requires implementing network security controls such as firewalls and IDS/IPS that protect against unauthorised access while preserving performance. When you incorporate continuous network monitoring and segmentation, you further enhance this equilibrium by ensuring security measures accommodate legitimate traffic while quickly identifying and addressing potential threats.

As you've seen throughout this chapter, ensuring availability requires a multifaceted, balanced approach. By integrating HA and DR strategies, adopting emerging technologies and continuously refining your practices, your organisation can maintain consistent availability despite an evolving threat landscape. The knowledge you've gained here provides a foundation for implementing availability measures that protect critical systems, maintain business operations and safeguard your organisation's reputation in an increasingly connected digital environment.

Revision Questions

1. What is availability, and why is it considered a core component of the CIA Triad in cybersecurity?

2. How can downtime impact businesses financially and reputationally? Provide examples.

3. Explain the difference between High Availability (HA) and Disaster Recovery (DR). How do they complement each other?

4. Identify and describe the main types of threats to availability.

5. Why is it essential to have a multi-layered defence approach to address the increasing complexity of threats to availability?

6. What role do insider threats play in disrupting availability, and how can organisations mitigate these threats?

7. Discuss the significance of network security controls like firewalls, IDS and IPS in maintaining network availability.

8. How does continuous network monitoring contribute to maintaining availability?

9. Explain the importance of Service Level Agreements (SLAs) in cloud availability and what factors should be considered when evaluating SLAs.

10. What emerging trends in technology are shaping the future of availability, and how should organisations adapt their strategies to these trends?

Chapter 8

Cybersecurity Hacks

Hack 1: Create a Simple Availability Dashboard Using Free Tools

Transform complex system monitoring into an easy-to-understand visual dashboard that consolidates all your critical availability metrics in one place. Use free tools like Google Sheets combined with StatusCake or UptimeRobot's free tiers to create a real-time availability dashboard that tracks your most important systems.

How to implement:

- Set up free monitoring accounts with services like UptimeRobot (monitors up to 50 websites for free)
- Create a Google Sheet with columns for: System Name, Current Status, Uptime %, Last Incident, Response Time
- Use Google Sheets' built-in functions to pull data from monitoring APIs or manually update weekly
- Add conditional formatting to highlight systems with uptime below 99% in red, 99-99.5% in yellow, and above 99.5% in green
- Share this dashboard with your team and update it during weekly meetings.

This hack provides a practical tool that demonstrates your understanding of availability metrics to supervisors and colleagues.

Hack 2: Implement the "3-2-1 Rule" with Free Cloud Storage

Master disaster recovery fundamentals by implementing the industry-standard 3-2-1 backup rule using free cloud storage services, ensuring your critical data survives any single point of failure without breaking the budget.

The 3-2-1 Rule:

- 3 copies of important data (1 original + 2 backups)
- 2 different storage media types (e.g., local drive + cloud storage)
- 1 copy stored off-site (cloud storage)

How to implement for free:

- Use your computer's built-in backup (Windows Backup or macOS Time Machine) for local backup
- Set up free cloud storage accounts: Google Drive (15GB), Dropbox (2GB), OneDrive (5GB)
- Rotate weekly backups between different cloud providers to maximize free storage
- Create a simple spreadsheet tracking what's backed up where and when
- Set monthly calendar reminders to verify backups can be restored.

This approach reinforces the concepts of redundancy, disaster recovery planning and business continuity while providing hands-on experience with backup strategies that can be scaled up in professional environments.

Chapter 9: Cyber Threat Intelligence

– Signals from the Shadows –

Objectives

By the end of this chapter, you will be able to:

- Define Cyber Threat Intelligence (CTI) and explain its role in shifting from reactive to proactive cybersecurity
- Describe the five stages of the intelligence cycle adapted for cyber
- Differentiate between the various types of CTI
- Explain how CTI enhances cybersecurity defence
- Identify common challenges in implementing CTI.

Knowledge is more than power—it is the key to survival. CTI transforms disparate data points into a meaningful narrative that illuminates adversaries' intentions, capabilities and tactics. Unlike raw data, intelligence in the cybersecurity context follows a systematic process of collection, analysis and dissemination that converts information into actionable insights, enabling organisations to anticipate and counter threats before they materialise into breaches.

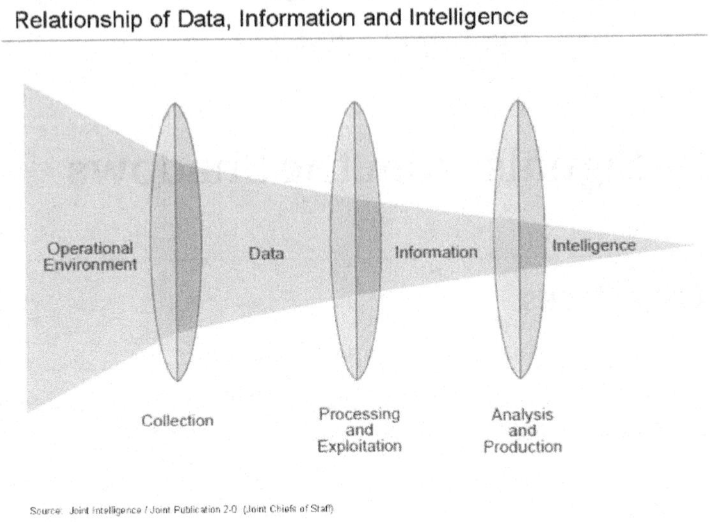

Source: By U.S. Joint Chiefs of Staff JP2-0.
https://commons.wikimedia.org/w/index.php?curid=47853614

The evolution of cybersecurity strategy mirrors warfare's historical progression, from reactive defences built around digital moats and walls to proactive operations that identify and neutralise threats at their source. This shift has occurred as cyber adversaries grow increasingly sophisticated, employing APTs, zero-day exploits and complex social engineering tactics that routinely bypass traditional security controls. The conventional model, detecting and responding to attacks after they occur, leaves organisations vulnerable to highly coordinated and technically advanced threats.

CTI bridges this gap by answering not just what is happening in a network, but also why and how it is occurring, and critically, what could happen next. This intelligence-driven approach enables security teams to:

- Detect early warning signs of cyberattacks before significant damage occurs
- Understand adversaries' Tactics, Techniques and Procedures (TTPs)
- Prioritise vulnerabilities based on real-world exploitation patterns

- Allocate security resources based on evidence rather than assumptions
- Shift from investigating past incidents to preventing future threats.

Take the 2017 WannaCry ransomware attack, which infected over 230,000 computers across 150 countries in just a few days, causing billions in damages. Organisations with mature CTI programs identified the malware's propagation method early, implemented targeted defences and significantly reduced their exposure.[75] Similarly, following the 2020 SolarWinds supply chain compromise, CTI was instrumental in identifying the sophisticated TTPs employed by the threat actors (later attributed to APT29 aka Cozy Bear), allowing organisations to detect indicators of compromise and mitigate further exploitation.[76]

Beyond its technical applications, CTI transforms cybersecurity from a tool-driven discipline to an intelligence-led approach focused on understanding cybercriminals and their behaviours. Instead of continually reacting to attacks, with CTI, you can anticipate threats, assess adversaries' strengths and weaknesses, and proactively implement security measures that disrupt cyberattacks before they succeed.

In this chapter, we will explore the intelligence cycle as it applies to cybersecurity, examine different types of CTI, demonstrate how to integrate intelligence into security frameworks and address the challenges organisations face when building an effective intelligence program. By mastering these concepts, you will gain the skills to transition from reactive cybersecurity to proactive threat hunting and strategic defence.

The Intelligence Cycle for Cyber Threats

The Intelligence Cycle is a foundational framework in CTI, providing a structured process for collecting, analysing and sharing intelligence on cyber threats. Adapted for CTI, it consists of five stages: Planning and Direction, Collection, Processing and Analysis, Dissemination and Collaboration, and Feedback. Each stage plays a vital role in ensuring intelligence remains actionable, timely and relevant to your organisation's cybersecurity posture.

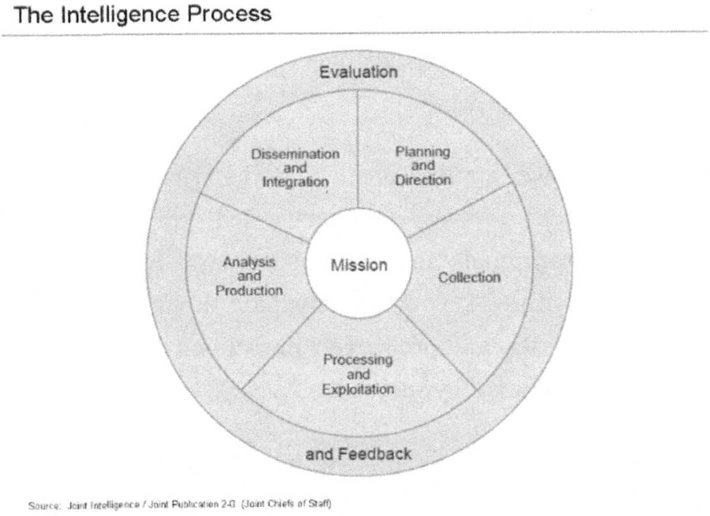

Source: By U.S. Joint Chiefs of Staff JP2-0.
https://commons.wikimedia.org/w/index.php?curid=47853466

Planning and Direction

The Planning and Direction stage lays the groundwork for the entire intelligence cycle. This phase defines intelligence collection requirements based on your organisation's specific needs and risk assessment. Cybersecurity teams identify key priorities, the threats to monitor, critical assets to protect and intelligence objectives. Strategic planning ensures intelligence activities align with your organisation's security goals. By setting clear directives, you can allocate resources efficiently, prevent data overload and generate intelligence that directly addresses your security concerns.

Collection

The Collection stage involves gathering cyber threat data from multiple sources, including internal logs, external threat feeds and Open-Source Intelligence (OSINT) such as news articles, social media and security forums. A diverse range of sources provides a comprehensive view of the threat landscape. Internal logs expose past incidents and system vulnerabilities, while external threat feeds deliver real-time updates on emerging threats and global trends. OSINT helps organisations uncover threat actors' activities and intentions using publicly available data. CTI partnership programs enhance intelligence-sharing, enabling organisations to strengthen their security collectively. This multi-source approach ensures more thorough and contextualised analysis.

Processing and Analysis

In the Processing and Analysis stage, raw data is refined into actionable intelligence. This involves sorting, filtering and interpreting collected information to identify patterns, Indicators of Compromise (IOCs) and potential attack vectors. Advanced analytical techniques, including ML and statistical analysis, help detect anomalies and predict emerging threats. The goal is to extract meaningful insights that inform decision-making and enhance security measures. For example, malware analysis may uncover novel attack methods, while correlating IOCs can map an adversary's TTPs, supporting more effective defensive strategies.

Dissemination and Collaboration

The Dissemination and Collaboration stage ensures processed intelligence reaches the right stakeholders at the right time, including security teams, incident responders and executive leadership. Effective dissemination strengthens decision-making and incident response. Collaboration with industry partners, Information Sharing and Analysis Centres (ISACs) and law enforcement agencies enhances intelligence quality and collective defence. By sharing intelligence beyond your organisation, you contribute to broader cybersecurity resilience.

Feedback

Feedback, often overlooked, is a valuable part of the intelligence cycle. This stage assesses CTI effectiveness and refines collection and analysis methods based on security operations (SECOPS) and stakeholder input. Continuous evaluation ensures intelligence remains adaptive to evolving cyber threats. Feedback mechanisms include post-incident reviews, intelligence accuracy assessments and stakeholder evaluations of the intelligence's practical value. This cyclical process drives ongoing optimisation of your CTI program.

Types of CTI

CTI comprises various intelligence categories, each serving distinct organisational needs. These categories offer insights ranging from high-level strategic perspectives to detailed technical data. Understanding their applications is essential for developing effective cybersecurity and incident response strategies. This section examines CTI types based on their sources and focus, highlighting their benefits and limitations for your SECOPS.

Strategic CTI

Strategic CTI delivers high-level intelligence on threat actors, their motivations, capabilities and potential targets. Senior management relies on this intelligence for cybersecurity decision-making, as it provides a comprehensive view of the threat landscape, assisting your organisation to anticipate emerging threats and align security resources accordingly. It includes analyses of geopolitical developments influencing cyber activities, such as nation-state actors targeting critical infrastructure. This intelligence enables your organisation to understand the broader context of threats and prepare for potential large-scale attacks. However, because of its broad focus, strategic CTI often lacks the specificity for immediate operational responses.

Tactical CTI

Tactical CTI concentrates on specific details about active threats, including malware samples, IOCs and mitigation techniques. This intelligence is critical for SECOPS teams requiring actionable information to defend against current threats. Tactical CTI helps you identify and respond to incidents quickly, reducing the impact of cyberattacks.

For instance, tactical CTI might provide a detailed analysis of a specific malware strain, including its behaviour patterns, command and control infrastructure, and associated IOCs. Security teams can apply this information to detect the malware within their networks and implement appropriate countermeasures. While highly actionable, tactical CTI can become challenging to manage due to the volume and velocity of incoming data.

Technical CTI

Technical CTI examines the detailed technical aspects of cyber threats, such as malware code, vulnerability exploits and attack methodologies. This intelligence is particularly valuable for incident responders and threat researchers who require in-depth technical knowledge to counter sophisticated threats. Technical CTI offers the detailed context necessary for advanced detection and mitigation strategies.

For instance, an analysis of a zero-day (aka 0-day) vulnerability might describe its exploitation methods and potential impact on affected systems. However, interpreting and implementing technical CTI effectively requires specialised expertise, which can be a limiting factor for some organisations.

Open-Source Intelligence (OSINT)

OSINT involves collecting publicly available information from sources like news articles, social media platforms and security forums. This intelligence type is accessible and often free, making it valuable for organisations with limited resources. OSINT provides insights into emerging threats, attacker tactics and vulnerabilities discussed in public domains, helping you stay informed about the latest cybersecurity trends. However, OSINT reliability varies considerably, often requiring verification from additional sources to ensure accuracy and relevance.

Commercial CTI

Commercial CTI encompasses intelligence feeds and analytical reports provided by specialised security vendors. These services deliver curated, high-quality intelligence on various threat actors and attack methods, frequently including insights not publicly available. Organisations seeking comprehensive and reliable intelligence to enhance their security posture typically use commercial CTI.

Security vendors provide detailed threat reports, including analysis of APTs, up-and-coming malware variants and threat actor profiles. These reports are invaluable for understanding and defending against sophisticated attacks. However, the primary limitation of commercial CTI is its cost, which may be prohibitive for smaller organisations with constrained security budgets.

Strategic Briefing: The 2020 SolarWinds Cyber Attack

In December 2020, investigators uncovered one of history's most significant and sophisticated cyberattacks, which targeted SolarWinds, a major IT management software provider. A nation-state actor exploited vulnerabilities in SolarWinds' Orion software platform, infiltrating government agencies, Fortune 500 companies and critical infrastructure operators.[77] This incident revealed the devastating potential of supply chain attacks and highlighted the essential role of CTI in modern cybersecurity defence.

Impact and Scope

The attack had far-reaching consequences, affecting nearly 18,000 SolarWinds customers who unknowingly installed the compromised Orion software update. Key impacts included:

- Compromise of U.S. government agencies, including the Departments of Treasury, Commerce and Homeland Security
- Infiltration of major private sector organisations, including Microsoft and cybersecurity firm FireEye
- Exfiltration of sensitive data and long-term network persistence by attackers
- Erosion of trust in supply chain security and third-party software vendors.

The breach, undetected for several months, highlighted the vulnerabilities inherent in complex IT ecosystems and the need for stronger supply chain security controls.

Attack Methodology

The attackers executed a highly sophisticated supply chain compromise by injecting malicious code into SolarWinds' software development environment.[78] This trojanised (maliciously altered) update, once deployed, created a backdoor that allowed adversaries to:

- Establish initial access through compromised Orion software
- Move laterally across victim networks while evading detection
- Exfiltrate sensitive data and maintain persistent access for months.

The stealth and precision of the attack demonstrated an advanced understanding of software supply chain vulnerabilities, making it one of the most concerning cyber incidents in history.

Response and Mitigation

The attack was first discovered by FireEye, which identified the breach within its own systems using advanced CTI tools. Upon investigation, FireEye uncovered the attack mechanisms, including specific IOCs and TTPs employed by the adversaries.[79] Immediate response actions included:

- Public disclosure of IOCs and TTPs by FireEye to enable widespread detection and response
- Rapid coordination between affected organisations and government agencies, including the Cybersecurity and Infrastructure Security Agency (CISA) and the FBI
- Patch deployment and forensic investigation by SolarWinds and impacted entities.

This collective response underscored the importance of timely information sharing in mitigating cyber threats.

Lessons Learned

The SolarWinds cyberattack revealed critical lessons for organisations reliant on third-party software and IT supply chains. It reinforced the need for:

- Enhanced supply chain security, including rigorous vetting of software updates and continuous monitoring of third-party vendors
- Proactive CTI strategies integrating threat intelligence feeds with security tools like SIEM and Endpoint Detection and Response (EDR) platforms
- Improved incident detection and response capabilities, emphasising threat hunting and behavioural analytics
- Strengthened collaboration between private and public sectors to share intelligence and coordinate defence efforts.

This incident demonstrated that sophisticated cyber threats demand equally advanced and coordinated defence strategies, reinforcing the indispensable role of CTI in cybersecurity resilience.

Using CTI in Cybersecurity Defence

CTI is a vital resource for enhancing your security posture. By applying this intelligence effectively, you can anticipate potential threats, improve incident response capabilities and systematically strengthen security measures. This section examines how CTI integrates into five critical aspects of cybersecurity defence: threat hunting, incident response, security awareness training, vulnerability management and security policy development.

Threat Hunting

Threat hunting employs searching for indicators of potential attacks within a network, using CTI as a guiding framework. This forward-looking approach enables organisations to identify and neutralise threats before they inflict significant damage. CTI provides essential context, including IOCs, attack patterns and threat actor profiles, allowing threat hunters to concentrate on the most critical and high-risk areas.

Consider a scenario where CTI reveals a new malware variant targeting financial institutions. Threat hunters within a bank can immediately search their network for IOCs specific to this malware. This preventative investigation can uncover malicious activity that might otherwise remain hidden, enabling timely intervention. Beyond immediate defence enhancement, threat hunting contributes significantly to developing a more extensive understanding of your specific threat landscape.

Other examples of IOCs include:

- **Unusual Outbound Traffic**—Unexpected data transfers to unfamiliar external IP addresses, especially large volumes of data leaving the network, may indicate data exfiltration or command-and-control (C2) communication.

- **Unexpected System Behaviour**—Unexplained system slowdowns, frequent crashes or disabled security tools can signal the presence of malware or an active attacker.

- **Suspicious File Hashes**—If a file's cryptographic hash (SHA-256) matches known malware signatures, it indicates the presence of malicious software.

- **Abnormal User Account Activity**—A spike in privileged account logins, logins from unusual locations or access attempts outside normal working hours could indicate compromised credentials or insider threats.

Incident Response

When a cyber incident occurs, a rapid and effective response is essential to limit damage and restore normal operations. CTI serves as a fundamental resource in incident response by providing specific information about the attacker's identity, motivations and methodologies. This intelligence empowers incident responders to make evidence-based decisions, establish clear priorities and deploy resources where they will have maximum impact.

During the 2017 WannaCry ransomware attack, organisations with access to relevant CTI quickly identified the ransomware's signature characteristics and implemented targeted defensive measures to contain the spread.[80] CTI helps responders accurately assess the scope of an attack, identify compromised systems and determine optimal mitigation strategies. Furthermore, post-incident analysis incorporating CTI can identify valuable lessons and improvement opportunities, strengthening future defence capabilities.

Security Awareness Training

One of the most effective methods for reducing cyber incident risk is educating employees about current attack methods based on contemporary CTI insights. Security awareness training programs that incorporate CTI provide employees with timely information about emerging threats and practical guidance on recognising and responding to them. Stay tuned for more of this topic in Chapter 18.

For example, if CTI indicates a surge in phishing attacks targeting a specific industry, training programs can incorporate authentic examples of these phishing emails alongside detailed identification techniques. This targeted approach enhances employee vigilance and improves defence against social engineering attacks. Regular training updates, informed by the latest CTI, ensure employees maintain awareness of evolving threats.

Vulnerability Management

Vulnerability management encompasses the identification, prioritisation, assessment and mitigation of security vulnerabilities within your systems (see Chapter 4). CTI delivers critical insights regarding which vulnerabilities threat actors are actively exploiting, allowing you to prioritise patching efforts according to current threat patterns.

For instance, CTI can identify specific vulnerabilities in widely used software that attackers are currently targeting. Resources, including Common Vulnerabilities and Exposures (CVEs) and Known Exploited Vulnerabilities (KEVs),[81] can help you locate such vulnerabilities. This intelligence enables you to prioritise addressing these specific vulnerabilities ahead of others. This risk-based approach ensures efficient allocation of limited resources to address the most significant threats first.

Security Policy Development

Is there value in creating cybersecurity policies that are not informed by your risk environment? The answer is unequivocally no. As examined in previous chapters, risk emerges from the combination of threats and vulnerabilities. Effective risk management requires governance activities, including the establishment of security policies. CTI informs security policy development by providing insights into the current threat landscape, helping you identify significant risks and implement appropriate policy controls.

If, for example, CTI indicates an increase in supply chain attacks, your organisation can develop specific policies requiring thorough vetting of third-party vendors and implement additional security measures to address this threat vector.

Integration with Other Security Tools and Processes

CTI derives from diverse sources requiring systematic integration and processing to generate actionable intelligence. The full benefits of CTI materialise only through integration with security tools and processes. This approach ensures CTI becomes an integral component of your organisation's defence strategy rather than an isolated information source.

For example, integrating CTI with SIEM systems enables automated correlation between threat intelligence and internal security events, facilitating faster detection and response. Additionally, incorporating CTI into automated incident response platforms streamlines workflows and reduces containment and remediation timeframes.

Challenges of Using CTI

Using CTI is not without challenges. Effectively leveraging CTI requires navigating issues such as:

- Data overload
- Converting raw data into actionable insights
- Managing the complexities of sharing and collaboration
- Balancing the costs associated with high-quality intelligence feeds.

Let's examine how to address these challenges to maximise the value of CTI.

Data Overload

One of the most pressing challenges of leveraging CTI is handling the enormous influx of data from multiple sources. Organisations receive intelligence from internal security logs, government agencies, commercial threat intelligence providers, industry groups and OSINT. This data varies widely in reliability, from high-confidence reports to unverified threat indicators, leading to information overload.

Security teams often struggle to filter through vast amounts of data to pinpoint relevant threats. This challenge intensifies when CTI feeds contain redundant, outdated or conflicting information. Critical threats may go unnoticed without effective processing mechanisms, increasing organisational risk exposure.

To combat data saturation, organisations must implement structured intelligence management strategies. Automated tools powered by Artificial Intelligence (AI) and Machine Learning (ML) refine CTI effectively. SIEM systems aggregate, correlate and prioritise threat intelligence by linking external data with internal network activity. SIEM platforms with CTI integration highlight patterns indicative of cyber threats, enabling security teams to focus on actionable intelligence rather than manually sifting through raw data.[82]

Additionally, organisations should establish clear intelligence prioritisation policies. Analysts need to define relevance criteria based on their specific threat landscape, focusing on intelligence aligning with their sector, infrastructure and risk profile. By setting predefined filters and thresholds, security teams can streamline data ingestion and target their defensive efforts at the most pressing threats.

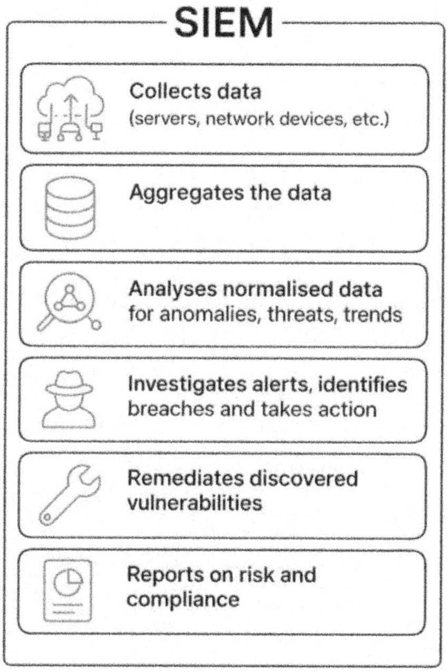

A SIEM solution brings together data from various sources, helping security teams manage workloads and enabling an effective data management strategy. Source: https://www.splunk.com/en_us/blog/learn/siem-security-information-event-management.html

Actionable Insights

Another major obstacle in using CTI effectively is the gap between raw intelligence and actionable insights. Collecting threat intelligence is insufficient; organisations must extract meaningful information from vast datasets and translate it into concrete defensive measures. Raw threat data often lacks useful context, making it difficult for security teams to understand its relevance and impact.

Cybersecurity professionals use structured frameworks to bridge this gap. The MITRE ATT&CK framework maps adversary tactics and techniques, helping analysts contextualise threats and understand attack methodologies. By categorising threats within a recognised framework, organisations can develop precise detection and response strategies tailored to specific attack methods.

Advanced analytical tools enhance intelligence interpretation. Threat intelligence platforms help security teams process large data volumes by automatically enriching raw indicators with contextual metadata. For example, a malicious IP address alone provides limited insight, but when cross-referenced with historical attack patterns, its role in a larger cyber campaign becomes clear.

Beyond technology, the human element remains critical. Analysts must possess expertise to recognise subtle threat indicators and correlate disparate intelligence sources. Regular training, simulated threat exercises and hands-on experience with threat-hunting methodologies equip cybersecurity professionals with the skills necessary to translate intelligence into meaningful security actions. When organisations prioritise skill development alongside technical solutions, they enhance their ability to convert intelligence into proactive defences.

Sharing and Collaboration

Threat intelligence sharing strengthens collective cybersecurity efforts, yet introduces significant challenges related to confidentiality, trust and legal constraints. Many organisations hesitate to share intelligence due to concerns about exposing sensitive information, risking competitive disadvantage or violating data protection regulations.

Despite these obstacles, structured intelligence-sharing frameworks facilitate secure collaboration. The ASD-ACSC supports initiatives such as the ACSC Partnership CTI Program, which enables organisations to receive and share intelligence while maintaining strict security protocols.

To standardise information exchange, organisations employ industry-accepted frameworks such as the Traffic Light Protocol (TLP),[83] Structured Threat Information eXpression (STIX)[84] and Trusted Automated eXchange of Intelligence Information (TAXII).[85] TLP categorises intelligence into different sensitivity levels, ensuring appropriate dissemination, while STIX provides a common language for describing threats. TAXII enables automated intelligence sharing, allowing security systems to exchange STIX-formatted reports seamlessly.

For example, an industry group tracking a phishing campaign might use STIX to structure intelligence about malicious domains and email headers. Security teams within the group can then access this data via TAXII and proactively block related threats before attackers gain a foothold. By leveraging structured sharing mechanisms, organisations enhance their defensive capabilities while preserving operational security and legal compliance.

Cost Considerations

The financial burden of acquiring and maintaining high-quality CTI feeds presents a formidable challenge, particularly for small to mid-sized organisations with limited cybersecurity budgets. Commercial intelligence services offer premium insights but come at a significant cost, often requiring annual subscriptions ranging from tens to hundreds of thousands of dollars. Additionally, the cost of integrating, analysing and acting on CTI can strain resources, particularly for teams with limited personnel.

Organisations should adopt a tiered intelligence approach to mitigate financial constraints. By combining free OSINT sources with selectively chosen commercial feeds, they can balance cost-effectiveness with threat awareness. Publicly available intelligence from government agencies, security blogs and industry forums provides valuable insights without the expense of proprietary feeds.

Furthermore, security teams must evaluate CTI investment returns based on tangible risk reduction. Consider a business debating whether to invest $50,000 in a CTI service. If the intelligence prevents a ransomware attack that could cost $200,000 in operational downtime and remediation, the investment proves highly cost-effective. Similarly, timely intelligence that enables rapid breach detection could prevent regulatory fines and reputational damage, further justifying the expenditure.

Automated threat intelligence integration optimises cost efficiency. Security Orchestration, Automation and Response (SOAR) platforms streamline intelligence processing, reducing manual labour and increasing response speed. Leveraging automation enables organisations to derive maximum value from their CTI investments while maintaining a sustainable financial model.

Overall, CTI is a vital component of modern cybersecurity, but its effective use requires overcoming key challenges such as data overload, extracting actionable insights, fostering secure collaboration and managing costs. Organisations can maximise the benefits of CTI by implementing effective data processing mechanisms, leveraging structured analysis frameworks, adopting secure intelligence-sharing protocols and optimising financial strategies. For cybersecurity newcomers, understanding these challenges and solutions lays a strong foundation for navigating the evolving threat landscape and contributing to effective cyber defence strategies.

Summary

CTI shifts cybersecurity from a reactive to a proactive approach, transforming raw data into actionable insights that help organisations anticipate and counter threats. This chapter examined how intelligence on adversaries and risks is systematically gathered, analysed and shared to enhance decision-making and strengthen defences.

The intelligence cycle structures CTI into five stages. Planning and Direction define intelligence needs based on risk assessments, ensuring efforts focus on critical threats. Collection compiles data from internal logs, external threat feeds and OSINT to create a comprehensive threat picture. Processing and Analysis refines raw data into intelligence by identifying patterns, IOCs and attack vectors. Dissemination ensures intelligence reaches the right stakeholders, while Feedback assesses effectiveness, refining future intelligence efforts.

CTI serves different organisational needs. Strategic CTI informs executives about threat actors, motivations and risks, guiding security investments. Tactical CTI provides time-sensitive intelligence for SECOPS teams responding to active threats. Technical CTI examines attack mechanisms, supporting incident responders and researchers. Organisations may also use OSINT, which draws from publicly available information, or Commercial Threat Intelligence, which provides curated insights from security vendors.

The 2020 SolarWinds cyber-attack demonstrates CTI's critical role. Attackers compromised 18,000 customers, including U.S. government agencies, via trojanised software updates, enabling unauthorised access. FireEye detected the breach using CTI tools and shared IOCs, allowing widespread threat mitigation. This case highlighted the importance of supply chain security, proactive intelligence strategies and stronger public-private collaboration.

CTI enhances cybersecurity in multiple areas. Threat hunting leverages IOCs and attack patterns to detect threats before they escalate. Incident response benefits from intelligence on adversaries' tactics, enabling faster, informed decision-making. Security awareness training uses real-world threats to educate employees, improving organisational vigilance. Vulnerability management prioritises patching efforts by focusing on actively exploited weaknesses. Security policy development ensures that security controls align with the most pressing risks in the shifting cyber landscape.

Despite its advantages, CTI presents challenges. The volume of threat data can overwhelm security teams, requiring advanced filtering and prioritisation. Effective intelligence analysis depends on skilled analysts and structured models like the MITRE ATT&CK framework. Confidentiality concerns often hinder information sharing, but protocols such as the TLP and formats like STIX/TAXII help standardise secure exchanges. Cost considerations, particularly for smaller organisations, necessitate balancing commercial intelligence feeds with open-source alternatives.

Addressing these challenges allows organisations to incorporate CTI into their cybersecurity strategy, transforming information gathering from a fragmented process into an intelligence-driven defence. This strategic shift empowers security teams to anticipate threats, respond effectively and strengthen resilience against increasingly sophisticated cyber adversaries.

Revision Questions

1. What is Cyber Threat Intelligence (CTI), and why is it critical in modern cybersecurity?

2. Explain the shift from reactive to proactive defence strategies in cybersecurity.

3. Describe the stages of the intelligence cycle adapted for CTI.

4. What are the different types of CTI based on source and focus?

5. How can organisations use CTI in threat hunting?

6. In what ways does CTI enhance incident response efforts?

7. What role does CTI play in security awareness training?

8. Discuss how CTI is integrated into vulnerability management.

9. Identify and explain two major challenges organisations face when implementing CTI.

10. Analyse how the SolarWinds cyberattack of 2020 demonstrates the importance of CTI in detecting and responding to sophisticated threats.

Cybersecurity Hacks

Hack 1: Build Your Personal CTI Dashboard Using Free Tools

Create a cost-effective threat intelligence monitoring system that consolidates multiple free sources into one accessible view.

How to Execute:

- Set up a simple dashboard using free tools like Google Sheets or Microsoft Excel combined with RSS feeds from reputable security sources (CISA alerts, security vendor blogs, CVE databases)
- Use Google Alerts to monitor specific keywords related to your industry or technology stack (e.g., "ransomware healthcare" or "phishing financial services")
- Create automated IFTTT (If This Then That) recipes to collect threat indicators from multiple sources into a single spreadsheet
- Schedule weekly reviews to identify patterns and prioritise threats relevant to your organisation.

This hack teaches you how to aggregate disparate intelligence sources without expensive commercial feeds while building practical skills in data organisation and pattern recognition.

Hack 2: Practice Threat Actor Profiling Through Open Source Research

Develop analytical skills by researching and profiling known threat actors using only publicly available information.

How to Execute:

- Choose a well-documented threat actor (e.g., APT29, Lazarus Group) and create a comprehensive profile using only OSINT sources
- Document their tactics, techniques, and procedures (TTPs) using the MITRE ATT&CK framework as your structure
- Track their evolution over time by analysing security research reports, court documents, and incident response case studies
- Create a one-page "threat actor card" summarising key attributes: motivation, capabilities, preferred targets and signature techniques
- Practice this monthly with different threat actors to build pattern recognition skills.

This exercise develops critical analytical thinking and research skills essential for CTI work.

Part II

Implementing Cybersecurity Measures

Chapter 10: Governance

– The Backbone of Cyber Resilience –

Objectives

By the end of this chapter, you will be able to:

- Define cybersecurity governance and explain its fundamental role in cybersecurity strategy
- Identify the key principles of cybersecurity governance structures
- Develop cybersecurity policies aligned with governance frameworks
- Learn how governance policies and security culture drive cybersecurity resilience
- Gain insights into governance effectiveness using appropriate metrics and maturity models.

In an era where cyber threats evolve faster than a mutating virus, cybersecurity governance isn't just another corporate buzzword—it's your organisation's survival manual. Let's demystify what this means, as we explore the first of the six cybersecurity functions outlined by the ISM and the NIST CSF and as described in Chapter 5.

At its core, cybersecurity governance is the framework guiding organisations in identifying, managing and overseeing cybersecurity. It encompasses activities from establishing strategies and policies to defining accountability and implementing oversight processes. This integrated set of practices ensures cybersecurity becomes embedded within an organisation's operations, culture and strategic decision-making.

Governance in cybersecurity extends beyond technical controls; it defines policies, roles and responsibilities that align security efforts with business objectives. Frameworks such as the NIST CSF and the Australian Signals Directorate Information Security Manual (ASD ISM) establish clear governance structures, ensuring accountability and compliance. The NIST CSF's Govern function sets cybersecurity policies, defines roles and allocates organisational responsibilities, while the ASD ISM outlines governance requirements specifically for Australian organisations.

Clear leadership and oversight are central to cybersecurity governance. This requires engaging key leadership roles, such as the Chief Information Security Officer (CISO), to provide strategic direction and align cybersecurity measures with your company's mission and stakeholder expectations. Such leadership is essential for integrating security risk management processes into the broader organisational risk management framework.

Cybersecurity governance functions as high-level oversight that guides decision-making, risk tolerance and policy enforcement. It helps you identify and assess the value of systems, applications and data, enabling more effective resource allocation and protection prioritisation. This process involves determining specific confidentiality, integrity and availability requirements for your assets, ensuring security measures address your organisation's particular needs.

Good governance fosters a proactive security approach. Instead of merely reacting to threats, organisations with strong governance structures anticipate potential risks and develop mitigation strategies beforehand. This includes implementing regular risk assessments to ensure security measures remain effective against evolving threats. Additionally, governance strengthens incident response capabilities by establishing clear protocols for detecting, responding to and recovering from cyber threats.

When governance is integrated into your cybersecurity strategy, you move from a reactive stance to a more strategic, risk-based approach. This elevates cybersecurity to a board-level concern and builds resilience in an increasingly unpredictable digital environment. As we proceed further through this chapter, we'll explore the key elements of cybersecurity governance and how to apply them to create a secure, adaptable organisation.

Principles of Cybersecurity Governance

Effective governance ensures that cybersecurity is an integrated aspect of business strategy, risk management and regulatory compliance. At its centre, governance establishes a structured framework that aligns cybersecurity with organisational objectives, promoting accountability, risk-informed decision-making and continuous oversight.

Executive Leadership and Commitment

Cybersecurity governance must originate at the highest levels of an organisation. Without leadership buy-in, cybersecurity risks remain confined within IT departments rather than being recognised as enterprise-wide concerns. Executive management must prioritise cybersecurity as a core business function, integrate it into the overall strategy and allocate sufficient resources.

This commitment becomes visible through establishing dedicated cybersecurity leadership, such as a CISO and cybersecurity steering committees. Security-conscious leadership builds stakeholder trust, strengthens operational resilience and safeguards long-term business viability. To achieve this, executives must stay informed about the rapidly developing threat landscape through regular briefings, scenario-based exercises and participation in cybersecurity training.

Governance Structure and Accountability

Regardless of size, all organisations need a governance structure that clearly defines the decision-making hierarchy, accountability structures and communication channels. This structure should establish governing bodies such as boards, committees and working groups, each with specific charters that articulate their purpose, roles and responsibilities for cybersecurity.

Clear ownership and accountability form the foundation of this structure. When you assign specific responsibilities to individuals or teams (through policies that people actually read), you ensure cybersecurity tasks are managed effectively. You'll also create clear escalation points for addressing issues. Your organisation should implement meaningful metrics and KPIs to measure cybersecurity effectiveness.

In well-designed governance models, cybersecurity functions extend across IT, legal, compliance and risk management teams, creating a multidisciplinary approach to security. This cross-functional collaboration enhances your organisation's ability to anticipate and address cyber threats that affect contractual agreements, supply chains, employee behaviour and business continuity planning.

Risk-Based Decision-Making

A practical cybersecurity governance framework relies on a risk-based approach, so that security measures align with the organisation's risk tolerance. Governance strategies must draw from risk assessments, allowing organisations to prioritise security investments based on the likelihood and impact of potential threats.

To guide risk-informed decision-making, you can adopt industry-recognised frameworks such as the NIST CSF, the NIST RMF, ISO 31000 and the ASD ISM, as discussed in Chapter 5. These methodologies help you define governance policies that address both technical and business risks, promoting a proactive rather than reactive approach to cybersecurity.

Policy Development and Regulatory Compliance

Cybersecurity governance depends on establishing policies that define security expectations and ensure regulatory compliance. These policies must be anchored in recognised industry standards, providing a structured approach to risk management, incident response and data protection.

Compliance with relevant sectoral regulations and industry best practices is non-negotiable in the current regulatory environment. Your organisation must adhere to your regionally specific legal obligations, such as:

- *The Australian Privacy Act 1988* and associated Australian Privacy Principles (APPs)

- *The General Data Protection Regulation* (GDPR)
- *The Australian Cyber Security Act* (CSA)
- Industry-specific guidelines like *The Health Insurance Portability and Accountability Act* (HIPAA)
- Sector-specific mandates like the Australian Prudential Regulation Authority CPS 234 (APRA CPS 234).

Regular security audits, compliance assessments and incident reviews confirm that policies are followed in practice, not just on paper. Governance frameworks should also incorporate a policy lifecycle approach, ensuring policies remain relevant through systematic reviews and updates.

Continuous Oversight and Adaptive Governance

Cybersecurity governance is dynamic; it must evolve in response to new threats, regulatory changes and business growth. Ongoing monitoring and assessment are vital to maintain the effectiveness of governance structures. A critical aspect of adaptive governance is resilience planning. Organisations must develop comprehensive incident response frameworks to effectively detect, respond to and recover from cyber incidents.

Finally, governance should cultivate a culture of continuous improvement. Lessons from security incidents, regulatory reviews and industry best practices should inform your framework evolution. Each incident should become a learning opportunity.

Integrating cybersecurity governance with your broader risk management strategy allows leaders to create a strong security posture that aligns with business objectives. Governance extends beyond policies and controls; it represents an ongoing commitment to security leadership, informed decision-making and continuous adaptation. That adaptability might just turn out to be your greatest asset in an ever-changing threat landscape.

Strategic Briefing: Updating Cyber Security Governance Principles

Background

In late 2024, the Australian Institute of Company Directors (AICD) and Cyber Security Cooperative Research Centre (CSCRC) successfully completed a significant update to the nation's Cyber Security Governance Principles.[86] This project, resulting in Version 2 of the Principles,[87] was a response to the rapidly evolving cyber threat landscape and increasing regulatory focus in Australia.

Objectives

The primary goals of this update were to address modern cyber threats and governance challenges, provide an updated framework for enhanced cyber resilience and offer practical guidance for proactive board oversight. These objectives reflected the growing concern about cyber vulnerabilities across Australian businesses and institutions.

Key Updates

Version 2 of the Principles incorporated several crucial emerging issues. First, it addressed digital supply chain risks, recognising the growing interdependence of businesses and their digital ecosystems. When companies rely on external vendors and partners for digital services, they inherit potential vulnerabilities that must be managed carefully.

Second, the update emphasised data governance, highlighting the importance of responsible data management and protection. As organisations collect and process increasing amounts of sensitive information, proper data handling becomes essential for maintaining trust and compliance.

Finally, the Principles provided enhanced guidance regarding effective cyber incident response and recovery, offering clear direction on preparedness and action during cyber crises to minimise damage and recovery time.

Implementation

The updated Principles include a practical implementation checklist emphasising actionable governance measures. Organisations are encouraged to establish clear roles and responsibilities for cybersecurity within their structures, ensuring accountability at all levels. The guidance recommended appointing a cyber 'champion' at the board level to advocate for cyber resilience and keep security concerns visible in strategic discussions.

Boards are also encouraged to transition from passive oversight to active engagement in cybersecurity matters. This involves regular improvements of security postures and meaningful participation in risk assessment processes. The checklist also offers guidance on third-party risk management, particularly regarding assessing digital service providers' security controls.

To enhance practical application, the project included case studies from corporate leaders and insights from recent major cybersecurity incidents, making the guidelines more relatable and applicable to real-world scenarios.

Significance and Impact

This update is particularly significant due to its alignment with the Australian government's 2023-2030 Cyber Security Strategy,[88] supporting the national goal of becoming a world leader in cyber security by 2030. The timing of the update meant it offered apt guidance for businesses during a period of rising cyber threats and evolving regulatory demands. Additionally, the framework provides adaptable best practices that could be implemented across various industries and organisation sizes, from small businesses to large corporations.

Outcomes and Future Directions

The successful update of the Cyber Security Governance Principles has increased awareness of cybersecurity as a critical governance issue at the board level. By elevating these concerns to strategic discussions, organisations can better anticipate and mitigate risks. The Principles provide a clear roadmap for organisations to enhance cyber resilience through systematic improvements and cultural changes. Perhaps most importantly, the update has fostered a more proactive approach to cybersecurity oversight among Australian directors, shifting from reactive crisis management to preventative strategy.

Moving forward, the AICD and CSCRC plan to continue monitoring the cyber landscape and provide regular updates to the Principles. They also aim to develop additional resources and training programs to support the implementation of these guidelines across Australian businesses.

This case study demonstrates how timely updates to governance frameworks can significantly contribute to national cybersecurity efforts and business resilience in an increasingly digital world. Understanding these principles and their practical applications can give you valuable insights into how organisations can protect themselves while navigating digital transformation.

Developing Cybersecurity Policies

In our increasingly complex and hostile cyber environment, well-defined cybersecurity policies form the backbone of governance, translating high-level principles into actionable security practices. To put it plainly—without a structured approach to risk management, organisations remain vulnerable to escalating threats, regulatory non-compliance and operational disruptions. These policies connect governance principles with practical security measures, ensuring risk management efforts are coherent, enforceable and aligned with broader business objectives.

Understanding Cybersecurity Governance

Governance involves overseeing, directing and establishing appropriate structures to support an organisation in achieving its mission and goals. In cybersecurity, the mission is to prevent security compromises while ensuring business continuity and minimising risk. This is typically achieved through a documentation hierarchy that guides an organisation at all levels, with cybersecurity policies at its core.

A cybersecurity policy framework establishes a unified security posture across an organisation. This ensures every department, from IT to human resources, operates under consistent security guidelines, reducing inconsistencies that could create exploitable vulnerabilities. These policies define the organisation's approach to security, provide clear models for employees, set standards for behaviour, outline responsibilities and establish procedures to follow.

Essential Cybersecurity Policies

Several foundational policies are critical to establishing effective security practices. These policies create a framework for implementing cybersecurity measures that manage risks and apply the CIA Triad. While not exhaustive, the following policies provide an excellent starting point:

Security Governance Policies

These policies define the roles of accountable authorities and management structures for security decisions. At the core is the Risk Management Policy, which aligns governance with an organisation's risk tolerance, defining clear risk appetite thresholds that guide security decision-making. Integrating risk assessment into business operations enables cybersecurity considerations to become an integral component of corporate governance rather than an afterthought.

Security governance policies should cover:

- Clear accountability through defined roles of authorities
- Management structures and responsibilities for security decisions
- Security planning and risk management processes
- Security maturity monitoring to adapt to emerging threats
- Governance for contracted goods and service providers
- Data breach notification protocols
- Incident response frameworks.

Organisations can effectively counter emerging threats by implementing continuous risk monitoring and aligning with industry frameworks such as ISO 27005, the NIST CFS, ISM, Essential Eight and APRA CPS 234.

Information Security Policies

These policies govern how information is classified and handled to reduce the risk of compromise. Key elements include Access Control and Acceptable Use Policies, which support a strong security culture by defining who can access systems and data. Role-Based Access Control (RBAC) and the principle of least privilege, limiting user permissions to only what is necessary for their responsibilities, reduce the likelihood of insider threats and accidental data exposure.

Essential information security policies include:

- Classification systems for protecting different types of information appropriately
- Access controls based on roles and responsibilities
- Safeguards against common and emerging cyber threats
- Requirements for secure ICT systems
- Acceptable use guidelines for ICT resources, including mobile devices
- Password policies for creating and maintaining strong credentials.

You'll also want to incorporate MFA and identity verification mechanisms to add security layers.

Incident Response and Business Continuity Policies

These policies put governance into action by ensuring organisations can detect, respond to and recover from cyber incidents effectively. They define incident classification and escalation procedures, allowing security teams to swiftly contain threats before they escalate. Predefined response plans facilitate rapid containment and mitigation, minimising downtime and financial losses.

Embedding business continuity and disaster recovery within cybersecurity policies helps organisations sustain operations during disruptions. Ongoing refinement through post-incident analysis enables them to strengthen resilience by adapting policies in response to real-world events.

Regulatory Compliance Policies

These policies ensure alignment with frameworks and legislation such as *The Australian Cyber Security Act* (CSA), ISM, Essential Eight, GDPR, APRA CPS 234 and Australia's *Security of Critical Infrastructure (SOCI) Act*. They outline an organisation's legal obligations, defining data protection and privacy measures that govern how sensitive information is handled. Compliance policies also establish protocols for mandatory reporting and regulatory audits, promoting transparency and accountability in cybersecurity operations.

Remember: Compliance isn't just about avoiding fines. It's about protecting your customers' trust and your organisation's reputation.

Personnel Security Policies

Critical for managing insider threats, these policies detail screening and vetting processes for potential employees, assessing their eligibility and suitability. They address ongoing assessments and the management of personnel separation, ensuring employees and contractors meet high standards of integrity and honesty.

Key personnel security policies include:

- Eligibility and suitability verification through screening processes
- Guidelines for the continuous evaluation of personnel
- Protocols for managing separation when staff leave the organisation.

Physical Security Policies

These policies define physical security measures and control mechanisms to safeguard an organisation's resources and information. They establish guidelines for providing a safe and secure physical environment for personnel, information and assets, covering aspects like building access controls and security pass management.

Ensuring Policy Effectiveness

To ensure cybersecurity policies remain effective, you should adopt these best practices for enforcement, accountability and periodic review:

- **Executive Buy-in**—Leadership support ensures cybersecurity policies are prioritised and integrated across all business units.
- **Clarity and Accessibility**—Policies must be clear, concise and readily accessible in straightforward language that all staff can understand, regardless of technical expertise.
- **Regular Training**—Employee awareness programs help reinforce security policies, reducing human error and insider threats while fostering a security-conscious culture.

- **Automated Monitoring**—Tools such as Security Information and Event Management (SIEM) systems provide real-time enforcement and alert mechanisms to detect policy violations.

- **Periodic Reviews**—Regular policy audits and updates allow organisations to stay ahead of new threats, regulatory changes and technological advancements.

Organisations strengthen cybersecurity by turning governance principles into practical security strategies and policies. A well-defined, risk-based cybersecurity policy serves as the foundation for a proactive approach that reduces threats, ensures regulatory compliance and supports ongoing business operations.

Fostering Cybersecurity Awareness

Your employees are your first line of defence against cyberattacks, particularly phishing and social engineering tactics. While governance frameworks remain essential, they're useless without genuine employee engagement. A strong security culture cultivates a shared responsibility (think of it as everyone watching each other's backs) that collectively protects your digital assets.

Importance of Cybersecurity Awareness

When awareness is lacking, employee approaches to security become inconsistent and chaotic, significantly increasing vulnerability. Human error continues to be a primary factor in security breaches. However, by educating staff and nurturing a security-conscious environment, you can dramatically reduce incident rates.

Cyber attackers methodically exploit human vulnerabilities through phishing schemes, weak password practices and social engineering tactics. A single employee clicking a malicious link can compromise an entire organisation's infrastructure. A well-established security culture not only prevents attacks but also strengthens incident response capabilities when breaches occur.

Security guidelines must transcend static documentation in employee handbooks to become integrated into daily workflows. This integration requires governance structures translating policies into practical, actionable behaviours through consistent engagement, education and appropriate enforcement mechanisms.

Key Elements of a Cybersecurity Culture

Leadership Commitment

Leadership commitment drives an effective cybersecurity culture. When executives visibly prioritise security initiatives, they signal its critical importance throughout your organisation. This commitment manifests through consistent communications, active participation in training programs and strategic resource allocation.

Executives must demonstrate security principles through their actions. When senior management diligently follows security protocols, they reinforce that security is non-negotiable rather than optional. Ideally, cybersecurity becomes embedded within your organisation's core values and integrated into strategic planning at every level.

Security Awareness Training

Effective training programs educate employees about emerging threats, best practices and proper reporting procedures. Generic approaches are largely ineffective; training should be tailored to specific roles and responsibilities. Comprehensive programs should address organisational policies, phishing identification, password management, safe internet practices and incident reporting.

Interactive methodologies, including gamification elements, simulated phishing exercises and hands-on workshops, significantly enhance learning outcomes and knowledge retention. However, organisations must carefully balance training frequency to prevent employee fatigue and disengagement.

Open Communication

Fostering open communication regarding security incidents enables swift detection and response. A 'no-blame' reporting culture empowers employees to report suspicious activities without fear of repercussions. Clear reporting channels, including anonymous options, create collaborative security environments where collective vigilance becomes the norm.

Shared Responsibility

Establishing cybersecurity as everyone's responsibility, not exclusively IT's domain, is fundamental to organisational security. All employees should clearly understand their role in maintaining security posture through regular reminders, explicit inclusion in job descriptions and recognition of exemplary practices. This shared responsibility extends to external service providers and contractors accessing organisational systems.

Incentives and Recognition

Implementing thoughtful incentives encourages consistent adherence to security protocols. Recognition of strong security practices motivates wider adoption across your organisation. Effective incentives range from formal acknowledgment to tangible rewards that reinforce positive behaviours.

Simultaneously, governance structures must establish clear consequences for security violations. Striking an appropriate balance between positive reinforcement and accountability measures is essential, as purely punitive approaches may generate fear rather than engagement. In contrast, a solely positive approach might fail to ensure necessary compliance.

Strategies for Promoting Cybersecurity Awareness

Different training methods support various learning preferences by combining traditional sessions, interactive online modules, collaborative workshops and gamified experiences. As cybersecurity threats constantly evolve, regular updates and refresher courses are needed to reinforce fundamental concepts while addressing emerging threats.

When incorporating real-world scenarios into training, including simulated phishing attacks, you must conduct these exercises ethically to avoid negative consequences such as employee humiliation. Security awareness programs should also address cognitive biases, as many employees incorrectly assume they're unlikely targets, leading to dangerous complacency.

Making security personally relevant by highlighting connections to employees' digital lives transforms cybersecurity from an abstract concept into a practical, relatable aspect of daily activities. By implementing these elements cohesively, you'll cultivate a culture of cybersecurity awareness that strengthens their overall security posture and empowers every employee to actively protect valuable digital assets.

Implementing and Maintaining Governance

So far, we've discussed the core elements and principles of cybersecurity governance necessary for success. Yet understanding these components alone doesn't provide the framework your organisation needs to identify, manage and oversee cybersecurity risks effectively. A well-designed governance model integrates these elements to create a practical, functioning framework. Think of it as your organisation's cybersecurity blueprint.

What is a Governance Model?

A governance model is a structured framework that defines the specific roles, responsibilities, processes and decision-making authorities within an organisation. This model ensures effective oversight, establishes clear accountability and aligns cybersecurity practices with your organisation's strategic objectives.[89] It's the organisational chart of your cybersecurity world.

The major components are:

This approach is commonly applied to corporate governance structures in most medium to large organisations, though smaller organisations can benefit from implementing a scaled-down version.

Applying this model to cybersecurity governance requires integrating security considerations within each component. For instance:

- The governance structure should include specific descriptions of executive management committees' cybersecurity responsibilities and conduct.
- The accountability component outlines the application of cybersecurity roles and responsibilities across all organisational levels
- Cultural dynamics encompass targeted activities and programs that foster a positive security culture
- Infrastructure incorporates cybersecurity policies as critical elements within the organisation's governance model.

Embedding cybersecurity principles into each aspect of the governance model allows organisations to establish a comprehensive approach to implementing and maintaining effective security oversight.

Implementing and maintaining effective cybersecurity governance also involves essential activities, including communication and awareness, training and education, monitoring and auditing, continuous improvement and cultural considerations. We will now examine these in more detail. Remember, effective governance requires commitment at all organisational levels and a proactive approach to adapting to evolving cyber threats.

Communication and Awareness

Effective communication forms the foundation of successful cybersecurity governance. It's like being the town crier for cybersecurity; you need to get the message out clearly and consistently. This involves disseminating information about the governance framework, policies and procedures to all employees, ensuring everyone understands their roles in maintaining security.

You should leverage multiple channels to share cybersecurity information: emails, intranet postings and meetings. Regular updates keep cybersecurity at the forefront of employees' minds. However, be cautious: overwhelming employees with excessive communication can lead them to tune out important security messages.

Meanwhile, awareness campaigns, including posters and newsletters, reinforce key messages and highlight cybersecurity's importance throughout the organisation.

Monitoring and Auditing

Regular monitoring and auditing are crucial for ensuring compliance with cybersecurity policies and identifying potential gaps. Continuous monitoring employs tools and techniques to track compliance and detect deviations from established security practices and policies.

Auditing involves periodic reviews and assessments of the organisation's cybersecurity posture and Information Security Management System (ISMS). These evaluations can be conducted internally or by external parties to objectively assess the organisation's security measures. For example, the Information Security Registered Assessors Program (IRAP) provides Australian ASD-endorsed cybersecurity professionals to perform assessment services in securing broader industry and Australian Government systems and data.[90]

Identifying and addressing non-compliance issues promptly maintains the integrity of the cybersecurity governance framework. Implementing automated monitoring tools can also enhance the efficiency and effectiveness of these oversight processes.

Continuous Improvement

You would have gathered by now that cybersecurity is a dynamic field, with new threats and technologies emerging regularly. What worked yesterday might not work tomorrow. Therefore, take care to review and update governance frameworks and policies regularly.

Continuous improvement involves assessing the effectiveness of current practices, learning from incidents and adapting to changes in the threat landscape and regulatory environment. Consider establishing processes for regularly collecting feedback from employees and stakeholders to identify areas for enhancement.[91]

Incorporating lessons learned from security incidents and industry best practices helps refine and strengthen governance measures. Staying informed about the latest developments in the field and participating in industry forums and training also support ongoing improvement efforts across the organisation.

Measuring Governance Effectiveness

Achieving successful cybersecurity governance requires continuous measurement and refinement beyond mere policy creation. Without clear metrics, organisations can't determine whether security efforts truly mitigate threats or simply satisfy compliance requirements. Meaningful governance metrics enable you to evaluate your cybersecurity posture and align it with both regulatory demands and business objectives.

Risk reduction assessment is fundamental. Track incident response by measuring Mean Time to Detect (MTTD) and Respond (MTTR) to security events, as well as metrics that reveal how quickly your organisation identifies and addresses threats. More on this later. Monitor security incidents over time to determine whether governance efforts yield tangible risk reductions. The percentage of threats mitigated before becoming critical incidents indicates proactive control effectiveness, while vulnerability remediation rates show how efficiently your team addresses weaknesses.

Compliance serves as a key indicator of cybersecurity effectiveness. Organisations should assess their security posture against recognised standards such as the ASD Essential Eight, ISO 27001, NIST CSF or the ASD ISM. Audit findings provide objective insights into adherence, with any gaps highlighting areas for improvement, while comparing implemented policies against required controls helps gauge operational effectiveness. Participation in employee security training also reflects awareness levels; high completion rates suggest a strong security culture, while low engagement may point to weaknesses in governance.

Employ Governance Maturity Models for systematic evaluation. The Cybersecurity Capability Maturity Model (C2M2) assesses governance from ad hoc to optimised practices. The Centre for Internet Security Risk Assessment Model (CISRAM) quantifies effectiveness by mapping security measures to risk exposure, while NIST CSF helps assess policies against industry standards. Control Objectives for Information and Related Technologies (COBIT) aligns cybersecurity with business goals, positioning security as a strategic enabler.

Governance assessments must keep pace with the rate of change in the threat landscape. Regular evaluation ensures alignment with current threats and regulatory requirements. Automated tools like SIEM platforms and Governance, Risk and Compliance (GRC) solutions deliver real-time insights that enhance decision-making.

Effective governance depends on structured feedback loops. Insights from security teams, executives and auditors should drive improvements, while incident lessons inform future strategies. Most crucially, governance must adapt continuously so that your organisation remains resilient.

Summary

Cybersecurity governance provides the strategic foundation for managing cyber risks and ensuring organisational resilience in an era of escalating threats. This chapter explored the principles, structures and practices that form effective governance frameworks, demonstrating how organisations can embed security into their operational DNA.

Cybersecurity governance at its core extends beyond technical controls to define policies, roles and responsibilities, aligning security with business objectives. This framework establishes leadership structures, allowing organisations to shift to proactive security postures. By integrating governance into their strategy, organisations can elevate security to boardroom priority, building resilience in an increasingly volatile landscape.

Effective cybersecurity governance rests on several fundamental principles. Security must originate at the highest organisational levels, with leadership demonstrating commitment through dedicated roles and regular engagement. This commitment flows through defined hierarchies with specific charters that articulate responsibilities. Governance frameworks must align with risk tolerance, drawing from comprehensive assessments to prioritise investments. Further, organisations must establish policies to define security expectations and ensure adherence to regulatory requirements, such as *The Australian Privacy Act*. Finally, security frameworks must evolve in response to new threats, regulatory changes and business growth.

Cybersecurity policies translate principles into actionable practices, forming the backbone of governance. These include Security Governance Policies that define risk management, Information Security Policies that guide data classification and Incident Response Policies that ensure effective threat mitigation. Additionally, Regulatory Compliance Policies align with legal frameworks, while Personnel and Physical Security Policies manage insider threats and safeguard resources.

Your employees constitute the first line of defence against cyberattacks; they're your human firewall. Creating a strong security culture requires leadership commitment, tailored awareness training, open communication, shared responsibility models and balanced incentive programs. When developing training, employ diverse methods, provide regular updates and incorporate real-world scenarios.

Implementing cybersecurity governance involves embedding security principles into every aspect of the organisational model. Effective governance requires consistent communication, regular monitoring, continuous improvement and positive cultural reinforcement.

You can measure governance effectiveness through several approaches. Risk reduction metrics provide quantitative assessments of your security posture. Compliance evaluations against frameworks like the Essential Eight establish your standing against industry standards. Governance maturity models help assess organisational progression from basic to advanced security practices. These assessments must keep up with the changing threat landscape, supported by feedback loops that drive strategy refinement.

This chapter provided the knowledge necessary to establish effective cybersecurity governance, equipping you to protect digital assets while maintaining operational efficiency and regulatory compliance in complex cyber environments.

Revision Questions

1. What is cybersecurity governance, and why is it critical in managing cyber risks for organisations?

2. How has the cybersecurity environment shifted in recent years, and what impact does this have on governance?

3. Explain the core principles for establishing effective cybersecurity governance.

4. What are the benefits of having a cybersecurity governance framework?

5. What are the key components of an effective cybersecurity governance model, and how do they interact?

6. How do cybersecurity policies translate governance principles into operational practices?

7. What types of cybersecurity policies are crucial for organisations, and what do they entail?

8. Why is fostering a culture of cybersecurity awareness important, and what are its key elements?

9. What steps are involved in the effective implementation and ongoing maintenance of cybersecurity governance?

10. Why is continuous improvement essential in cybersecurity governance, and how can organisations achieve it?

Cybersecurity Hacks

Hack 1: Create a Cybersecurity Awareness Calendar:

Establishing a comprehensive cybersecurity awareness calendar represents a valuable strategic approach for organisations seeking employee engagement. By having monthly themes, such as phishing prevention, password security and data protection, companies maintain security as an ongoing priority rather than a one-time initiative. Each theme can be supported through educational materials, focused training sessions and interactive activities reinforcing key concepts. This methodical approach ensures cybersecurity remains at the forefront of employee consciousness while gradually building a security-focused organisational culture throughout the year.

Hack 2: Establish "Security Champion" Networks:

Implement a grassroots "Security Champion" program by identifying enthusiastic employees across different departments who can serve as cybersecurity advocates and first points of contact. These champions can receive slightly enhanced security training and act as liaisons between their teams and the IT security department. This approach leverages existing social networks within the organisation to amplify security awareness without requiring massive resource investments. Champions can help identify department-specific security concerns, assist with policy interpretation, and provide peer-to-peer support during security incidents. This hack creates a distributed governance model that strengthens the security culture from within, making cybersecurity feel less like an external imposition and more like a shared responsibility.

Chapter 11: Identification

– Mapping Assets, Risks and Responsibilities –

Objectives

By the end of this chapter, you will be able to:

- Explain the role of the Identify principle in cybersecurity
- Understand how to develop an effective asset inventory covering hardware, software, data and personnel
- Apply risk identification techniques to uncover security gaps before they escalate into vulnerabilities
- Establish classification frameworks to prioritise cybersecurity efforts based on asset value and risk
- Recognise a scalable identification strategy tailored for organisations of different sizes.

In the last chapter, we covered the Govern function, the first of the six cybersecurity functions outlined by ISM and the NIST CSF (Govern, Identify, Protect, Detect, Respond and Recover). We now move onto the second function, Identify, which is the foundation for an effective cybersecurity posture.

Identification means systematically recognising and cataloguing all assets in your organisation's digital environment. It's like taking stock in a warehouse; you need to know what's there before you can protect anything. The Identify function offers visibility into your technology landscape and its risks, forming the basis for all other security measures.

At its core, identification encompasses several essential components. Asset management forms the foundation by maintaining an inventory of hardware, software, data and network components. This inventory feeds into risk assessment, where potential threats and vulnerabilities are evaluated. Governance then defines policies that align cybersecurity with business objectives. Finally, supply chain security manages risks within third-party relationships, recognising that modern organisations rarely operate in isolation.

Without knowing what assets you possess, securing them is effectively impossible. Poor visibility into IT infrastructure creates blind spots, leaving vulnerabilities exposed. Attackers specifically target unmanaged systems, outdated software and unpatched vulnerabilities that remain invisible to security teams.

The WannaCry ransomware outbreak of 2017 exemplifies the consequences of identification failures.[92] This global attack exploited unpatched Windows systems that organisations failed to identify, causing billions in damages across multiple sectors. This incident is a clear reminder that poor identification isn't just risky—it's costly. Being able to prevent such a breach is a valuable professional skill.

Effective identification reveals your attack surface (all points where unauthorised access or data loss could occur). In a small business, this might comprise a dozen computers and a server. But the challenge becomes far more complex in large organisations, especially in government. Overlooking asset management is often what causes the weakest links in your defences.

The identification process involves discovering assets, documenting their details and applying governance to monitor for changes. Organisations implement identification strategies through both automated tools and manual verification processes.

Standards like ISO 27001 and the ASD's Essential Eight offer structured asset management methods. These frameworks are especially important in Australia, where laws like *The Security of Critical Infrastructure (SOCI) Act* (2018) require detailed identification practices across sectors such as energy and telecommunications.

The Identify principle also supports the other cybersecurity functions. Protection can only work when security teams know what to protect. Detection capabilities rely on establishing accurate baselines of normal system behaviour. Similarly, Response and Recovery efforts require all-inclusive asset information to prioritise actions.

This chapter gives you practical strategies for implementing identification, helping organisations manage risk before problems emerge. Strengthening this key skill builds resilience against modern cyber threats, an ability that will serve you well throughout your career.

Asset Management: A Bedrock of Cybersecurity

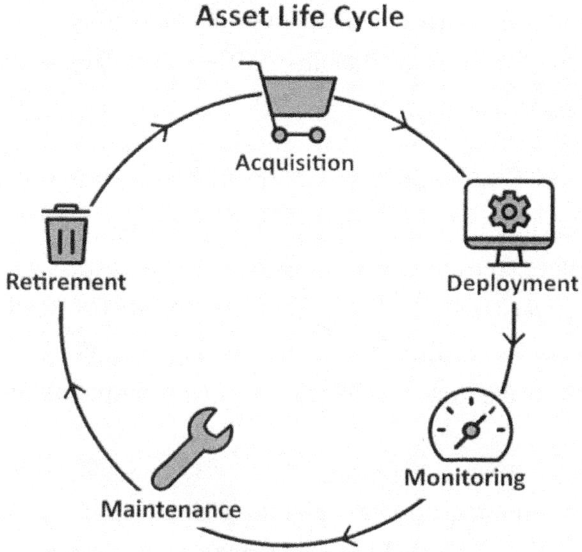

Fundamentals of Cyber Asset Management

Organisations now rely on an ever-expanding network of interconnected assets, ranging from physical hardware to cloud-based services and critical data repositories, including Large Language Models (LLMs). Cyber asset management forms the foundation upon which all security strategies are built, ensuring visibility, control and protection of these essential resources. The ISO 55000:2024 standard defines asset management as the "coordinated activity to realise value from assets",[93] highlighting its role in maximising efficiency and mitigating risks.

This concept of asset management is formalised within cybersecurity through two key frameworks: the ASD's ISM and the NIST CSF. These frameworks provide you with structured approaches to categorising and assessing security risks across your organisation's digital landscape.

When using the ISM, you will rely on identification principles categorised into three key areas. These principles help assess and document the business importance, confidentiality, integrity and availability requirements of systems, applications and data. This documentation serves as a foundation for implementing security controls that align with your organisation's operational needs.

Complementing this approach, the NIST framework's Identify function emphasises practical asset management through six interconnected components. These components establish systematic practices for identifying physical and software assets, determining appropriate ownership, prioritising critical resources according to their business value and maintaining continuous monitoring systems that alert you to potential security issues.

Your cyber asset management strategy must integrate several interconnected elements. Strategic planning aligns asset activities with your organisational goals, while lifecycle management oversees assets from acquisition to disposal. Accurate data management maintains inventories and condition assessments, working alongside financial management practices like lifecycle costing. These elements function within risk management processes that help identify and mitigate asset-related vulnerabilities, creating a continuous improvement framework.

Integrating cyber asset management with broader enterprise risk management enables your organisation to enhance regulatory compliance, streamline operations and strengthen its security posture. In our current threat environment, maintaining clear visibility into your organisational assets isn't merely advantageous, it's essential for building and maintaining resilience against increasingly sophisticated attacks.

Building a Comprehensive Asset Inventory

Creating and maintaining a complete inventory stands as one of the most critical steps in asset management (seriously, you can't protect what you don't know exists). This process involves developing a detailed, systematic record that catalogues all hardware, software and network components within your organisation. An effective inventory encompasses hardware, software, cloud services, data assets and personnel, ensuring every digital and physical component receives proper accounting.

Hardware assets include traditional endpoints like servers, workstations and network devices but also extend to mobile devices, Internet of Things (IoT) endpoints and operational technology (OT) systems. Software assets span from operating systems and enterprise applications to unlicensed or legacy software, which frequently introduce security risks. With growing reliance on cloud infrastructure, tracking SaaS, PaaS and IaaS offerings becomes equally important.

A comprehensive inventory requires several key components working in concert. Each item needs a unique identifier for tracking, accompanied by detailed descriptions of software and hardware versions and patching histories. Support information must document end-of-support and end-of-life dates, while condition tracking monitors age and status for maintenance planning. Assets require categorisation based on criticality, sensitivity and security requirements, supported by appropriate documentation including security plans and manuals that respect classification requirements.

Maintaining accurate inventories presents significant challenges, yet modern cybersecurity solutions offer automated asset discovery and continuous monitoring capabilities. Tools like Tanium and CrowdStrike provide endpoint detection, while vulnerability scanners such as Qualys and Nessus enhance visibility. IT Asset Management solutions like System Centre Configuration Manager (SCCM) and ServiceNow consolidate records, while cloud-native tools, including AWS Config and Azure Defender, deliver real-time insights into cloud environments.

Without a thorough understanding of existing assets, securing them becomes impossible. A well-managed inventory not only strengthens security but enables efficient resource alignment, preventing overlooked vulnerabilities and compliance failures.

Classification and Categorisation for Risk-Based Management

Once you've established an inventory, the next step involves classifying and categorising assets based on their criticality, sensitivity and security requirements. Not all assets hold equal value; some prove indispensable to business continuity, while others contain sensitive data requiring stringent protection. Effective security depends on thoughtful classification and categorisation, ensuring resources receive appropriate safeguarding based on their risk profile.

Classification determines each asset's importance to your organisation and the potential impact if compromised. Assets are typically evaluated based on their CIA impact. High-value systems, often called 'The Crown Jewels', warrant heightened security measures due to their operational significance or regulatory exposure. The ASD Essential Eight provides a structured approach for prioritising asset protection, emphasising patch management, MFA and access control for these critical systems.

Business Impact Analysis (BIA) plays a crucial role in identifying critical assets and establishing Recovery Time Objectives (RTOs) and Recovery Point Objectives (RPOs), which will guide incident response and disaster recovery planning. This analysis forms the foundation of risk-based asset management, ensuring protection measures align with business priorities.

Categorisation involves grouping assets with similar characteristics or security needs, facilitating more efficient management. Logical diagrams of the organisation's ICT infrastructure assist in grouping digital assets, supporting targeted security controls for specific asset groups, optimised resource allocation and streamlined compliance with regulatory requirements.

Security classifications also intersect with compliance mandates. ISO 27001 requires asset inventories within information security management systems, while the ASD ISM describes classification based on CIA principles. Similarly, the GDPR mandates detailed processing records, emphasising the necessity of meticulous asset tracking in today's regulatory landscape.

Ownership, Roles and Responsibilities in Asset Identification

Good asset management extends beyond technical functions. It requires clear ownership, accountability and cross-organisational collaboration. Location tracking is a principal component, involving knowledge of hardware and software component placement and assigning mobile assets like laptops and phones that may contain sensitive data. This physical tracking works alongside virtual monitoring of data movement across networks and cloud environments to create comprehensive asset visibility.

IT teams shoulder responsibility for provisioning and decommissioning assets, ensuring infrastructure remains secure and current. Security teams play a complementary role by identifying vulnerabilities, enforcing controls and monitoring threats. Business units contribute by aligning assets with operational objectives and regulatory requirements. This tripartite structure creates checks and balances that strengthen overall asset governance.

A well-defined governance model fosters coordination between these teams. Configuration Management Databases (CMDBs) constitute central repositories tracking ownership, changes and dependencies. However, maintaining accurate CMDBs presents persistent challenges due to outdated records and incomplete data. Automation significantly enhances accuracy by integrating real-time platforms like ServiceNow with discovery tools to update records, reducing manual errors and improving visibility.

Beyond IT governance, data stewardship ensures proper classification and protection of sensitive assets. Data stewards enforce security policies, manage access rights and maintain data integrity across your organisation. A common risk to data integrity emerges through shadow IT (unauthorised applications, devices or cloud services introduced without security oversight). People often resort to shadow IT when official systems move at the speed of a glacier; they simply want tools that meet their needs and want them now. But these unmanaged technologies can create dangerous blind spots, significantly increasing the risk of breaches through unmonitored endpoints.

You must implement strategies to detect and mitigate shadow IT risks through Network Traffic Analysis (NTA) and Cloud Access Security Brokers (CASBs). Several enterprises have experienced this danger when unapproved cloud storage buckets exposed sensitive customer data.[94] Proper asset governance could have prevented such exposures by ensuring visibility and access controls across all digital resources. Effective location tracking enables quick identification and remediation of security incidents while keeping all assets properly secured and accounted for.

Dynamic Asset Discovery and Inventory Management

Asset inventories must be continuously updated to reflect an organisation's evolving infrastructure. This ongoing monitoring ensures awareness of the current asset landscape, facilitating proactive security measures in response to changes. As you develop your cybersecurity practices, you'll encounter various asset management approaches, each offering distinct advantages and challenges.

Automated discovery tools provide invaluable capabilities for scanning networks and identifying assets in real-time. These sophisticated tools employ algorithms to detect and catalogue hardware and software, delivering continuous updates to asset inventories. Automation ensures inventories remain current even in rapidly changing environments. These solutions also excel at identifying unauthorised infrastructure additions and flagging potential security risks for investigation.

Despite automation's advantages, manual inventory processes remain necessary, particularly for identifying non-standard or less visible assets. These processes incorporate physical audits and verification checks to ensure all assets receive proper accounting, including those that automated tools might miss. Traditional manual tracking can introduce challenges through human error, which is why many organisations implement a hybrid approach combining technological efficiency with human verification.

Your selection of management tools depends on organisational scale. Large enterprises benefit from comprehensive platforms like Qualys and ServiceNow ITAM, while mid-sized companies may leverage SCCM or Ivanti. Small businesses can achieve appropriate visibility through cost-effective options such as Spiceworks or Open-AudIT.

Integrating asset management with complementary security practices like vulnerability management and access control strengthens your overall security posture. This creates a foundation for cybersecurity resilience that adapts to changing threats while minimising potential security gaps.

Risk Identification and Cybersecurity Resilience

Principles of Risk Identification

Risk identification is essential to effective cybersecurity risk management, serving as a precursor to risk assessment. While risk assessment evaluates the likelihood and impact of threats, as discussed in Chapter 4, risk identification focuses on systematically discovering potential risks *before* they escalate into security incidents. Even the most thorough assessments can overlook critical threats without identification processes, resulting in flawed mitigation strategies and unaddressed vulnerabilities.

Cybersecurity depends on identifying vulnerabilities before adversaries can exploit them, a challenging yet necessary activity. A proactive approach to risk mapping assists in this process by helping organisations anticipate threats before they materialise. When you leverage intelligence sources such as industry reports, historical incident data and threat intelligence feeds, you'll more effectively predict and prepare for emerging risks. The frameworks described in Chapter 5 also provide valuable structured methodologies for mapping vulnerabilities and risk systematically, ensuring organisations maintain a consistent and repeatable identification process.

Cybersecurity risks span multiple categories: operational, financial, reputational, regulatory and technological. Identifying risks within these categories allows you to align security priorities with business objectives, appropriately safeguarding critical systems and sensitive data.

Continuous monitoring plays a pivotal role in risk identification by providing real-time visibility into security threats. Automated tools such as SIEM systems, intrusion detection systems and vulnerability scanners enhance your ability to detect threats early. However, technology alone can't replace human judgment. Expert analysis remains essential for interpreting contextual risks that automated tools might miss. Integrating technology with expert analysis, you'll develop a risk identification process that strengthens your organisation's cybersecurity resilience.

Identifying Weaknesses in the Organisation

Picture trying to secure a house when you don't know where all the doors and windows are. That's the challenge security teams face with gaps in asset visibility. Cybersecurity resilience hinges on your ability to identify and address weaknesses before they become exploitable vulnerabilities.

As briefly mentioned already, one of the most significant blind spots is shadow IT, which refers to unauthorised or unmonitored devices, applications and cloud services operating outside security teams' supervision. Legacy systems with outdated software introduce security risks due to unpatched vulnerabilities, while misconfigurations in cloud services, identity management and network controls further expose assets to potential threats. Without maintaining a centralised asset inventory, security teams struggle to control an organisation's digital footprint, significantly increasing the likelihood of security incidents.

Numerous breaches illustrate the consequences of inadequate risk identification. The 2017 Equifax breach (detailed in Chapter 7) demonstrates how failing to patch a known Apache Struts vulnerability resulted in exposing 147 million records. More recently, the 2020 SolarWinds attack shows how undetected supply chain compromises allowed threat actors to infiltrate government and corporate networks worldwide. These incidents emphasise why risk identification and IT asset visibility must be organisational priorities.

To address these risks, you should employ structured identification methods, including security audits, penetration testing and attack surface management. Regular security awareness training enables employees to recognise social engineering attempts and potential insider threats, fostering a vigilance culture throughout your organisation. Applying the principle of least privilege allows you to significantly reduce risk exposure by ensuring staff access only to the resources necessary for their specific roles.

When organisations integrate risk identification into their broader business risk management framework, cybersecurity evolves from a technical consideration to a strategic priority. This strategic alignment enhances the overall security posture while securing leadership support and appropriate budget allocation for essential security initiatives that protect critical assets.

The next step involves assessing cybersecurity risks through a comprehensive process to identify asset vulnerabilities, validating and recording them. These vulnerabilities must be evaluated against the latest Cyber Threat Intelligence (CTI) from various sources, as described in Chapter 9.

Supply Chain and Third-Party Risk Identification

Modern cybersecurity risks extend well beyond your organisation's internal infrastructure. Vendors, suppliers and service providers can introduce vulnerabilities that threaten your security posture, making supply chain risk identification vital for cybersecurity resilience.

Third-party risks manifest in multiple forms, including data exposure, weak security practices and software vulnerabilities. Understanding these risks requires mapping your entire supply chain and identifying interdependencies that could amplify security threats. Regulations such as Australia's *Security of Critical Infrastructure SOCI Act* emphasise the securing of supply chains, particularly for organisations operating in critical industries where breaches could have far-reaching consequences.

Assessing third-party cyber risks begins with conducting rigorous due diligence before onboarding external vendors. Frameworks such as the ASD's Supply Chain Security Guidance and NIST's Cyber Supply Chain Risk

Management provide structured approaches for evaluating third-party risks. To maintain visibility and control over supply chain security, you should implement security audits, establish contractual security obligations and continuously monitor third-party access and activity across your systems.

Significant breaches illustrate the profound impact of supply chain vulnerabilities. During the 2020 SolarWinds attack, threat actors inserted malware into a widely used IT management tool. In 2021, the Kaseya VSA attack exploited a remote monitoring tool to distribute ransomware to managed service providers and their clients.[95] These incidents demonstrate why you need stringent third-party security vetting, continuous monitoring and robust incident response preparedness.

To strengthen supply chain security further, you should assess hardware and software authenticity and integrity before acquisition and use. Additionally, evaluating critical suppliers before procurement helps maintain your organisation's supply chain integrity and reliability, reducing compromise risk through trusted relationships.

Establishing a Risk Register

A risk register serves as a central repository for tracking identified risks, assigning accountability and prioritising mitigation efforts. By documenting risks systematically, you ensure informed decision-making when allocating security resources to critical threats.

Structure your risk register with key components: clear risk descriptions defining potential impacts, severity assessments using likelihood and impact ratings (see Chapter 4 – qualitative assessments, risk matrix) and criticality assessments mapping risks to business objectives. Document existing and planned mitigation controls and assign specific stakeholders as risk owners responsible for managing each identified risk.

Differentiate between inherent risk (before controls) and residual risk (after controls) to assess your security measures' effectiveness. By evaluating vulnerabilities against current threat intelligence, you can document threats with their potential impacts and exploitation likelihood, enabling prioritised risk responses and appropriate management strategies.

Governance, Risk and Compliance (GRC) platforms streamline risk tracking and reporting, while color-coded risk heatmaps help visualise priorities for executive decision-making. Meanwhile, regular updates and periodic stakeholder reviews ensure alignment with enterprise risk management strategies and changing business operations.

Integrating risk register insights with incident response planning improves your ability to anticipate and mitigate threats. AI-driven analytics can help predict emerging risks and prioritise mitigation efforts based on business impact and exploitation likelihood.

Cybersecurity Identification in Action: Industry Applications

Cybersecurity identification isn't one-size-fits-all. It must be tailored to each industry's unique needs. Different sectors possess distinct assets, vulnerabilities and regulatory requirements that shape their identification strategies. Effective cybersecurity begins with understanding what requires protection, potential threats and who's responsible for what. Let's explore how asset identification, risk assessment and responsibility mapping operate across government, finance and healthcare sectors.

Identifying Critical Assets in Different Sectors

Understanding what constitutes a critical asset is the basis of any cybersecurity strategy. Industries define and prioritise assets based on regulatory mandates, operational importance and exposure to cyber threats.

Government and Defence: National Security Data Identification Strategies

For government and defence organisations, protecting national security data represents the highest priority. Information classification plays a central role, with data categorised into unclassified, classified and top-secret tiers, each requiring stringent access controls. Defence agencies must account for supply chain security as third-party contractors often handle sensitive information.

Integrating CTI forms an essential aspect of cybersecurity identification in this sector. By mapping assets against known adversarial tactics, governments can prioritise defensive measures against espionage, sabotage and cyber warfare. Regulatory frameworks like the ASD's Essential Eight, the ISM and the Defence Industry Security Program (DISP) direct security measures.

Financial Sector: Asset Management in Regulated Environments

The financial sector operates in a heavily regulated environment where asset identification connects directly to risk management and compliance. Due to their economic significance, core banking systems, payment networks and trading platforms attract cybercriminals. Identifying these high-value assets enables institutions to implement strong security controls.

Personally Identifiable Information (PII), such as full name, date of birth and driver's license number, requires stringent protection as another key asset. Financial institutions must map and secure PII to maintain customer trust and comply with regulations such as APRA's CPS 234. Third-party risk identification proves vital as banks increasingly rely on cloud providers and fintech partners.

This sector necessitates a dynamic approach to risk assessment. Continuous monitoring ensures financial institutions can quickly respond to new threats, fraudulent activities and evolving regulatory requirements.

Healthcare: Identifying Patient Data and Medical Device Security Risks

The healthcare industry faces unique cybersecurity challenges due to patient data sensitivity and connected medical devices. Electronic Health Records contain confidential patient information, making them valuable targets for cybercriminals. Protecting these records is mandated by Australia's *My Health Record Act* and international frameworks like HIPAA.

Medical devices introduce additional security concerns. Pacemakers, infusion pumps and other internet-connected devices require ongoing monitoring to prevent unauthorised access (imagine someone remotely tampering with life-support equipment—it has happened!). The security challenge grows with healthcare's reliance on third-party providers, insurers and research institutions.

Hospitals and emergency services are classified as critical infrastructure under national cybersecurity frameworks, ensuring they receive prioritised protection against cyber incidents that could disrupt patient care.

Continuous Identification and Adaptation

Cybersecurity identification isn't a "set and forget" process. It must evolve and adapt as new technologies, threats, and business changes emerge. To maintain resilience, organisations must implement continuous asset discovery. But what does that look like in practice?

Business transformations, such as cloud adoption, BYOD policies, and mergers or acquisitions, often introduce unmanaged assets and shadow IT. These blind spots can pose significant risks, making real-time visibility not just helpful, but vital. At the same time, regulatory landscapes are anything but static. As new compliance mandates arise, identification strategies must adapt to ensure ongoing legal and operational alignment.

In response, emerging technologies are reshaping how organisations approach identification. The Zero Trust model, for example, requires constant verification of users, devices, and workloads—fundamentally shifting how trust is granted within networks. Complementing this, AI and machine learning tools detect rogue devices and anomalous behaviour automatically, removing the burden of manual oversight. They also enable risk-based prioritisation, helping organisations focus resources where they're needed most.

As the cyber threat landscape evolves, an adaptive approach to asset identification ensures organisations remain resilient to future challenges.

Summary

This chapter explored identification as a core cybersecurity principle, highlighting that security starts with knowing what to protect. Organisations must systematically document all assets in their digital ecosystem before implementing security measures.

Effective identification consists of four key components: asset management (maintaining inventories), risk assessment (evaluating threats and vulnerabilities), governance (aligning security with business objectives) and supply chain security (managing third-party risks). Without a structured approach to identification, organisations create dangerous blind spots that attackers can exploit.

Asset management's role in cybersecurity represents the central theme throughout this chapter. Organisations must develop detailed inventories covering hardware, software, cloud services, data assets and personnel. Each asset should have unique identifiers, descriptions and classifications based on criticality and sensitivity. While maintaining accurate inventories presents challenges, automated tools dramatically improve visibility across complex environments.

Asset classification is another critical aspect of identification. Some systems, often called "The Crown Jewels", require heightened protection due to their operational importance. This chapter emphasised evaluating assets based on confidentiality, integrity and availability, using BIA to establish recovery objectives and frameworks like the ASD's Essential Eight to prioritise security measures.

Beyond technical considerations, identification involves ownership and accountability. IT teams handle provisioning; security teams assess vulnerabilities; and business units align assets with organisational goals. We also examined shadow IT as a significant risk, requiring network analysis and CASBs for detection.

Risk identification, a precursor to risk assessment, focuses on uncovering threats before they escalate into incidents. Common weaknesses include poor asset visibility, shadow IT and misconfigurations, all creating security gaps.

We noted that supply chain security is an increasingly urgent concern, as third-party vendors can introduce vulnerabilities through data exposure and weak security practices. Recent incidents highlight the importance of thorough due diligence and continuous monitoring of third-party access.

Finally, we discussed how identification strategies must evolve to address industry-specific challenges, emerging technologies and shifting threat landscapes. By applying the concepts covered in this chapter, you can improve asset visibility, strengthen cybersecurity posture and take a more proactive approach to managing security risks in a complex digital environment. The goal isn't perfection—it's visibility, understanding and systematic improvement of your security foundation.

Revision Questions

1. What does the Identify principle refer to in cybersecurity, and why is it fundamental?
2. What are the four essential components of the Identify principle?
3. What was the WannaCry ransomware outbreak of 2017 an example of?
4. What major categories should be included in a comprehensive asset inventory?
5. What is the CIA Triad used for in asset classification?
6. What roles do different teams play in asset management?
7. What is shadow IT and why is it a significant risk?
8. How does risk identification differ from risk assessment?
9. What major cybersecurity incidents demonstrate the importance of supply chain security?
10. How are Zero Trust and AI transforming asset discovery?

Cybersecurity Hacks

Hack 1: Implement the "Asset Criticality Traffic Light System"

Develop a visual asset classification system that simplifies complex risk assessment concepts for your organisation. Create a three-tier classification system using traffic light colours: Red (Critical/Crown Jewels), Yellow (Important), and Green (Standard). For each asset, ask three simple questions: "Would losing this asset stop business operations?" (Red), "Would losing this asset significantly impact business operations?" (Yellow), or "Would losing this asset cause minor inconvenience?" (Green).

Create simple visual tags or labels for physical assets and use colour-coded spreadsheet systems for digital assets. This system should align with our earlier CIA triad discussion - Red assets typically have high confidentiality, integrity, and availability requirements, while Green assets may only need basic protection. Start small by categorising 5 - 10 key assets in your immediate work area, then expand gradually. This approach makes the complex concepts of Business Impact Analysis and asset prioritisation tangible and actionable for beginners.

Hack 2: Create a Simple Shadow IT Detection System

Build a basic shadow IT monitoring system using readily available tools to identify unauthorised devices and applications in your environment. Start by enabling basic network monitoring through your router's admin panel or free tools like Wireshark to capture network traffic during off-peak hours. Create a simple spreadsheet template with columns for device MAC addresses, IP addresses, device names, and approval status. Schedule monthly "device discovery walks" where you physically inspect workspaces for unauthorised devices like personal phones, tablets, or USB drives connected to corporate systems. Ensure an approved policy has been implemented and staff are aware of discovery walks.

For cloud-based shadow IT, use your organisation's firewall logs or free tools like Cloudflare's DNS filtering to identify unusual web traffic patterns pointing to unauthorised cloud services. Set up simple alerts for common shadow IT services like personal Dropbox, Google Drive, or unauthorised messaging apps. Document your findings in a risk register format with three categories: "Approved," "Needs Review," and "Prohibited." This approach helps junior professionals understand the scope of shadow IT while building practical identification skills.

Chapter 12: Protection

— Securing Systems, Strengthening Defences —

Objectives

By the end of this chapter, you will be able to:

- Explain the Protection principle and its role within cybersecurity
- Apply the CIA Triad (Confidentiality, Integrity and Availability) to evaluate and implement appropriate security controls
- Understand Identity, Access and Privilege Management (IAM) strategies using the principle of least privilege
- Recognise data protection strategies incorporating classification, encryption and monitoring systems
- Appreciate resilient network and infrastructure security measures.

As the third core principle in both the NIST CSF and the ASD frameworks, Protection transforms abstract security strategies into concrete defences. The Protection principle acts as your organisation's digital fortress. It enables organisations to maintain operations even when confronted with sophisticated cyber threats by implementing appropriate safeguards for critical infrastructure services.

The Protection function operates cohesively alongside the other six complementary framework components—Govern, Identify, Protect, Detect, Respond and Recover—to create a comprehensive security posture. While the Identify function maps an organisation's security landscape, Protection implements specific controls to shield identified assets from threats. These controls are precisely calibrated according to asset criticality and threat severity, creating a tailored security approach.

At the heart of the Protection principle lies the CIA Triad, which provides the essential framework for evaluating and implementing security controls. The CIA Triad guides organisations in prioritising which assets require the strongest protection measures and determines how these measures should be balanced. When properly applied, this framework ensures that sensitive information remains accessible only to authorised users through encryption and access controls, that data can be trusted through verification mechanisms and that systems remain operational. Security teams use the triad as an analytical lens to analyse vulnerabilities and allocate resources where they matter most, without sacrificing business functionality.

Defence-in-depth forms the architectural foundation of the Protection principle through its establishment of multiple security barriers. This approach ensures that secondary and tertiary defences remain intact if one protective layer fails. Today's strategies go beyond perimeter-based defences, evolving into Zero-Trust Architecture, which operates on the principle of "never trust, always verify". This evolution requires continuous validation for every user and device attempting to access organisational resources. More on this shortly.

Strong protection requires proactive measures, including systematic software patching, scheduled vulnerability assessments and comprehensive data encryption. As cyber threats grow, forward-thinking organisations turn to automation and AI. Systems like Security Orchestration, Automation and Response (SOAR) help accelerate incident response, reduce human error and free up security teams to focus on high-level threats requiring specialised attention.

This chapter provides detailed strategies and practical techniques for implementing the Protection principle within your organisation, helping you construct strong defences and develop resilience in an environment where threat vectors constantly evolve.

Identity, Access and Privilege Management

Protecting organisational assets requires more than perimeter defences. It's about answering the fundamental question: "Who are you, and by what authority do you have access?" Identity, Access and Privilege Management (IAM) is your critical security foundation, ensuring only authorised individuals can access specific resources while maintaining effective security controls.

Modern IAM systems use the Authentication, Authorisation and Accounting (AAA) model. Authentication verifies user identity through various mechanisms; Authorisation determines what actions authenticated users may perform; and Accounting tracks user activities for compliance and security monitoring. Organisations must maintain strict governance throughout the identity lifecycle (from creation through modification to eventual deactivation) to prevent unauthorised access and potential data breaches.

Identity proofing forms the first critical step. Before granting system access, organisations must verify an individual's claimed identity through evidence such as physical document verification, biometric data or knowledge-based verification. This verified identity is then bound to credentials, ensuring the person using those credentials is genuinely who they claim to be. Financial institutions, for example, often employ the 100-point check before allowing access to sensitive financial data.

Authentication methods continue to evolve beyond traditional password-based systems, which remain vulnerable to credential stuffing, brute force attacks and social engineering. MFA enhances security by combining independent credentials across different categories: something you know (password), something you have (security token) and something you are (biometric). The growing adoption of passwordless authentication solutions like FIDO2 and passkeys represents a significant advancement, using cryptographic key pairs rather than shared secrets.

The principle of least privilege functions as a foundational security concept within IAM frameworks. This approach restricts access permissions to the minimum levels necessary for users to perform their job functions, thereby reducing the potential attack surface. Consider a marketing employee; they shouldn't be able to access financial records unless it is essential for their role. When implemented alongside separation of duties policies (which distribute responsibilities across multiple individuals), organisations significantly mitigate risks associated with privilege abuse. The 2020 Twitter breach illustrates this principle's importance: attackers compromised internal administrative tools, but proper privilege restrictions could have limited their ability to hijack high-profile accounts.

Zero Trust Architecture represents a paradigm shift from traditional perimeter-based security to an identity-centric approach, summarised by "never trust, always verify". This model assumes that threats exist both outside and inside the network, requiring continuous validation of every access request regardless of origin. Context-aware access controls evaluate multiple risk factors (user behaviour, device health, location, time) to make dynamic authorisation decisions. As organisations increasingly adopt cloud services and support remote work, Zero Trust principles become paramount for securing distributed environments where traditional network boundaries have dissolved.

Privileged Access Management (PAM) focuses on securing administrative accounts with elevated system rights. These high-value targets require additional protections, including credential vaulting, just-in-time (JIT) access provisioning and session monitoring. Implementing automated credential rotation and enforcing time-limited access helps minimise exposure windows for privileged credentials.

As cyber threats evolve in sophistication, you must recognise that identity has become the security perimeter. Carefully managing digital identities throughout their lifecycle and enforcing principle-based access controls enables your organisation to establish a strong foundation for its overall security posture.

Data Protection and Privacy

Data has become the new oil; arguably, it's more valuable than the black stuff we pump from the ground. This escalating value has transformed data protection from a 'nice-to-have' into a fundamental business imperative. You can't just slap a password on a folder and call it secure anymore.

Effective data protection requires a comprehensive approach that incorporates proper classification, sophisticated encryption strategies and vigilant monitoring systems. These elements work together to safeguard information throughout its lifecycle.

Data Classification, Handling and Lifecycle Management

Protecting data begins with understanding what you're securing. Through the Identify principle introduced earlier and using data classification frameworks, information is sorted into distinct categories, typically ranging from public material to highly confidential secrets. This foundational work enables organisations to apply proportionate security controls, ensuring safeguards match the sensitivity of the data.

When implementing classification models, consistency becomes crucial. A document marked 'confidential' must receive identical protection regardless of location. These classifications inform data handling procedures, dictating how information should be stored, transmitted and eventually destroyed.

Data Lifecycle Management (DLM) provides a structured approach to information governance from creation through disposal. Each phase presents unique security challenges that require different approaches:

- Data creation—Access controls must be immediately established.
- Data active use—Encryption and access monitoring become essential.
- Data archival or deletion—Organisations must ensure complete data sanitisation to prevent unauthorised recovery.

Cryptographic erasure, where encryption keys are securely deleted rather than the data itself, offers efficient methods for rendering sensitive information permanently inaccessible. Role-based access policies operationalise the principle of least privilege, ensuring employees access only the data necessary for their responsibilities. A financial analyst requires access to accounting records but not necessarily customer service databases. These policies must be regularly reviewed during personnel changes to prevent privilege creep, the gradual accumulation of unnecessary access rights over time.

Comprehensive auditing mechanisms create accountability by maintaining tamper-evident logs of all access attempts, modifications and potential security violations. Think of it as having security cameras that can't be turned off.

Encryption Strategies

Data protection depends significantly on encryption across three distinct states: at rest, in transit and, increasingly, in use.

Encryption at rest protects stored data using algorithms like AES-256, implemented through full-disk encryption solutions such as BitLocker for Windows or Linux Unified Key Setup (LUKS) for Linux. Database-level encryption provides additional granularity, allowing organisations to encrypt specific tables or columns containing sensitive information while maintaining performance for less critical data.

Data in transit faces unique vulnerabilities as it traverses networks potentially exposed to interception. Transport Layer Security (TLS) 1.3 represents the current standard for securing web communications. It eliminates vulnerable legacy protocols while providing forward secrecy, so that any possible compromise of current encryption keys cannot expose previously captured traffic.

End-to-end encryption provides additional assurance for highly sensitive communications by encrypting data on the sender's device and decrypting only on the recipient's end, preventing service providers from accessing message contents.

Protection of data in use represents the newest frontier in encryption technology. Homomorphic encryption allows computations on encrypted data without decryption, while trusted execution environments create isolated processing spaces resistant to tampering. These emerging technologies address the traditional vulnerability gap when data must be decrypted for processing.

Key management practices determine overall security effectiveness for all encryption implementations. Hardware Security Modules (HSMs) provide tamper-resistant environments for cryptographic operations, while key rotation policies limit the window of opportunity should keys become compromised. Effective implementations separate key management duties across multiple individuals, preventing any administrator from having complete control over cryptographic systems.

Data Loss Prevention (DLP)

DLP systems are the defensive perimeter against unauthorised data exfiltration through automated policy enforcement. These solutions operate across endpoints, networks and cloud environments, identifying and blocking suspicious data transfers based on content patterns, metadata and contextual information.

When properly implemented, DLP tools can prevent common data leakage scenarios such as uploading sensitive files to personal cloud storage, forwarding confidential emails to external addresses or transferring controlled information to unauthorised removable media.

Compliance Consideration

Australian organisations face increasingly stringent regulatory requirements, including *The Privacy Act's* Notifiable Data Breaches scheme, *The Security of Critical Infrastructure (SOCI) Act 2018* and *The Cyber Security Act 2024*. These frameworks establish mandatory security standards and breach notification obligations, with significant penalties for non-compliance.

Multinational organisations that interact with the EU, for example, must also navigate international regulations like the GDPR, which imposes strict requirements on data handling and protection practices regardless of where the company is headquartered. The regulatory landscape continues evolving, making compliance a moving target that requires constant attention.

Strategic Briefing: The 2022 Optus Data Breach

In September 2022, Optus, Australia's second-largest telecommunications company, experienced a major data breach that exposed the personal information of approximately 9.8 million customers.[96] The breach, which involved sensitive personal identifiers including passport and driver's licence numbers, demonstrated how inadequate Application API security controls could allow attackers to bypass perimeter defences and access customer data directly.[97]

The incident highlighted critical security failures in data protection practices and triggered significant regulatory responses from the Australian government. It serves as a powerful case study of the importance of comprehensive security protocols, particularly for organisations handling sensitive customer information.

Impact and Scope

The data breach had widespread implications affecting both customers and organisations. Key impacts included:

- Exposure of approximately 9.8 million customer records
- Compromise of sensitive Personally Identifiable Information (PII), including passport numbers
- Compromise of driver's licence numbers and other identity documents
- Risk of identity theft for affected customers
- Need for credit monitoring services for impacted individuals
- Requirement for document replacement for many customers.

The incident triggered immediate regulatory scrutiny and forced emergency regulatory changes by the Australian government. It also led to a comprehensive review of *The Privacy Act 1988*, resulting in increased penalties for serious or repeated breaches, jumping from the previous $2.1 million to up to $50 million.[98]

Attack Methodology

The attackers gained access through an exploitable vulnerability in Optus's digital infrastructure. The breach methodology involved:

- Exploitation of an unauthenticated API endpoint
- Complete bypass of perimeter defences
- Direct access to customer data without triggering security alerts
- Exploitation of insufficient data encryption protections
- Taking advantage of inadequate access limitations to sensitive data.

This breach highlighted how attackers could exploit technical vulnerabilities to gain unauthorised access to vast amounts of customer information, particularly when proper authentication mechanisms were not in place.

Response and Mitigation

Upon discovering the breach, Optus implemented several response measures to address the situation. The company notified affected customers about the data compromise and engaged with regulatory authorities, including the Australian government and the privacy commissioner.

Optus provided support services to affected customers, including credit monitoring and assistance with document replacement processes. The company also thoroughly reviewed its security infrastructure and implemented enhanced protection measures for API endpoints and customer data.

The Australian government responded by implementing emergency regulatory changes to mitigate the impact on affected individuals and prevent similar incidents in the future.

Lessons Learned

The Optus data breach revealed critical lessons for telecommunications companies and other organisations handling sensitive customer data:

- **API Security Controls**—Organisations must implement robust authentication and authorisation mechanisms for all API endpoints and conduct regular security testing to identify vulnerabilities.

- **Data Minimisation Practices**—Retaining sensitive identity documents long after their business purpose had expired unnecessarily amplified the breach's impact. Organisations should implement strict data minimisation policies, retaining sensitive information only as long as necessary for legitimate business purposes.

- **Comprehensive Encryption**—Sensitive customer data requires end-to-end encryption with proper key management practices to ensure protection even if perimeter defences are breached.

- **Defence-in-Depth Strategies**—Relying solely on perimeter security is insufficient. Organisations need layered security controls, including robust authentication, access controls, encryption and continuous monitoring.

- **Regulatory Compliance**—Privacy failures can have severe consequences, including regulatory penalties and reputational damage.

Organisations must stay current with evolving privacy regulations and implement effective compliance programs.

The Optus breach reminds us that even well-funded companies can fail if they overlook core security principles. Learning from these failures builds stronger, more resilient defences. If you're taking notes, you're already ahead of those who learn the hard way.

Building Resilience: Network and Infrastructure Security

You've probably heard the word 'resilience' as it applies to individuals, meaning to stay strong when life throws curveballs your way. Technology infrastructure resilience, on the other hand, refers to a system's ability to withstand, adapt to and recover from disruptions while maintaining continuous availability. As adversaries continuously evolve their techniques, a comprehensive, layered security approach incorporating defence-in-depth becomes essential for protecting assets across your digital infrastructure.

One of the key strategies to build such resilience is network segmentation. Network segmentation functions as a fundamental security principle, dividing large networks into smaller, isolated sections to contain breaches and restrict lateral movement. By implementing logical segmentation through Virtual Local Area Networks (VLANs) and subnets, plus physical separation for highly sensitive systems, organisations establish security boundaries that prevent attackers from easily traversing the network. Micro-segmentation enhances this approach by applying fine-grained security policies at the workload level. The 2021 Colonial Pipeline attack illustrates segmentation's vital importance. Proper isolation between Operational Technology (OT) and Information Technology (IT) networks would have prevented attackers from simultaneously compromising billing systems and pipeline operations.

Protecting networks from unauthorised access requires multiple defensive layers working harmoniously. Firewalls remain key components. They've evolved from simple packet filters to sophisticated next-generation solutions that integrate deep packet inspection, intrusion prevention capabilities and threat intelligence feeds.

However, firewall effectiveness hinges on meticulous configuration. You must regularly audit rule sets to prevent dangerous misconfigurations. Intrusion Detection and Prevention Systems (IDPSs) augment firewalls by analysing network traffic for suspicious patterns, generating alerts or automatically blocking threats. With remote work becoming standard, secure access solutions like VPNs now safeguard data in transit by creating encrypted tunnels across public networks.

Cloud environments present distinct security challenges as traditional network boundaries disappear in multi-cloud architectures. Consider implementing cloud-native security controls, including security groups, network Access Control Lists (ACLs) and cloud workload protection platforms. Comprehensive encryption strategies safeguard information at rest, in transit and during processing with confidential computing technologies. Ongoing monitoring through Cloud Security Posture Management (CSPM)[99] tools preserves visibility and maintains compliance with security standards across dynamic cloud infrastructures.

As we've discussed previously, Zero Trust security model marks a significant shift from traditional perimeter-based approaches to an identity-centric strategy. Zero Trust Network Access (ZTNA) continuously verifies every access request, regardless of origin, by assessing user identity, device health and the context of the request before granting only the minimum necessary permissions. When paired with Secure Access Service Edge (SASE) frameworks, which unify networking and security functions as cloud-delivered services, this approach provides effective, consistent protection that follows users and data across all environments.

Environmental threats, including natural disasters and power disruptions, represent significant risks to infrastructure resilience. Strategic data centre location decisions must consider geographical hazards; you wouldn't build a house on a floodplain, right? As such, when locating data centres, avoid flood zones and regions susceptible to extreme weather events. Uninterruptible Power Supplies (UPS) and backup generators maintain critical systems during outages while protecting sensitive equipment from power fluctuations. Additionally, environmental monitoring systems track temperature, humidity and air quality, and detect potential issues before they develop into service-impacting events.

Effective resilience develops from thorough planning for both routine operations and adverse scenarios. You should implement redundancy across multiple levels, from network connections and power sources to entire data centres, establishing failover mechanisms that ensure service availability during component failures. Disaster recovery planning creates clear protocols for responding to major incidents, specifying Recovery Time Objectives (RTOs) and restoration priorities for critical systems. Regular drills and simulations develop organisational 'muscle memory', equipping teams to execute recovery procedures effectively during emergencies.

Ensuring sufficient resource capacity constitutes a key element of infrastructure resilience. Proactive capacity planning helps you anticipate growth requirements before resource limitations affect availability. This involves analysing usage patterns and forecasting future demands, essential for both traditional data centres and cloud environments, where capacity directly influences technology budgets. Contemporary architectures employ auto-scaling capabilities to adjust resources dynamically based on demand variations, while continuous performance monitoring detects potential bottlenecks before they impact users.

By implementing these protection strategies and resilience measures, your organisation can establish durable technology infrastructures capable of withstanding evolving threats while maintaining continuous availability of critical systems and services.

Empowering Users: Awareness and Training

Throughout this book, we keep returning to the importance of training and awareness. The reason? Cybersecurity is as much a people problem as a business or technology problem. People are your frontline and should be empowered with knowledge and skills to recognise and respond to potential threats. In Chapter 18, we'll explore this in greater detail. For now, it's important to know that awareness and training are fundamental components of the Protection principle in cybersecurity.

Security awareness and system hardening constitute the foundations of effective cybersecurity defence. While sophisticated technical controls provide valuable protection, the human element remains simultaneously your greatest vulnerability and most adaptable defence mechanism. Creating resilient systems necessitates a multi-layered approach that addresses both human and technical vulnerabilities through comprehensive awareness programs and systematic hardening protocols.

Human susceptibility to manipulation constitutes a significant attack vector that technical controls alone can't mitigate. Think of it like teaching someone to spot a con artist. Once they know the tricks, they're much harder to fool. Cybersecurity awareness training transforms employees from potential security liabilities into vigilant human firewalls capable of identifying and responding to emerging threats. Effective programs educate staff on recognising social engineering tactics such as phishing emails with urgent requests, pretexting schemes that fabricate scenarios to extract information and tailgating attempts to gain unauthorised physical access. You'll want to cultivate a security-first mindset that embeds protective behaviours into daily operations, so that password hygiene, data handling protocols and physical security measures become instinctive rather than burdensome.

Training effectiveness increases significantly when you deploy phishing simulations and gamified learning experiences that transform abstract concepts into practical skills. Forward-thinking organisations implement role-specific training that addresses distinct risks faced by various stakeholders. For example, ICT professionals require training to implement and manage security measures, including firewalls, intrusion detection systems and encryption protocols. Conversely, executives must be guided on how to make strategic decisions about cybersecurity investments, risk assessment and policy development. These specialised approaches ensure security knowledge aligns precisely with job responsibilities.

Of course, building a security-aware culture requires more than just one-off or periodic training. It requires continuous reinforcement in daily operations. Leaders play a critical role by setting an example and actively promoting cybersecurity best practices. By fostering a security-first mindset, you can strengthen their defences, adapt as different threats arise and create a resilient work environment.

Infrastructure (Platform) Security

Don't be misled by security companies' glossy marketing claims about comprehensive platform solutions. If such all-encompassing security platforms truly existed, we would not witness the continuing surge in cybersecurity incidents worldwide.

It's also worth noting here that, in the context of the Protection principle, 'platform' refers to an organisation's ICT infrastructure rather than vendor-marketed security software products. This dual interpretation has created industry confusion, so this chapter will use the more precise term 'infrastructure' instead.

Securing ICT infrastructure requires an all-inclusive strategy that protects both physical and virtual components against cyber threats. This approach safeguards hardware, software and services to establish a resilient foundation for all digital operations.

Configuration Management

Many security incidents stem from misconfigurations and inadequately trained ICT staff. Effective configuration management forms the foundation for maintaining secure system settings, ensuring consistent application and monitoring of hardware and software configurations across the ICT infrastructure. Security strategies, policies, staff training and compliance auditing provide little protection if security protocols are not correctly applied to infrastructure component configurations.

This practice requires defining baseline configurations for systems and networks, documenting all changes and regularly reviewing configurations to identify and address deviations. For example, the 2013 Target data breach, which exposed over 40 million customers' personal information, was partially attributed to misconfigured systems that allowed unauthorised access.[100] Proper configuration management could have mitigated this risk by ensuring correctly applied security settings with ongoing compliance monitoring.

System Hardening

System hardening transforms vulnerable default configurations into strengthened defensive positions. This process begins by reducing the attack surface through eliminating unnecessary services, removing default credentials and disabling unused features that might contain exploitable vulnerabilities. Organisations reinforce this foundation by implementing secure baselines aligned with industry standards such as CIS Benchmarks or frameworks like NIST and ASD guidance.

While hardening strengthens systems at their core, modern threats also require advanced protection at the endpoint. Endpoint Detection and Response (EDR) and Extended Detection and Response (XDR) solutions employ behavioural analysis and ML to identify sophisticated attack patterns that evade signature-based detection. When threats emerge, automated response mechanisms quickly isolate infected devices, preventing attackers from expanding their access while security teams investigate and respond.

Software Maintenance, Patching and Updates

Regular software maintenance, patching and timely updates represent fundamental practices for maintaining ICT infrastructure security and integrity within the Protection Principle. Unaddressed software vulnerabilities create entry points for cyber attackers who exploit these weaknesses to gain unauthorised access to systems and sensitive data. Regularly updating and patching software closes these security gaps.

While Chapter 3 provides detailed guidance on vulnerability and patch management strategies, routine software upkeep remains vital to protecting systems from exploitation. Timely updates are not only about functionality; they are a core protection mechanism. Delayed patching expands the attack surface, exposing systems to preventable compromise. Neglecting this aspect can result in significant financial, operational and reputational consequences following a breach.

The urgency of prompt patching is demonstrated by recent incidents such as the February 2024 ConnectWise ScreenConnect vulnerabilities (CVE-2024-1708 and CVE-2024-1709). These critical flaws, one of which received the highest possible CVSS score of 10.0, allowed authentication bypass and remote code execution. Attackers rapidly exploited these weaknesses in unpatched systems running the widely used remote management tool, compromising IT service providers and downstream clients alike. [101]

Secure Hardware Lifecycle Management

Consider what happens when equipment fails or when organisations acquire second-hand devices. Without proper controls, these situations could introduce malware, enabling attackers to establish footholds in your ICT infrastructure and extract your data.

Secure hardware lifecycle management ensures appropriate handling of hardware from acquisition through disposal. This includes securing hardware during transportation, ensuring proper device configuration and maintenance, and securely decommissioning and disposing of hardware to prevent data leakage. If sensitive information remains on devices, improper hardware disposal can lead to data breaches. Organisations should implement strict protocols for secure data destruction on retired hardware to mitigate this risk.

The Australian Signals Directorate Information Security Manual (ASD ISM) provides relevant guidance: Before releasing ICT equipment into the public domain, it must undergo sanitisation, destruction or declassification. Even after these processes, equipment may still pose security risks, so its release requires formal authorisation from the appropriate organisational authority. Additionally, the disposal process must include removing all labels and markings indicating ownership, sensitivity, classification or other information linking the equipment to its previous use, ensuring the equipment does not attract unwarranted attention after disposal.[102]

Log Management Practices

Modern operating systems, applications and network devices produce log files documenting their operation. These files contain detailed records of events, processes and transactions essential for monitoring system performance, troubleshooting issues and ensuring security through historical activity documentation. The volume of this information presents significant management challenges. Imagine trying to read every diary entry from a thousand people simultaneously.

Log management systematically addresses these large volumes of information. Effective approaches involve defining appropriate logging content, establishing retention policies, centralising logs into SIEM systems and employing automated analysis tools to identify suspicious activity patterns.

Log analysis helps identify behavioural patterns indicating potential insider threats or external attacks and improves security incident response effectiveness. Numerous resources, including free information from log management service providers, can assist in determining and applying effective log management. Before establishing policies or procedures, review the NIST Guide to Computer Security Log Management[103] and the ISM Guidelines for System Monitoring to establish fundamental practices.

Secure Software Development Practices

Early in IT's evolution, bug hunting focused on ensuring software functionality and requirements compliance, with little consideration for cyber threat vulnerabilities. This approach has evolved considerably, particularly given the emergence of software zero-day vulnerabilities.

Implementing secure software development practices should be central to application development rather than an afterthought. This involves integrating security throughout the software development lifecycle and establishing a Secure Software Development Lifecycle (S-SDLC), building upon the SDLC practices described in Chapter 3.

Beyond infrastructure-level controls, secure software development plays a critical role in reducing risk before systems are even deployed. Secure coding standards, peer code reviews and automated security testing are key elements of the Protection principle. These practices help identify and eliminate vulnerabilities early in the development lifecycle. Integrating security checks into Continuous Integration (CI) and Continuous Delivery (CD)[104] pipelines ensures these safeguards are applied consistently across all builds and deployments.

Securing hardware, software and services through infrastructure security measures represents an important aspect of the Protection principle. Effective configuration management, timely software maintenance and patching, secure hardware lifecycle management, comprehensive log management and secure software development practices all contribute to creating a secure and resilient ICT infrastructure.

Summary

Chapter 12 examined the Protection principle. As the third core principle in both NIST and ASD frameworks, Protection transforms abstract security strategies into concrete defences that keep your organisation running when cybercriminals come knocking.

The CIA Triad forms the essential framework for evaluating security controls (like a three-legged stool—remove one leg and everything collapses). Proper application ensures only authorised users access sensitive information, verification mechanisms establish data trustworthiness and systems remain operational without compromising functionality. Protection works alongside six complementary components—Govern, Identify, Protect, Detect, Respond and Recover—creating a complete security posture.

Identity, Access and Privilege Management (IAM) provides a critical security foundation through the AAA model. Organisations have evolved beyond traditional passwords to implement MFA, strengthening defences against credential attacks. Least privilege serves as a central security doctrine by restricting access permissions to only what users need for specific job functions, significantly reducing attack surfaces.

Effective data protection combines proper classification, encryption and active monitoring. Classification frameworks define clear sensitivity categories. Encryption protects information in three states: at rest, in transit and increasingly, in use. DLP systems create defensive barriers against unauthorised exfiltration through automated policy enforcement.

Network and infrastructure security builds resilience through strategies like network segmentation, which divides networks into isolated sections to contain breaches. It's similar to compartmentalising a submarine: if one section floods, others remain watertight. Zero Trust represents a significant advancement from traditional perimeter-based approaches to an identity-centric strategy following "never trust, always verify". This approach requires continuous validation of every user and device attempting access.

Security awareness training converts employees from vulnerabilities into vigilant human firewalls. Effective infrastructure security requires systematic configuration management, system hardening, secure hardware lifecycle processes and thorough log management. Regular software maintenance and patching eliminate security gaps that attackers exploit.

The 2022 Optus data breach illustrated how inadequate API security allowed attackers to bypass defences and access personal information of 9.8 million customers. This incident revealed serious security failures and triggered significant regulatory responses from the Australian government. It serves as a compelling case study demonstrating why thorough security protocols matter for organisations handling sensitive customer information.

Revision Questions

1. What is the Protection principle in cybersecurity frameworks, and how does it function?

2. What is the CIA Triad, and why is it important to the Protection principle?

3. How has the Protection strategy evolved from traditional approaches?

4. What is Identity and Access and Privileged Access Management (IAM and PAM), and how does it enhance security?

5. What are the three states in which data requires encryption protection?

6. Why is network segmentation important in security architecture?

7. What primary security failure led to the 2022 Optus data breach?

8. How does system hardening contribute to security?

9. What role do humans play in the Protection principle?

10. Why is secure hardware lifecycle management necessary?

Cybersecurity Hacks

Hack 1: Establish a "Data Classification Quick Reference" System

Create a simple visual guide that helps colleagues instantly identify data sensitivity levels and corresponding protection requirements. Design a one-page reference card or digital poster that uses colour coding and clear examples to distinguish between public, internal, confidential, and restricted data. Include specific handling instructions for each classification level, such as "Green = Public (can be shared freely)," "Yellow = Internal (company employees only)," and "Red = Confidential (encrypted storage required)."

Place these reference guides at workstations, in meeting rooms, and as digital wallpapers. When employees can quickly identify data sensitivity, they naturally apply appropriate protection measures without lengthy decision-making processes.

Hack 2: Implement the "Three-Touch Rule" for Password Security

Develop a streamlined approach to password management that combines security best practices with user convenience. The "Three-Touch Rule" requires employees to use three distinct elements: a password manager (first touch), multi-factor authentication (second touch), and a unique recovery method (third touch). This creates a simple mental framework that ensures comprehensive credential protection without overwhelming users with complex security procedures.

Chapter 13: Detection

– Catching Threats Before They Strike –

Objectives

By the end of this chapter, you will be able to:

- Explain the Detection principle and its critical role in the cybersecurity framework
- Identify the three fundamental pillars of detection: monitoring, analysis and alerting
- Distinguish between different monitoring approaches for networks, endpoints, physical environments and supply chains
- Understand the concept of Indicators of Compromise (IoCs) and their role in threat detection
- Evaluate detection effectiveness and address common challenges.

Detect, the fourth component of the six cybersecurity functions (Govern, Identify, Protect, Detect, Respond and Recover) we are exploring in Part 2 of this book, is a core cybersecurity principle focused on identifying threats before they escalate into major incidents. While Protect aims to prevent compromise, Detection accepts a difficult truth: some threats will inevitably bypass even the most advanced defences. Think of Detection as your cybersecurity watchdog, a vigilant sentinel positioned between protective controls and incident response, forming a vital link in your security chain.

Early identification of threats dramatically limits potential damage. Detecting a breach within minutes, rather than months, allows for swift containment before sensitive data is stolen or systems are compromised. This transforms cybersecurity from a static, preventative function into a responsive, adaptive one, ready to meet evolving threats head-on.

Modern attackers are constantly devising ways to evade traditional controls. Techniques like Living-off-the-Land (LotL) attacks (which repurpose legitimate tools), fileless malware (which operates entirely in memory) and APTs make detection challenging. IBM reports that organisations take an average of 204 days to identify a breach.[105] In that time, attackers can entrench themselves, elevate privileges and extract sensitive data unnoticed. The SolarWinds supply chain attack highlighted this risk, with threat actors maintaining undetected access for nearly nine months.

Implementing effective detection means continuously monitoring across networks, physical environments and human activity. Technologies such as Intrusion Detection Systems (IDSs), Security Information and Event Management (SIEM) and Network Traffic Analysis (NTA) provide visibility into your infrastructure and surface irregularities. At its core, detection involves analysing event logs and other telemetry. It's digital detective work, leveraging threat intelligence and correlating data from multiple sources to distinguish normal operations from malicious activity.

Detection strengthens every other cybersecurity function. It builds on Identify through targeted oversight of assets. It enhances Protection by revealing gaps in preventative defences. It enables incident Response by triggering timely actions. It also informs Recovery by clarifying how breaches occur.

Without strong Detection capabilities, you're operating in the dark, unaware of threats until damage becomes visible. In an environment where digital assets rank among an organisation's most critical resources, Detection serves as a primary element of an effective cybersecurity strategy.

This chapter outlines essential methods and tools for detecting cyber threats in today's computing environments. You'll explore how comprehensive monitoring creates actionable threat awareness across systems, networks, user behaviours and third-party access. Each domain adds a unique layer to your organisation's detection capabilities, assisting you to recognise and respond to threats before they inflict serious harm.

The Core Components of Cybersecurity Detection

Effective cybersecurity detection functions as your organisation's immune system, actively scanning for threats while distinguishing between normal operations and potential attacks. This critical security function relies on three fundamental pillars: monitoring, analysis and alerting. Each provides vital capabilities that work together like a well-orchestrated defence team.

Monitoring is your vigilant sentry, maintaining continuous surveillance across your organisation's infrastructure. Security teams deploy technologies such as SIEM systems, IDS and EDR solutions to capture comprehensive views of network activities. Effective monitoring involves more than simply deploying the right technologies. It requires a thoughtful integration of logs from multiple sources, including firewall traffic, endpoint activity, cloud resource usage and authentication records. Just as crucial is striking the right balance between monitoring for immediate threat detection and scheduled log reviews that uncover subtle patterns developing over time.

Analysis converts extensive collected data into actionable security insights. Security operations use sophisticated correlation techniques to link seemingly unrelated events across different systems, exposing attack patterns that remain hidden when examined individually. Behavioural analytics (the science of establishing baseline models of normal system and user activities) flags deviations that might indicate compromise. This analytical capability becomes more powerful when enhanced with threat intelligence, such as contextual information about attack techniques, compromise indicators and adversary behaviours observed throughout the global threat landscape. Despite technological progress, skilled human analysts remain essential. They apply critical judgment to differentiate genuine threats from harmless anomalies and investigate suspicious activities that automated systems identify.

Alerting completes the Detection triad by ensuring identified threats receive appropriate attention. An effective alerting strategy balances thorough coverage against alert fatigue, where security teams become desensitised to notifications due to excessive false positives. Organisations address this challenge through risk-based prioritisation frameworks that assign severity levels based on potential impact, affected assets and detection confidence. The MITRE ATT&CK framework offers a common language for classifying alerts according to adversary tactics and techniques, facilitating precise threat communication. SOAR platforms improve alerting by automating initial response actions, enriching alerts with context and directing them to appropriate personnel using predefined playbooks.

Your organisation's current detection capabilities could range anywhere from basic to advanced implementations. Organisations at the basic level function with limited visibility, depending on default security configurations and responding to incidents after damage occurs (hopefully you're not here!). Those with intermediate maturity implement dedicated security tools with some correlation capabilities but lack appropriate threat hunting processes. Advanced detection operations incorporate AI-enhanced analytics, fully integrated threat intelligence and automated response workflows managed by dedicated Security Operations Centres (SOCs).

As cyber threats become increasingly sophisticated, detection capabilities must evolve from isolated security tools to integrated systems that provide visibility across all technology assets. By strengthening each Detection pillar—monitoring, analysis and alerting—your organisation establishes the situational awareness necessary to identify and neutralise threats before they achieve their objectives.

Network and Infrastructure Monitoring

Just as traffic authorities monitor highways to ensure safety, network monitoring serves as a vigilant sentinel guarding your organisation's digital highways. In our interconnected landscape, comprehensive monitoring of networks and infrastructure forms the foundation of effective threat detection, providing visibility across your entire technology stack while enabling rapid response to potential security incidents.

Network-based detection tools function as your first line of defence. IDSs continuously analyse network traffic patterns for suspicious activities, comparing behaviours against known threat signatures and anomalies to identify potential intrusions in real-time. Their more advanced counterparts, Intrusion Prevention Systems (IPSs), extend this capability by actively blocking malicious traffic. While signature-based detection excels at identifying known threats, anomaly-based approaches prove invaluable for detecting novel attack vectors. The 2020 SolarWinds breach (see Chapter 9) demonstrated how sophisticated attackers can bypass signature-based systems, highlighting the critical need for behavioural analysis capabilities.

SIEM platforms transform disparate security data into actionable intelligence. These systems collect, aggregate and centralise log data across your organisation's infrastructure, including network devices, servers, applications and endpoints. By correlating seemingly unrelated events from multiple sources, SIEM systems reveal coordinated attack patterns, unusual login activities and other indicators of compromise that might otherwise remain hidden when examining individual systems in isolation. Modern SIEM solutions increasingly incorporate Machine Learning (ML) algorithms that establish baseline behaviour patterns and flag anomalous deviations, reducing alert fatigue while improving detection accuracy.

The evolution toward SOAR platforms represents a significant advancement in SECOPS. SOAR tools automate routine investigation tasks and orchestrate responses across multiple security controls, dramatically reducing incident response times from hours to minutes. This automation proves particularly valuable during critical incidents when every minute counts.

NTA provides deeper insights into threat activities through techniques like Deep Packet Inspection (DPI) and flow analysis. DPI examines packet contents beyond simple header information, revealing malicious payloads, protocol violations and command-and-control (C2) communications hidden within legitimate traffic. Complementing this approach, flow analysis focuses on metadata, examining traffic patterns, communication relationships and data volume changes to identify suspicious activities like data exfiltration attempts or unusual lateral movement. This method effectively detects stealthy attacks that traditional monitoring techniques might miss.

Cloud environments introduce unique monitoring challenges that conventional tools struggle to address. The abstracted nature of cloud infrastructure limits visibility into underlying network traffic, while shared responsibility models create potential monitoring gaps. Cloud service providers offer native monitoring solutions that provide essential visibility into infrastructure changes, access patterns and potential misconfigurations. However, multi-cloud and hybrid environments significantly complicate monitoring efforts due to inconsistent logging formats and fragmented visibility.

Rapid threat detection requires not just the right tools but also integration of these capabilities into a cohesive security monitoring strategy. Combining IDS/IPS, SIEM, SOAR and traffic analysis capabilities assists organisations to create a defence-in-depth that transforms potential crises into manageable incidents. As attack techniques continue to evolve, your monitoring capabilities must similarly advance, incorporating AI, behavioural analytics and automated response mechanisms to match the speed and sophistication of modern cyber threats.

System and Endpoint Monitoring

In our digital battlefield, threats evolve relentlessly and endpoint monitoring acts as your organisation's watchful guardian. It is similar to having security cameras throughout your house, except these digital sentries watch every device, workstation, server and mobile phone—the endpoints where attackers love to get their first foothold.

Modern cybersecurity relies on effective visibility across all endpoints. EDR solutions play the starring role in this defence strategy. They're continuously analysing device behaviour through signature-based detection and sophisticated behavioural analytics. When suspicious activities occur, these systems automatically isolate compromised systems, terminate malicious processes and kick off incident response protocols before threats spread laterally through your network.

What is the essence of effective monitoring? Systematic log collection and analysis. Windows Event Logs document critical operating system and application activities. Syslog captures key data from network devices and Unix/Linux systems. Security-specific audit logs record authentication attempts, permission changes and administrative actions, all potential indicators of compromise when anomalies appear.

You'll need to centralise these diverse log sources through SIEM platforms. This establishes a correlation between seemingly disparate events that might collectively signal an attack in progress. Without proper retention policies and real-time alerting mechanisms, valuable security intelligence remains buried within overwhelming volumes of log data.

File Integrity Monitoring (FIM) delivers another critical layer of defence. It establishes cryptographic baselines of essential system files and configurations, like taking a fingerprint of every important file. When unauthorised modifications occur (from malware infection or a malicious insider), FIM solutions immediately flag these deviations from the established baseline.

This capability is valuable for protecting web servers from defacement, securing industrial control system configurations and safeguarding financial transaction records from tampering. You'll need to balance monitoring scope with performance considerations carefully. Most organisations implement agent-based solutions for critical systems and agentless monitoring for broader coverage.

Application behaviour analysis extends monitoring beyond static files to dynamic application activities. Security teams can identify anomalous patterns by establishing baselines of normal application behaviour through ML models. These include unexpected API calls, excessive data transfers or suspicious privilege escalations.

User and Entity Behaviour Analytics (UEBA) enhances this approach by correlating endpoint activities with user behaviour patterns. This enables organisations to detect compromised accounts and insider threats that might otherwise appear legitimate.

A well-structured monitoring strategy transforms raw data into actionable intelligence. It enables security teams to detect, analyse and respond to threats with precision. By integrating endpoint telemetry, log analysis, FIM and application behaviour analytics, you'll create a multi-layered defence capable of identifying both known attack signatures and emerging threats.

However, effective monitoring isn't just about deploying the right tools. It requires continuous fine-tuning, well-defined alerting mechanisms and skilled analysts who can distinguish genuine threats from false positives. As attackers refine their techniques, organisations must remain equally agile, ensuring that any unusual activity from new attack methods is promptly detected and triggers an immediate response.

Monitoring the Physical Environment

It might surprise you to learn that not knowing who can access physical assets like servers and network equipment can lead to severe security breaches. Effective monitoring of the physical environment ensures that unauthorised attempts to access, tamper with or damage these assets are detected and addressed promptly.

Access control systems represent the most cost-effective method for restricting physical access to sensitive areas. These systems regulate entry to specific zones within a facility using keycards, biometrics or PIN codes. They simultaneously log all entry attempts, both successful and unsuccessful. This logging functionality proves invaluable during forensic analysis after security incidents, providing detailed records of who was present at specific times.

Security cameras complement access control systems as an essential component of physical security management. Position these cameras strategically to monitor all entry and exit points, as well as sensitive areas like server rooms. Modern security systems featuring motion detection and night vision capabilities enable real-time monitoring and can immediately alert security personnel to suspicious activities.

Environmental monitoring plays an equally important role in maintaining physical asset integrity. By tracking temperature, humidity and other environmental factors, your organisation ensures that conditions remain within safe operational parameters. Sudden environmental changes may indicate potential problems such as overheating or water leaks, which can damage hardware and cause data loss.

Specialised sensors can trigger alerts when conditions deviate from predefined thresholds, enabling immediate corrective action. Think of them as smoke detectors or thermometers for your IT infrastructure. Hopefully, you'll never need them, but you'll be grateful they're there if something goes wrong.

Integrating these physical monitoring methods into your overall security strategy will significantly strengthen your security posture. This comprehensive approach not only protects against cyber threats but also preserves the integrity and availability of critical physical assets. A secure physical environment forms the foundation of an effective cybersecurity strategy and directly contributes to your organisation's operational resilience.

User and Behavioural Analytics in Detection

The approach to threat detection has evolved beyond traditional methods, with User and Behavioural Analytics (UBA) transforming how we monitor security. Instead of relying solely on static rules, UBA focuses on human behaviour, the most unpredictable element in any system. Using advanced algorithms, UBA establishes normal activity patterns, detects anomalies and intercepts potential threats before they escalate.

Your organisation faces significant risks from insider threats, which originate from current employees, former staff or contractors with legitimate access credentials. These individuals can cause data breaches, financial losses and reputational damage, whether through malicious intent or simple negligence. Consider a disgruntled employee exfiltrating confidential information. Such security breaches often have far-reaching consequences, making early detection critical for protection.

Behavioural baselines constitute the foundation of effective threat detection. Security teams meticulously track login times, access locations, system interactions and data usage patterns to create a digital fingerprint of normal activity. These baselines serve as reference points, helping analysts identify deviations that warrant investigation. However, defining 'normal' presents considerable challenges. Work habits vary among individuals, roles evolve and organisational restructuring frequently alters access requirements.

User Activity Monitoring (UAM) systems identify potential security risks by flagging unusual patterns. Personnel accessing sensitive files after hours, connecting from unfamiliar locations or transferring large volumes of data trigger immediate alerts. Additional warning indicators include unauthorised software installations, security setting modifications or attempts to reach restricted network segments. For instance, when the credentials of a finance employee who typically works business hours are suddenly used to download confidential records at 3 AM from an overseas location, behavioural analytics immediately flags this deviation.

Traditional security measures (firewalls and access controls) prove ineffective against insider threats because these attackers already possess valid credentials. UBA addresses this vulnerability by continuously analysing user behaviour beyond basic authentication. You can identify warning signs like privilege escalation, repeated access failures or unusual data movements, allowing detection of compromised accounts or malicious insiders before substantial damage occurs.

AI has significantly enhanced behavioural analytics capabilities. ML algorithms analyse extensive datasets to refine baselines and detect subtle anomalies that human analysts might overlook. Integration with SIEM platforms enables automated responses, including access restriction, account lockdown or real-time security team alerts. These automated defences dramatically reduce incident response times, preventing threats from developing into major security incidents.

Beyond technological solutions, employee awareness and training form crucial components of comprehensive cybersecurity. Educating staff about potential risks, monitoring practices and security protocols effectively reduces accidental and deliberate security breaches. Employees who understand cybersecurity's importance become active participants in maintaining your organisation's security posture; they're your first line of defence, not just potential vulnerabilities.

Despite its effectiveness, behavioural monitoring raises significant ethical and privacy considerations. Organisations must navigate legal regulations while maintaining transparency in data collection and analysis practices. The ethical implementation of UBA requires clear policies and governance frameworks that balance effective security measures with respect for individual privacy. When implemented thoughtfully, UBA provides powerful threat detection capabilities while fostering a workplace culture of trust and security awareness.

Supply Chain and Third-Party Risk Detection

Organisations operate within complex, interconnected networks that rely on numerous third-party relationships. This expanded digital footprint increases exposure to cyber threats, as attackers recognise that breaching a single supplier can provide access to multiple organisations. The 2020 SolarWinds attack demonstrated this risk on a massive scale. In a single campaign, attackers compromised a software vendor's update process and infiltrated over 18,000 organisations.

Traditional security measures focused on protecting an organisation's internal network are insufficient in our current environment. Threats can originate from within when trusted vendors directly access critical systems. Effective detection requires continuous monitoring across your entire vendor ecosystem, examining direct suppliers and subcontractors. Most organisations work with hundreds of third parties, and each relationship presents a potential entry point for sophisticated attackers.

Securing the supply chain begins with clearly defined contractual obligations. Service Level Agreements (SLAs) should specify security expectations, data protection policies and incident response procedures. These agreements ensure vendors align with your security standards rather than applying their own, potentially weaker measures. Regular assessments through security audits, penetration tests and policy reviews help you identify vulnerabilities and enforce compliance with requirements.

Behavioural analytics serves as a critical tool in detecting supply chain breaches before they escalate. By establishing a baseline for normal vendor activity, you can quickly spot anomalies that might signal compromise. Consider a scenario where a third-party maintenance provider accesses customer databases outside regular business hours. Similarly, when you monitor activity logs from external services, you can identify unauthorised access attempts or unusual data transfer patterns that often precede major breaches.

A Zero-Trust Approach strengthens supply chain security by assuming breaches will occur and requiring verification for all third-party interactions. This means you never automatically trust any vendor, regardless of relationship history. You can limit potential damage from suppliers by implementing JIT access provisioning and network segmentation. Every third-party connection must undergo strict authentication, authorisation and behavioural validation before accessing sensitive systems.

Integrating threat intelligence transforms supply chain security from reactive to proactive protection. Industry-specific Information Sharing and

Analysis Centres (ISACs) provide early warnings about threats targeting particular supply chain components. Dark web monitoring can reveal stolen vendor credentials before attackers exploit them. Advanced ML models can analyse these intelligence sources to assign dynamic risk scores to each supplier, allowing your security team to focus monitoring efforts where they matter most.

As supply chains grow increasingly complex, automated threat hunting becomes essential for identifying sophisticated attackers who deliberately evade standard detection methods. When you proactively search for signs of compromise in third-party networks and regularly test supplier access points, you can discover and address vulnerabilities before cybercriminals exploit them. Stay vigilant—your security is only as strong as your weakest supplier.

Tactical Briefing: Indicators of Compromise (IoCs)

Understanding IoCs

Cyberattacks typically leave digital evidence, IoCs, enabling security teams to detect and respond to threats efficiently. These telltale signs, including suspicious file hashes, unusual network traffic patterns or unauthorised system modifications, function as early warning signals of malicious activity. When identified in real-time, IoCs allow organisations to detect ongoing attacks and address security risks before they escalate into major incidents. Since cybercriminals constantly evolve their tactics, security professionals must regularly validate and update their IoC databases to maintain effective protection. The primary purpose of monitoring IoCs is to identify potential security threats and malicious activities within an organisation's systems and networks.

How to Identify and Classify IoCs

Effective security monitoring starts with recognising anomalous behaviour that could indicate a breach. IoCs fall into several categories, each offering insight into different types of threats:

- File-based IoCs include file hashes, names and sizes identifying known malware.

- Network-based IoCs cover suspicious IP addresses, domain names and URLs linked to malicious activity

- Behavioural IoCs involve unusual system activity, such as repeated login failures or unexpected data transfers

- Artifact-based IoCs include registry keys, configuration changes and log modifications that may indicate an attack.

Threat actors attempt to evade detection by altering malware signatures or using encrypted communication channels, making it crucial to analyse IoCs in context. A single IoC alone may not confirm an attack, but when combined with other indicators, it can provide a clearer picture of a security incident.

Documenting and Validating IoCs

To be useful, IoCs must be accurately documented in a structured format. Security teams commonly use:

- YARA rules to identify malware families based on file patterns - https://yara.readthedocs.io/en/v3.4.0/writingrules.html

- STIX (Structured Threat Information eXpression) for sharing threat intelligence - https://oasis-open.github.io/cti-documentation/

- OpenIOC for exchanging standardised IoC data - https://cloud.google.com/blog/topics/threat-intelligence/openioc-basics/
- Text-based formats (CSV, JSON) for easy integration into security tools.

Each IoC should include key details such as file hashes (e.g., SHA-256), associated malicious IP addresses, domain names and specific behavioural patterns. However, inaccurate IoCs can lead to false positives, wasting resources and causing unnecessary alarms. To prevent this, teams should cross-reference findings with trusted threat intelligence sources and test IoCs in controlled environments before deployment.

Sharing and Implementing IoCs Effectively

The value of IoCs increases significantly when shared across organisations, strengthening collective defences against common threats. Security teams typically integrate these indicators into various detection tools, including SIEM systems, IDS and IPS, and firewalls. By automating the ingestion of IoCs through specialised threat intelligence platforms, organisations improve their response times while reducing manual workload for security staff. Given the ever-changing nature of cyber threats, security teams must regularly review and update their IoC databases to maintain their effectiveness against new attack techniques.

Investigating Security Incidents Through Forensic Analysis

When monitoring systems detect an attack, forensic analysis becomes necessary for tracing its origin and assessing its impact. However, forensic analysis requires specialised skills and strict chain-of-custody requirements, and it should only be undertaken by qualified staff rather than amateur or enthusiast IT personnel. Security teams with proper forensic training meticulously examine file systems, memory dumps and network logs to reconstruct the sequence of events during the incident. This forensic evidence may later support legal proceedings against attackers, necessitating strict adherence to data privacy regulations and chain-of-custody protocols throughout the investigation.

Strengthening Detection Capabilities in Cybersecurity

The combination of IoC monitoring and forensic analysis creates a powerful framework for threat detection. As cyber threats continue to grow in sophistication, security professionals must constantly refine their detection strategies and adapt their techniques to ensure organisational resilience against evolving attack methods.

Evaluating and Improving Detection Effectiveness

Deploying detection mechanisms is only half the challenge. Their effectiveness determines whether threats stay hidden or are swiftly exposed.

Mean Time to Detect (MTTD) is a key performance metric that measures how quickly threats are identified after entering your environment. Lower MTTD values reflect more responsive and capable detection systems. When organisations begin tracking this metric, they often uncover dangerous gaps between compromise and discovery, sometimes finding that advanced threats persist undetected for weeks or even months.

False positive rates and alert accuracy serve as complementary evaluation measures. When legitimate activities trigger security alerts, security teams experience alert fatigue, increasing the risk of overlooking genuine threats amid excessive noise. Consider Target's high-profile 2013 data breach. A U.S. Senate report found that Target's security solution issued multiple automated warnings of potential intrusion, yet these alerts were ignored. Subsequent analyses suggest that frequent false alarms contributed to the failure to act on genuine warnings.[106]

The fine balance between sensitivity and specificity presents a crucial calibration challenge. Increased sensitivity detects more potential threats but typically generates more false positives. Conversely, higher specificity reduces false alarms but risks missing subtle attack indicators. False negatives prove more dangerous than false positives; they allow threats to operate undetected within your environment.

Fine-tuning detection rules requires methodical precision. Establishing an accurate baseline behaviour enables meaningful anomaly detection (what constitutes 'normal' varies significantly across organisational environments). Adjusting detection thresholds based on this baseline helps reduce noise while maintaining vigilance for genuine threats. Rather than relying on generic detection parameters, organisations must tailor algorithms to match their specific network characteristics, application behaviours and business processes.

As we've seen repeatedly throughout this book, cybersecurity's dynamic nature necessitates continuous improvement cycles. This is no different when working on Detection. Threat landscape monitoring ensures detection mechanisms evolve alongside adversary tactics. Regular audits and assessments stress-test these systems, revealing blind spots before attackers exploit them. Creating feedback loops between detection and Incident Response Teams (IRTs) facilitates valuable intelligence exchanges, where responders' frontline experiences inform detection rule modifications.

As threats grow in sophistication, manual tuning alone proves inadequate. Progressive organisations increasingly employ ML for continuous analysis and automatic rule adaptation, reducing human intervention while improving accuracy. Through careful measurement, systematic refinement and cross-functional collaboration, organisations transform detection from a static shield into adaptive armour that evolves with each encounter.

Overcoming Common Detection Challenges

Our current threat landscape presents formidable obstacles for even sophisticated detection systems. As attackers continuously refine their techniques to evade discovery, security teams face an ongoing battle to maintain visibility across increasingly complex environments. Understanding these detection blind spots, the shadowy corners where adversaries operate undetected, is vital for developing effective monitoring strategies.

Sophisticated adversaries employ a variety of evasion techniques to remain undetected. One particularly effective tactic is Living off the Land (LotL) attacks, where legitimate system tools (such as PowerShell or Windows Management Instrumentation (WMI)) are weaponised to perform malicious actions. These tools blend in with routine operations, making them difficult for signature-based defences to flag. Attackers also use code obfuscation and packing methods to disguise malware signatures, while encrypted communications and steganography help conceal command and control traffic within ordinary data streams.

Threat actors increasingly turn to legitimate cloud platforms and SaaS applications to host their infrastructure. By aligning malicious traffic with standard business patterns, they reduce their chances of detection, much like a pickpocket blending into a crowd.

The sheer volume of security alerts presents a formidable challenge for detection teams. SIEM platforms often generate thousands of alerts daily, increasing the risk that serious threats will be lost amid a flood of false positives. This makes risk-based alerting strategies essential. By prioritising alerts based on asset criticality, anomalous user behaviour and real-time threat intelligence, teams can focus on what matters most.

ML offers valuable support in filtering alerts. Algorithms can correlate diverse events to reveal genuine attack patterns while suppressing benign anomalies. In effect, it's like employing a discerning assistant who learns to differentiate between emergencies and distractions.

Improving log management further sharpens detection. Rather than collecting all data indiscriminately, mature strategies focus on high-value log sources (such as DNS queries, authentication events and endpoint telemetry) while filtering out low-utility information. Regularly tuning detection rules, alert thresholds and SIEM settings ensures that security teams remain focused on real threats instead of becoming overwhelmed by noise.

Even as detection technologies advance, one core challenge endures: maintaining full visibility while accurately separating signal from noise. By methodically addressing blind spots, understanding adversary tactics and implementing intelligent alert strategies, organisations can substantially improve their ability to identify and stop threats before they succeed.

Summary

This chapter examined how Detection forms the crucial bridge between protective controls and incident response, creating an essential link in your cybersecurity chain.

Early threat identification dramatically reduces damage (hopefully before your CEO asks uncomfortable questions). Breaches detected within minutes, rather than months, prevent data exfiltration and system compromise. However, modern threats like LotL attacks, fileless malware and APTs are specifically designed to evade detection. IBM reports show organisations take an average of 204 days to identify breaches. Imagine a burglar living in your house for nearly seven months before you notice them!

Effective detection relies on three core pillars: monitoring, analysis and alerting. Security tools like SIEM, IDS and EDR continuously track infrastructure activity. Behavioural analytics and threat intelligence refine detection accuracy, while network monitoring identifies suspicious traffic through techniques like DPI and flow analysis. SOAR platforms automate response workflows, and endpoint monitoring covers workstations, servers and mobile devices using FIM and UBA to detect anomalies.

Physical security plays a vital role in detection by monitoring access to server rooms and network equipment. Supply chain security is equally vital. Attacks like SolarWinds demonstrate how a single compromised vendor can expose multiple organisations. A Zero-Trust Model, which assumes breaches will occur, mitigates these risks.

Understanding IoCs, telltale signs of malicious activity, enables security teams to detect and respond to threats efficiently. These include file-based, network-based, behavioural and artifact-based indicators that function as early warning signals. Accurate documentation and validation of IoCs and sharing across organisations strengthen collective defences against common threats.

Evaluating detection effectiveness requires monitoring metrics like MTTD and false positive rates. The delicate balance between sensitivity and specificity presents a crucial calibration challenge. Increased sensitivity detects more potential threats but generates more false positives, while higher specificity reduces false alarms but risks missing subtle attack indicators.

Common detection challenges include unmonitored network segments, insider threats, encrypted traffic, supply chain compromises and various evasion techniques employed by sophisticated adversaries. The overwhelming volume of security alerts creates perhaps the most challenging obstacle, making risk-based alerting strategies and ML algorithms vital for effective threat detection.

This chapter covered Detection's role in cybersecurity, highlighting its function as the bridge between prevention and response. It explored key detection technologies, strategies for identifying threats and challenges such as alert fatigue and evasion techniques. Continuously refining detection capabilities enables organisations to stay ahead of evolving threats and strengthen their overall security posture.

Revision Questions

1. What is the Detection principle in cybersecurity and why is it important?
2. What are the three fundamental pillars of effective cybersecurity detection?
3. According to IBM security reports, how long do organisations typically take to identify breaches?
4. What is Network Traffic Analysis (NTA) and what techniques does it use?
5. What are Indicators of Compromise (IoCs) and what categories do they fall into?
6. What detection challenge did the 2013 Target data breach illustrate?
7. What is Living off the Land (LotL) and why is it challenging for detection systems?
8. How does User and Behavioural Analytics (UBA) enhance detection capabilities?
9. Why is physical environment monitoring important for cybersecurity?
10. What approach to supply chain security assumes breaches will occur and requires verification for all third-party interactions?

Cybersecurity Hacks

Hack 1: Implement Deception Technology

Deception technology helps detect cyber threats early by deploying decoy assets, known as honeypots, within a network. These honeypots mimic real systems, tricking attackers into engaging with them. When an attacker interacts with a decoy, it reveals their presence and tactics, allowing security teams to gather intelligence on potential threats without exposing actual systems to risk. This approach strengthens security by identifying malicious activity before it causes damage.

Hack 2: Use Threat Hunting

Threat hunting is a cybersecurity practice that involves actively searching a network for signs of hidden threats that bypass traditional security measures. Instead of waiting for automated alerts, security teams analyse system activity, logs and data patterns to detect unusual behaviour. Regular threat-hunting exercises help identify vulnerabilities before they can be exploited, reducing security risks. For beginners, building a simple threat-hunting playbook, a structured guide outlining key steps and techniques, can make this process more effective and manageable.

Building a Basic Threat-Hunting Playbook

Step 1: Define Objectives and Scope

Key Questions:

- What threats are you targeting? (e.g., insider threats, APTs, ransomware)
- Which assets and systems are in scope?
- What outcomes do you expect?

Action: Establish clear, measurable objectives to guide your hunt.

Step 2: Gather Intelligence and Form Hypotheses

Threat intelligence provides context, while hypotheses focus the hunt.

Sources:

- Open-source intelligence (OSINT)
- Threat intelligence feeds (e.g., MITRE ATT&CK, VirusTotal)
- Internal logs and past incidents.

Example Hypotheses:

- "If phishing is used, there will be unusual logins from unknown IPs".
- "If an attacker deploys malware, endpoint logs may show unexpected process executions".
- "If lateral movement occurs, admin credentials may be misused".

Action: Analyse intelligence and develop testable hypotheses.

Step 3: Identify and Analyse Data

To validate hypotheses, collect relevant data.

Common Sources:

- SIEM logs
- EDR data
- Firewall logs

- Network traffic captures.

Action: Ensure visibility into key data sources and begin analysis.

Step 4: Investigate and Respond

Confirm whether anomalies indicate real threats and take action.

Investigation Steps:

1. Correlate findings with other data.
2. Determine incident scope.
3. Validate findings through additional queries or manual inspection.
4. Define response actions based on threat severity.

Action: Confirm or refute threats, document findings and escalate if necessary.

Next Steps:

- Implement and tailor the playbook to your organisation
- Refine it through feedback and evolving threats.

Chapter 14: Response

– From Detection to Resolution –

Objectives

By the end of this chapter, you will be able to:

- Understand the foundational processes involved in responding to cybersecurity incidents
- Understand how to develop a comprehensive Incident Response Plan (IRP)
- Apply effective techniques for incident triage, validation, categorisation and prioritisation
- Understand appropriate containment and mitigation strategies during cyber incidents
- Conduct incident analysis and establish processes for continuous improvement.

In cybersecurity, Detection is only half the battle. An organisation's true resilience is revealed in how swiftly and decisively it can act once a threat emerges. That's where the Respond principle, our fifth core tenet, takes centre stage. It's about the actions that follow Detection—how you contain threats, limit damage and ensure your plans are put into motion effectively. If you neglect Respond, it's like having smoke detectors but no fire brigade.

Responding requires more than reacting in the moment. It involves building and rehearsing incident response strategies. You hope you'll never need them, but those rehearsed steps become vital in a crisis. These strategies define how to isolate affected systems, eliminate malicious elements, communicate clearly and restore disrupted services. A timely, coordinated response helps reduce impact, accelerate recovery and strengthen defences for the future.

The Respond principle spans several interconnected efforts critical to incident management. It begins with prompt internal and external communication using established channels and notification protocols. From there, it covers analysis, containment, eradication and recovery. These activities depend on well-crafted response plans, business continuity measures and disaster recovery procedures, all designed to restore operations quickly and minimise disruption.

Key components include incident management: executing response plans, triaging events, verifying legitimacy, categorising incidents and prioritising based on urgency. Then comes incident analysis: uncovering root causes, evaluating impact, preserving evidence and supporting forensic work. Coordination with third parties, such as law enforcement and cybersecurity experts, is often essential. Regular training and simulated drills ensure your teams stay sharp and processes evolve with emerging threats. Ultimately, the Respond principle is about reducing harm, speeding recovery and reinforcing your overall security stance.

This chapter will give you the knowledge and tools to confidently navigate cyber incidents. You'll explore the foundational elements of an Incident Response Plan (IRP) and learn how to triage and validate incidents effectively. We'll also look at methods for classifying threats based on severity and potential fallout. Finally, we'll walk through escalation paths, recovery protocols and the crucial steps of post-incident analysis. Mastering these practices will assist you to safeguard your organisation and enhance the resilience of your digital environment against ever-changing cyber risks.

Building the Foundation for Effective Response

Incident Management

Effective cyber incident response requires a structured incident management framework. This methodical approach transforms chaotic crisis into coordinated responses that minimise organisational impact and accelerate recovery. While security tools play important roles, it's the framework, processes, people and practices, that determines response effectiveness.

Incident management follows a defined lifecycle spanning preparation through resolution, much like a hospital emergency room. Beginning with preparation, organisations establish the necessary infrastructure, tools and training before incidents occur. When anomalies arise, detection mechanisms identify potential security events, triggering analysis to validate legitimate incidents and assess scope. Once confirmed, containment isolates affected systems to prevent further spread, followed by eradication to remove threat actors and their persistence mechanisms.

Effective categorisation and prioritisation form the framework's foundation. Incidents range from minor events with limited impact to enterprise-wide breaches threatening critical operations. By classifying incidents based on severity, impact scope and service disruption, organisations allocate resources appropriately and escalate responses per predefined criteria. A ransomware attack encrypting financial systems, for example, requires immediate executive involvement and comprehensive mobilisation, a more significant response when compared to what is needed for an isolated malware detection on a non-critical workstation.

Clear stakeholder roles eliminate confusion during high-pressure situations. The Incident Commander maintains overall response coordination, while Security Operations Centre (SOC) analysts provide technical detection and investigation. IT responders implement containment and recovery actions, supported by legal and communications teams managing regulatory obligations and stakeholder messaging. This approach prevents duplicate efforts and ensures comprehensive coverage.

Response playbooks and Standard Operating Procedures (SOPs) transform abstract plans into actionable workflows. These predefined, documented response sequences for common incident types (data breaches, malware outbreaks, unauthorised access) provide responders with step-by-step guidance during stressful situations. Modern organisations increasingly incorporate automation to accelerate initial response actions, enabling human responders to focus on complex analysis and decision-making rather than repetitive tasks.

The framework must also include comprehensive measurement mechanisms. Key Performance Indicators (KPIs) like Mean Time to Detect (MTTD) and Mean Time to Respond (MTTR) provide quantitative assessments of response capabilities. Regular testing through tabletop exercises and simulated breaches validates response plans against practical scenarios, identifying improvement opportunities before real incidents occur.

Implementing an effective incident management framework empowers your organisation to transform security incidents from overwhelming crises into manageable events with predictable processes and outcomes.

The Incident Response Plan (IRP)

Think of a cyber incident as a house fire. You wouldn't stand around debating who calls the fire brigade. A well-defined IRP provides your organisation with a structured, methodical approach to managing and mitigating cyber incidents. When cyber threats (like those described in Chapter 2) strike, they can cause significant damage if you don't handle them efficiently. An effective IRP outlines specific procedures, assigns responsibilities and ensures swift, coordinated responses that minimise disruption and reduce overall impact.

Importance of an IRP

An IRP serves as a guide during the chaotic and high-pressure period following a cyber incident. Without a structured plan, your organisation's response often becomes disorganised, slow and ineffective, leading to prolonged downtime, loss of sensitive data, regulatory penalties and reputational damage. A well-prepared IRP ensures all team members understand their specific roles, communication channels remain open and predefined steps are followed to contain, mitigate and recover from incidents.

Components of a Practical IRP

A comprehensive IRP comprises several essential components that must be clearly defined to ensure a seamless response when an incident occurs:

Roles and Responsibilities

Defining the roles and responsibilities of all stakeholders is needed for an efficient response. This includes appointing an Incident Response Team (IRT) leader who oversees response efforts and ensures timely decision-making. Other key roles typically include security analysts, IT personnel, legal advisors and public relations officers. Each IRT member should have immediate access to their role-specific responsibilities, separate from the full IRP document. This enables them to act promptly without navigating through extensive documentation during an incident.

Communication Plan

Communication is frequently compromised during a cyber incident. The IRP should establish clear communication protocols detailing how and when to engage with internal stakeholders (executives, employees and IT teams) and external parties (customers, partners, regulatory bodies and law enforcement when necessary). A predefined communication strategy prevents misinformation, facilitates timely updates and maintains stakeholder confidence throughout the crisis.

Escalation Procedures

Since incidents vary in severity, an IRP should include a tiered escalation procedure. This framework defines when incidents should be escalated to senior management, legal teams or external cybersecurity experts. Establishing clear escalation criteria ensures that appropriate personnel make critical decisions at the right time, avoiding unnecessary delays in incident resolution.

Recovery Steps

After containing an incident, restoring normal business operations becomes the priority. The IRP should include a detailed recovery plan outlining procedures for system restoration, data recovery and forensic analysis to determine the incident's root cause. A thorough post-incident review helps identify vulnerabilities and strengthen security measures, creating continuous improvement cycles in your security posture.

Developing an Effective IRP

When creating and refining your IRP, consider using resources such as the Cyber Incident Response Guidance from the ASD-ACSC.[107] This guidance provides valuable insights into best practices for incident management tailored to the Australian context. Regular testing through simulation exercises and updates to your IRP ensures it remains relevant and effective against evolving cybersecurity threats.

Implementing a well-structured IRP significantly enhances your organisation's ability to respond to cyber threats, minimise damage and ensure swift recovery, ultimately protecting business operations, reputation and stakeholders.

The Response Lifecycle in Action

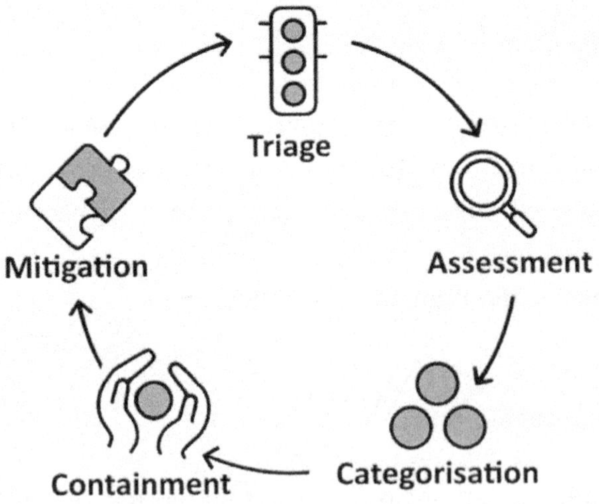

Initial Response: Detection to Triage

A cybersecurity incident response hinges on the swift transition from threat detection to active containment. When security systems identify a potential threat, organisations must rapidly validate its authenticity, assess severity and implement appropriate countermeasures to minimise damage.

The response process begins with incident triage, a methodical approach that ensures security teams focus their limited resources on genuine threats. Security analysts first identify potential incidents by monitoring multiple information sources, including network logs, Security Information and Event Management (SIEM) systems and user-reported anomalies. Automated tools and threat intelligence feeds enhance this detection capability by flagging patterns that may indicate malicious activity.

Not all security alerts represent equal threats, however. Once identified, incidents must be prioritised based on several critical factors: the sensitivity of affected data, the extent of system compromise and the potential impact on business operations. High-priority incidents, for example, should trigger immediate action, often activating formal escalation procedures within your organisation's IRP. This approach determines not only the response urgency, but also which teams and resources must be deployed.

Validation serves as an important checkpoint in the response workflow. Security teams must confirm that a flagged threat represents a genuine incident rather than a false positive (like crying wolf when there's just a dog). This verification typically involves detailed analysis of system logs, correlation of events with known attack patterns and consultation with subject matter experts. Effective validation prevents unnecessary resource expenditure while ensuring real threats receive prompt attention.

The first 60 minutes following incident confirmation often prove decisive in limiting damage. Response teams must make rapid decisions about containment strategies, whether to isolate affected systems immediately or temporarily monitor attacker movements to gather intelligence. Simultaneously, they must preserve critical evidence through the proper collection of volatile data while maintaining a chain of custody to support potential legal proceedings.

Communication flows in parallel with technical response efforts. Key stakeholders, including technical teams, executive leadership, legal counsel and affected partners, require timely notification with appropriate levels of detail. Throughout this process, the Incident Response Team (IRT) continuously evaluates business impact, carefully balancing operational continuity against security requirements.

The effectiveness of this initial response phase typically determines whether an incident remains a contained security event or escalates into a major organisational crisis requiring prolonged recovery efforts.

Assessment and Categorisation

When security incidents occur, swift classification becomes vital for effective response. Assessment and categorisation convert raw alerts into actionable intelligence, allowing organisations to deploy resources efficiently while methodically addressing threats.

Industry-standard frameworks such as NIST SP 800-61 and European Union Agency for Cybersecurity (ENISA) Reference Incident Classification Taxonomy offer structured approaches to incident categorisation. These frameworks establish defined taxonomies (malware infections, data breaches, denial-of-service attacks), enabling responders to recognise patterns and implement proven remediation strategies.

Different incident categories require specific response approaches. Malware infections typically require system isolation, while data breaches require forensic investigation and stakeholder notification. Although automated tools expedite classification for common scenarios, human analysts are essential for interpreting sophisticated attacks evading standard categorisation.

You must evaluate incident severity through multiple dimensions, especially impacts on confidentiality, integrity and availability. Organisations now use quantitative models like the Common Vulnerability Scoring System (CVSS) to assess impact objectively, converting technical indicators into business risk metrics that combine vulnerability data with threat intelligence.

When handling multiple incidents simultaneously, prioritisation becomes critical. A risk-based strategy ensures threats to mission-critical systems receive urgent attention based on potential damage to sensitive data, the extent of system compromise and overall organisational risk. Regardless of technical severity, regulatory requirements often prioritise specific incidents, particularly those involving personally identifiable information.

Well-defined escalation procedures come to the fore when incidents surpass the initial team's capabilities. Your IRP should specify the previously mentioned areas (command structure, escalation triggers, communication protocols, etc.) to ensure incidents are managed at the appropriate organisational level, maintaining coordinated response while decreasing MTTR for complex security events.

Containment and Mitigation Strategies

Once a cyber threat is detected, rapid yet strategic containment is essential to limiting damage while preserving critical forensic evidence. Security teams must balance neutralising the immediate risk and gathering intelligence to understand the attack's scope. A well-executed containment plan prevents further compromise without unnecessarily disrupting operations.

The containment approach depends on the attack's nature and the response objectives. Immediate containment, such as isolating an infected system during a ransomware outbreak, halts the spread but may alert the attacker, prompting evasive actions or escalation. Conversely, delaying containment is sometimes preferable for persistent threats, such as Advanced

Persistent Threats (APTs), where continued observation can expose attacker tactics and infrastructure.

Containment operates at multiple levels: at the network layer, measures like firewall rules and VLAN (Virtual Local Area Network) segmentation prevent lateral movement, while at the endpoint level, actions such as disabling compromised accounts and quarantining hosts limit exposure while preserving critical forensic data.

Effective isolation techniques prevent further compromise while allowing security teams to analyse the threat. Physically disconnecting a system, such as removing it from the network, may be necessary to contain destructive malware. However, logical isolation, achieved through network segmentation and access restrictions, enables ongoing monitoring without permitting further exploitation. A well-executed isolation strategy also considers operational continuity. Rather than shutting down access for an entire department, organisations may impose targeted restrictions on affected users or systems, ensuring critical functions remain operational while mitigating the threat.

Mitigation strategies must align with the type of cyber incident. In malware and ransomware attacks, containment involves disconnecting infected hosts, blocking malicious domains and securing forensic data by capturing memory and disk images before rebooting systems. Insider threats and data breaches require a different approach, restricting access to suspicious accounts while enhancing monitoring can prevent further damage without prematurely alerting the insider. Data Loss Prevention (DLP) tools help contain breaches by preventing unauthorised data transfers and tracking exfiltration attempts.

Executing mitigation actions with precision preserves evidence integrity and ensures an effective post-incident response. Selecting the right containment strategy requires assessing the incident's severity, scope and intelligence value. Security teams must weigh the urgency of threat suppression against the benefits of continued monitoring. As cyber threats evolve, real-time intelligence should inform containment efforts, ensuring responses remain adaptable and aligned with the attack's progression.

Investigation and Analysis

Evidence Collection and Preservation

An effective cyber incident response depends on careful collection and preservation of digital evidence. Think of it like a crime scene: you've got one chance to capture everything before it's gone. A core principle of digital forensics is evidence volatility, which describes how quickly certain data changes or disappears if you don't capture it promptly.

Volatile data must be collected first because it's lost when systems power down. This includes RAM contents, active network connections and running processes. Network packet captures are particularly valuable as evidence during this initial collection phase. In contrast, non-volatile data (disk images, system logs and archived files) remains intact longer but still requires secure preservation. Responders must act swiftly and systematically to prevent losing critical evidence.

Legal and compliance requirements shape how you'll handle evidence. Digital evidence must be collected and stored according to legal frameworks to ensure that it's admissible in court and compliant with cybersecurity regulations. In Australia, this includes *The Privacy Act 1988*, ASD-ACSC guidelines and relevant international standards. Failure to follow these protocols can compromise investigations, render evidence inadmissible or result in regulatory penalties.

Maintaining the chain of custody is a non-negotiable for preserving evidence integrity. Every interaction with evidence must be documented, including timestamps, personnel involved and reasons for access. This detailed record establishes the evidence's authenticity and ensures reliability in legal or investigative contexts.

Secure handling measures further protect evidence integrity. Cryptographic hashing verifies that data remains unchanged, while write-blocking prevents unintentional or malicious modifications. Secure storage and restricted access to forensic images ensure that only authorised personnel handle critical evidence.

Meanwhile, forensic investigators use specialised tools to capture and analyse digital evidence. FTK Imager and Autopsy create forensic disk images, while Volatility extracts memory dumps for investigation. Wireshark and SIEM systems assist in Network Traffic Analysis (NTA), helping to identify suspicious activity and security breaches.

Applying write-blockers during disk imaging prevents alterations, preserving original evidence integrity. Hashing techniques confirm that data remains unmodified. Overall, adhering to established forensic methodologies, such as the NIST guidelines and International Organization on Computer Evidence (IOCE) principles, ensures that evidence collection meets industry standards and withstands legal scrutiny.

Analytical Approaches

When responding to cyber incidents, thoroughly analysing the incident is essential to identify its root causes, determine the extent of the damage and understand the methods employed by the attackers. Root Cause Analysis (RCA) and systematic event correlation are vital to this process, assisting cybersecurity teams to reconstruct the incident and identify its origin. The analysis begins with collecting forensic evidence, such as system logs, network traffic and endpoint data, to determine the initial compromise point. This evidence is then cross-checked with threat intelligence sources, such as Indicators of Compromise (IoCs) and Tactics, Techniques and Procedures

(TTPs). Analysts can then gain insights into the threat actor's methods and motives by comparing the findings with known attack patterns.

RCA often relies on structured frameworks, like the Five Whys, Fishbone Diagrams or Fault Tree Analysis. These approaches help trace the incident's origin, whether it be a system vulnerability, a human error or a combination of both. For example, a malware infection might be traced to an unpatched system or weak password protocols, highlighting the need for stronger preventative measures. Forensic analysis is crucial for understanding the technical aspects of the breach and maintaining data integrity, which is often vital in legal or compliance investigations.

Once you've identified the root cause, constructing a detailed timeline of events is critical. By correlating data from various sources, such as SIEM logs, cloud logs and endpoint detection systems, analysts can establish a precise sequence of the attack. When mapped against models like the Kill Chain or MITRE ATT&CK, this timeline offers vital insights into the attacker's tactics, allowing teams to pinpoint detection and response gaps.

Source: Cyber Kill Chain®, Lockheed Martin. https://www.lockheedmartin.com/en-us/capabilities/cyber/cyber-kill-chain.html

In complex incidents, traditional methods may fall short. The following advanced techniques provide deeper visibility and support more effective response strategies:

- Behavioural and Anomaly-Based Analysis—Uses machine learning and statistical methods to detect subtle deviations from normal activity—capturing threats that signature-based systems may overlook.

- Memory and Disk Forensics—Analyses system memory and storage to uncover evidence of persistence mechanisms, lateral movement, and hidden malicious artefacts.

- Reverse Engineering Malware—Deconstructs malware samples to reveal command-and-control (C2) infrastructure, functionality, and potential vulnerabilities—enabling tailored detection signatures.

Together, these techniques offer a comprehensive view of the incident, enabling more effective containment, eradication, and recovery.

Stakeholder Management During Incidents

Internal Communication Strategies

Consider again the analogy of a cybersecurity incident being like a house fire. Without clear communication, firefighters will bump into each other instead of coordinating their efforts. Similarly, during a cybersecurity incident, effective internal communication ensures all stakeholders remain informed, aligned and capable of responding swiftly. A structured approach prevents miscommunication from creating delays, confusion or misguided decisions that amplify the crisis.

Establishing Clear Communication Protocols

Your IRT needs structured communication protocols to deliver timely updates and support informed decision-making. Regular briefings keep executives informed with concise summaries tailored to their strategic concerns (operational impact, regulatory obligations and resource requirements). These briefings should follow a consistent schedule, balancing transparency with operational efficiency.

Executives will be worried about financial, legal and reputational implications. Their concerns must be addressed proactively, so that their leadership decisions align with both broader business objectives and the technical priorities you will highlight using your expertise. Remember, executives don't need to understand the technical intricacies of a SQL injection—they just need to know what it means for the business.

Cross-Functional Team Coordination

Cyber incidents require seamless collaboration between IT security, legal, compliance and public relations teams. Clearly defined roles and responsibilities minimise ambiguity and prevent duplicate efforts, allowing each team to focus on its specific function.

Real-time collaboration tools facilitate information sharing, ensuring that containment measures, legal considerations and external communications work together. Maintaining consistency in public messaging prevents contradictory narratives that could undermine stakeholder trust and damage your organisation's reputation.

Employee Communication During Incidents

Precise, timely communication with employees alleviates uncertainty and reinforces security protocols. Regular updates provide actionable guidance, such as identifying phishing attempts connected to the incident.

Dedicated support channels, including help desks or response teams, enable employees to seek clarification and report concerns. This approach enhances your organisation's overall resilience by turning every employee into an additional layer of defence.

External Communication and Reporting

Effective external communication during a cyber incident is vital for meeting regulatory obligations, preserving stakeholder confidence and protecting an organisation's reputation. A structured approach ensures legal requirements are fulfilled while addressing customer concerns and media inquiries in a clear and controlled manner.

Regulatory Reporting Requirements

Complying with national and international reporting laws is an unavoidable part of legal compliance and risk mitigation. Regulations such as the GDPR in Europe and Australia's Notifiable Data Breaches (NDB) scheme impose specific timelines and formats for reporting cyber incidents. Non-compliance results in financial penalties and legal consequences that no one wants to explain to the board.

Certain industries face more stringent reporting standards. Financial institutions, healthcare providers and critical infrastructure operators must fulfil additional obligations to safeguard sensitive data and maintain public trust. Transparent, well-documented reporting satisfies legal requirements and strengthens organisational accountability.

Customer and Public Communication

Timely, transparent communication with affected customers minimises reputational damage and maintains trust. You must provide clear details about the incident, its impact and recommended protective actions. For example, instruct customers to reset passwords, monitor financial transactions or implement additional security measures.

Beyond immediate responses, open and honest messaging builds long-term confidence. Companies that acknowledge the issue, explain their remediation efforts and outline future security improvements can maintain customer loyalty even after a breach. It's counterintuitive, but transparency often strengthens relationships.

Media Relations During Cybersecurity Incidents

A well-prepared spokesperson and carefully crafted statements ensure consistency in public messaging and limit misinformation. Actively managing media engagement allows organisations to provide accurate updates, reducing speculation and uncertainty. Following an incident, media strategies should highlight recovery efforts and planned security enhancements to reassure stakeholders and rebuild confidence.

Summary

The Respond principle, the fifth core element in cybersecurity, bridges the crucial gap between detection and resolution when security incidents occur. This principle emphasises your organisation's ability to act swiftly and effectively after identifying a threat through incident response strategies.

Effective response encompasses several interconnected activities: isolating affected systems, eliminating malicious elements, communicating with stakeholders and restoring compromised services. These coordinated actions minimise damage, accelerate recovery timeframes and strengthen your defences against subsequent attacks.

Throughout this chapter, we've explored the multifaceted nature of incident response, from establishing incident management frameworks and developing detailed response plans to conducting proper triage, analysis and stakeholder communication. A well-executed response process significantly reduces downtime, financial losses and reputational damage while demonstrating your commitment to regulatory compliance and security best practices.

Incident response effectiveness depends on your organisation's preparedness and proactive planning. This includes investing in effective IRPs, establishing clear communication protocols and defining precise roles for all team members. When these elements work together, you can face cyber incidents confidently, protecting critical assets and maintaining operational continuity even under attack.

The dynamic nature of cybersecurity threats means you can't rely on static approaches to incident response. Regularly refining your response strategies based on lessons learned, evolving threat intelligence and technological changes ensures your organisation remains prepared. This ongoing improvement process, updating plans, enhancing communication methods and conducting realistic simulation exercises, keeps your response capabilities aligned with current threats.

As the digital landscape continues to evolve, effective incident response capabilities become vital for organisations across all sectors. The Respond principle represents an essential component of your cybersecurity framework, connecting preparation with action and enabling your organisation to protect its digital assets effectively against increasingly sophisticated threats.

Revision Questions

1. What is the Respond principle in cybersecurity and why is it important?

2. What are the key components of an effective incident management framework?

3. What essential components should be included in an Incident Response Plan (IRP)?

4. How should organisations approach incident triage?

5. What factors should be considered when categorising and prioritising incidents?

6. What containment strategies can be employed during cyber incidents?

7. How should digital evidence be collected and preserved during incident response?

8. What analytical approaches help identify the root causes of cyber incidents?

9. What internal communication strategies are essential during a cybersecurity incident?

10. What external communication responsibilities do organisations have during cyber incidents?

Cybersecurity Hacks

Hack 1: Create Incident Response Playbooks for Common Scenarios

Developing playbooks for frequently encountered incidents provides you and your organisation with valuable, ready-to-deploy response strategies. These step-by-step guides should address common scenarios such as ransomware attacks, phishing campaigns and data breaches, incorporating specific containment actions, communication templates and recovery procedures tailored to each situation. When implemented effectively, these playbooks eliminate decision paralysis during high-pressure incidents, ensure consistent response quality regardless of which team members are responding and measurably reduce your Mean Time to Respond (MTTR) to security events.

Hack 2: Establish a "Digital First Aid Kit" for Immediate Response

Create a readily accessible digital toolkit containing essential resources for the first 30 minutes of any incident response. This kit should include: pre-written communication templates for different stakeholder groups (executives, employees, customers), a quick reference card listing critical system administrator contacts and escalation phone numbers, screenshots of normal system baselines for rapid comparison during incidents, and a simple incident classification flowchart that helps determine severity levels. Store this toolkit in multiple locations (printed copies, cloud storage, mobile devices) to ensure availability even if primary systems are compromised.

Chapter 15: Recovery

– Rebuilding After the Storm –

Objectives

By the end of this chapter, you will be able to:

- Define Recovery and explain its role as the final principle in the cybersecurity framework
- Understand how to conduct a Business Impact Analysis (BIA)
- Implement effective data protection and backup strategies for cyber incident recovery
- Measure and improve recovery capabilities through metrics and continuous learning
- Navigate common challenges in executing recovery operations.

The aftermath of a cybersecurity incident is like the wake of a cyclone; your systems lie damaged, operations are disrupted and the road ahead is murky with uncertainty. Recovery, the sixth foundational principle in the cybersecurity framework, completes the process outlined by the ISM and the NIST CSF that progresses from Govern, to Identify, Protect, Detect and Respond. Recovery marks the shift from managing a crisis to restoring normal operations. While Respond is about containing threats during an incident, Recovery focuses on rebuilding once the immediate danger has passed.

Recovery involves more than technical fixes; it's akin to restoring a home after cyclone damage. It reflects your organisation's ability to bounce back from disruption, a strategic strength that must be prepared for in advance, put into action during crises and refined once the dust has settled. This principle rounds off the cybersecurity lifecycle while laying the groundwork for the next phase of security evolution. Doing so creates a continuous improvement loop that strengthens organisational resilience.

In today's complex digital environment, cybersecurity incidents are inevitable—even with advanced preventive tools in place. The NotPetya ransomware attack of 2017, which caused over $10 billion in damages,[108] shows just how severe the fallout can be when recovery planning falls short. A well-designed recovery process acts as your safety net when defences fail, assisting to protect your organisation's reputation and maintain customer trust.

The Recovery principle plays several connected roles in a mature cybersecurity program. It ensures continuity of operations and reduces costly downtime through solid Business Continuity and Disaster Recovery (BCDR) strategies. It also helps restore services, meet governance obligations and turn setbacks into opportunities for learning, like turning lemons into lemonade.

Modern recovery efforts operate in a web of technical interdependencies, legal requirements and stakeholder demands. Effective recovery strikes a balance between quick service restoration and careful incident investigation, legal compliance and lasting improvements. Achieving this balance calls for detailed planning, clear accountability and constant re-evaluation of recovery readiness.

This chapter explores the many dimensions of recovery, from key concepts and practical implementation to performance measurement and common challenges. By mastering the Recover principle, you can turn inevitable incidents into opportunities for growth and elevate your organisation's overall security posture.

Understanding the Recovery Principle

Recovery in cybersecurity encompasses the coordinated activities your organisation undertakes to restore systems, data and services following a security incident. This process extends beyond technical restoration to include rebuilding stakeholder trust, addressing vulnerabilities and strengthening your overall security posture.

Recovery serves as a bridge between reactive incident management and proactive security enhancement. It's where you conclude one incident while preparing for future challenges, turning a crisis into an opportunity for improvement.

Unlike disaster recovery (which addresses service availability regardless of cause), cybersecurity recovery explicitly addresses the aftermath of malicious activities. This distinction matters because when recovering from security incidents, you must consider evidence preservation, attacker persistence, public disclosure requirements and potential legal ramifications that don't typically apply to natural disasters or technical failures.

Recovery timelines vary significantly based on incident severity, organisational preparedness and available resources. A minor malware infection might require only hours to remediate, while sophisticated data breaches could demand weeks or months of coordinated effort. This variability underscores why you need realistic Recovery Time Objectives (RTOs) for critical systems during planning. By categorising systems according to business impact, you'll allocate resources appropriately during high-pressure situations.

Effective recovery requires mastery of several interconnected activities:

Recovery Planning

Recovery planning establishes the foundation for all post-incident responses. This forward-looking process identifies critical systems, documents restoration procedures, assigns responsibilities and establishes decision-making frameworks before incidents occur. Think of it as creating an emergency evacuation plan; you don't want to figure out the exits when the building is on fire.

Your plans must be specific to your organisation and industry. For example, a hospital might prioritise patient management systems over administrative systems, ensuring critical care continues during an active incident. Without comprehensive planning, your recovery efforts become reactive and disorganised, potentially extending downtime and increasing financial and reputational damage.

Post-Incident Improvements

Post-incident improvements transform security events from losses into valuable learning opportunities. This reflective process examines what occurred, why defences failed and how your organisation responded. The resulting insights drive targeted enhancements to policies, architectures, training programs and detection capabilities, creating a positive feedback loop that strengthens your entire security program.

Recovery Communications

Recovery communications maintain stakeholder trust and fulfil compliance obligations during restoration. This involves sharing timely, accurate information with internal teams, executives, customers, regulators and potentially the public. Effective recovery communication balances transparency against security requirements, provides actionable guidance to affected parties and demonstrates your organisation's competence despite the incident.

Building Recovery Resilience

Building recovery resilience requires ongoing investment in both technology and people. Regular testing through tabletop exercises and simulated recovery scenarios helps identify gaps before real incidents occur.

These exercises should include stakeholders from across your organisation, not just IT and security teams.

Documenting lessons learned from each recovery effort creates an institutional knowledge base that improves future response capabilities and assists new team members to understand past challenges and solutions.

Business Continuity and Recovery Objectives

Two critical metrics guide your recovery planning and execution: Recovery Time Objective (RTO) and Recovery Point Objective (RPO). These measurements help you quantify and communicate recovery expectations across your organisation.

Recovery Time Objective (RTO)

RTO specifies the maximum acceptable downtime for a system or service, essentially answering the question, "How quickly must we restore functionality?" This timeframe directly impacts business operations and customer experience. For example, financial institutions typically implement stringent RTOs for transaction systems, often requiring near-immediate restoration to prevent significant financial losses and maintain regulatory compliance.

When establishing RTOs, you must consider both technical feasibility and the business impact of extended outages.

Recovery Point Objective (RPO)

RPO defines the maximum acceptable data loss measured in time, answering the question, "How much recent data can we afford to lose?" This metric determines your backup frequency and replication strategy. For example, a healthcare provider might set an RPO of zero for patient records, requiring continuous replication to prevent any data loss, while accepting a longer RPO for marketing materials.

You must balance recovery costs against potential losses, as near-zero RPOs require significant investment in redundant systems and real-time replication technologies.

Why These Objectives Matter

These objectives serve multiple crucial purposes within your recovery framework. They establish clear expectations for stakeholders about service restoration timeframes, guide investment decisions for backup systems and redundant infrastructure, and provide objective benchmarks for evaluating recovery performance.

Following a ransomware incident, you can compare actual recovery times against established RTOs to identify improvement opportunities. Without clearly defined RTOs and RPOs, your organisation will struggle to allocate recovery resources effectively and cannot objectively determine whether recovery capabilities meet business requirements.

Building a Recovery Framework

Business Impact Analysis (BIA): Foundation of Recovery Planning

BIA is the vital first step in recovery planning, like a triage system for your digital assets. At its core, BIA gives your organisation a clear understanding of how disruptions affect critical business functions, helping you prioritise recovery efforts based on operational impact. Rather than treating all systems equally, which would be like giving a paper cut the same attention as a heart attack, this process enables informed decisions about resource allocation during crisis situations.

When conducting a BIA, you first identify your organisation's critical functions, those essential activities that support core business objectives. For a financial institution, this might include payment processing and customer authentication systems, while a healthcare provider might prioritise patient record access and emergency communications. Once the critical functions are identified, you must evaluate potential consequences across multiple dimensions.

Financial impacts include direct revenue loss, emergency remediation costs and potential regulatory penalties. During the 2020 SolarWinds supply chain attack, for example, affected organisations faced significant expenses for incident investigation, system rebuilding and legal consultation services.

Operational impacts affect your ability to maintain business continuity. The 2021 Colonial Pipeline ransomware attack demonstrates these consequences.[109] The six-day operational shutdown caused widespread fuel shortages and supply chain disruptions throughout the eastern United States, affecting millions of consumers.

Reputational impacts can outlast technical recovery, as the 2017 Equifax breach illustrates. Despite eventually restoring systems, the company faced sustained public criticism, customer distrust and brand damage that persisted long after technical operations resumed.[110]

Integrating Recovery into Risk Management

Recovery planning begins with thorough integration into your broader risk management strategy, establishing the foundation for effective response capabilities. You must align recovery priorities with enterprise risk assessments to ensure critical systems receive appropriate protection and restoration precedence. This risk-based approach directs resources toward business-critical assets instead of attempting to protect everything equally, creating realistic recovery objectives that acknowledge your organisational constraints.

Well-implemented risk mitigation strategies enhance your recovery capabilities by reducing both the likelihood and impact of disruptive events. When you implement defence-in-depth security controls, including network segmentation, access controls and endpoint protection, you prevent minor security events from escalating into major incidents requiring full recovery activation. For example, properly configured network segregation can contain ransomware to a limited segment of your infrastructure, significantly reducing recovery scope and time compared to an organisation-wide infection.

Regulatory frameworks significantly shape recovery planning requirements across different sectors. If you operate within the financial services industry, you must demonstrate recovery capabilities with specific testing requirements and documentation standards. Similarly, healthcare organisations face strict requirements regarding patient data protection and availability under privacy legislation. You must integrate these compliance requirements into your recovery frameworks not only to avoid regulatory penalties but also to strengthen your operational resilience. For instance, the Australian Prudential Regulation Authority (APRA) requires financial institutions to maintain comprehensive business continuity plans with regular testing and board-level oversight.[111]

Data Protection and Backup Strategies

The 3-2-1 Backup Strategy provides fundamental data protection for your organisation by maintaining three copies of critical information on two different storage media, with one copy stored offsite. This approach creates multiple recovery options when your primary systems become compromised. You can select the most appropriate restoration source based on incident circumstances. For instance, if your primary data centre experiences flooding, you can recover from off-site backups without incurring extensive downtime. Modern implementations often extend this concept to include cloud-based tertiary backups that provide geographic dispersion beyond traditional off-site storage models.

Immutable and air-gapped backups serve as essential defences against ransomware variants that specifically target backup infrastructure. Immutable storage prevents data modification or deletion even with administrative credentials, while air-gapped solutions maintain physical or logical separation from your production networks to prevent compromise. These approaches ensure clean recovery sources remain available even when your primary backups become encrypted or corrupted through advanced persistent threats.

Recovery validation testing verifies backup integrity and restoration procedures through regular automated checks supplemented by complete recovery simulations. Your automated validation should focus on data completeness and corruption detection, while periodic full restoration exercises validate recovery timeframes and identify procedural gaps. They also verify system functionality following recovery operations. Many organisations discover critical shortcomings only during actual incidents, which you can avoid by implementing testing protocols that simulate realistic failure scenarios. Regular testing also aids your technical teams in maintaining the skills necessary for high-pressure recovery situations.

Disaster Recovery (DR) Site Options

Recovery site selection represents a strategic decision balancing cost constraints against recovery objectives. When planning your cybersecurity infrastructure, you must evaluate several options that offer different compromises between recovery speed and implementation costs.

Hot sites maintain fully configured environments with real-time data replication, enabling near-immediate failover but requiring substantial investment in duplicate infrastructure. It's like having a fully furnished backup office ready at a moment's notice. Major financial institutions and Defence capabilities often implement hot sites to ensure continuous availability, accepting higher costs to minimise disruption impact. These environments mirror your production systems completely, allowing rapid operations redirection with minimal data loss.

Warm sites provide pre-configured hardware requiring only data restoration and final configuration, balancing moderate costs with recovery times measured in hours rather than minutes. This middle-ground approach suits many organisations with moderate recovery objectives, as demonstrated by Australian retail companies maintaining warm recovery environments for e-commerce platforms. Your recovery teams only need to restore recent data and activate standby systems.

Cold sites furnish basic infrastructure without pre-installed systems, offering economical solutions when recovery timeframes of days are acceptable. These facilities provide essentials like power, connectivity and physical security but require your team to ship, install and configure systems during recovery. Government agencies with non-critical administrative functions often select this option to meet budgetary constraints while maintaining recovery capabilities.

Your selection criteria must incorporate multiple factors beyond cost considerations, including geographic separation, regulatory requirements and scalability needs during prolonged recovery operations. Geographic separation protects against regional disasters, while scalability ensures your recovery environment can support extended operations if primary facilities remain unavailable.

Hybrid recovery models combine traditional infrastructure with cloud-based capabilities to optimise recovery efficiency and operational costs. These approaches leverage cloud platforms for non-critical workloads while maintaining dedicated infrastructure for sensitive applications with specific compliance requirements. Many Australian organisations implement hybrid models to balance data sovereignty requirements with cloud recovery flexibility.

Recovery Environments: On-Premises vs. Cloud

Traditional on-premises approaches present considerable challenges when your organisation requires rapid restoration after an incident. Physical infrastructure rebuilding demands extensive time, from hardware procurement and configuration to system deployment and data restoration, potentially extending business disruption by days or weeks.

Cloud-based recovery environments offer compelling alternatives through inherent flexibility and scalability. Instead of maintaining idle physical systems, you can dynamically provision resources on demand, eliminating procurement delays and accelerating restoration timelines. This allows you to match recovery resources to actual needs rather than investing in maximum-capacity infrastructure that remains unused during normal operations. However, cloud solutions introduce specific considerations within the shared responsibility model, where providers secure underlying infrastructure while you retain accountability for data security, access management and compliance requirements.

Disaster Recovery as a Service (DRaaS) exemplifies recovery environment evolution, offering automated, orchestrated failover capabilities that minimise manual intervention during crisis situations. These solutions continuously replicate your critical systems to standby environments activated within minutes of a declared disaster. Ongoing validation built into DRaaS platforms ensures your recovery procedures remain effective as systems and threat landscapes evolve. Many Australian healthcare providers implement DRaaS for patient management systems, balancing strict data sovereignty requirements with rapid recovery capabilities.

Communication and Human Elements in Recovery

Effective communication is essential for successful cyber incident recovery. Even technically excellent recovery efforts may collapse amidst confusion and misinformation without well-structured protocols. You'll need communication strategies spanning three critical phases to ensure a coordinated response and maintain stakeholder trust.

Pre-incident communication planning creates the framework for controlled information flow during crises. This preparation starts with clearly defined roles and responsibilities, specifying who communicates what to which audiences. Develop pre-approved templates for various incident scenarios (data breaches, ransomware attacks, service disruptions) to deploy consistent messaging when urgency mounts. Australian financial institutions often maintain specific templates to meet regulatory requirements and customer expectations; it's like having emergency scripts ready when stress levels spike.

During active incidents, communication becomes a vital operational function. Real-time updates via centralised incident management platforms help all responders maintain situational awareness while documenting key decisions. Your internal stakeholders need clear briefings balancing thorough disclosure with security requirements. For external communications, trained spokespersons must manage media interactions, preventing information gaps that invite speculation.

Post-incident communication centres on rebuilding trust and capturing lessons learned. Debriefing sessions with internal teams document communication strengths and weaknesses, guiding improvements. External stakeholders (customers, partners and regulators) need thoughtful reassurance through formal statements acknowledging the incident, explaining its impact and detailing preventative measures you've implemented.

Throughout each phase, navigate specific legal and regulatory reporting requirements. Privacy laws and industry regulations mandate notification timeframes and content requirements. *The Australian Privacy Act* and Notifiable Data Breaches (NDB) scheme require specific disclosures within strict timeframes. Early legal consultation ensures communications meet these obligations while limiting liability exposure.

The Human Element in Recovery

The human dimension represents the most critical, yet frequently overlooked, aspect of cyber incident recovery. While technical systems can be restored from backups (relatively straightforward), psychological, operational and organisational human elements require deliberate cultivation for crisis resilience.

A culture of preparedness forms the foundation of effective human-centred recovery. You must weave security consciousness into your organisational framework, with leadership actively championing resilience initiatives rather than treating them as isolated IT responsibilities. This elevates recovery from emergency response to a strategic capability embedded within daily operations.

Resilience-focused training equips personnel for inevitable cyber incident disruption. Scenario-based exercises simulating realistic attack conditions develop instinctive crisis responses, advancing beyond theoretical knowledge to practical application. These simulations should address role-specific responsibilities, with executives practising strategic decision-making under pressure, IT teams executing technical recovery procedures and communications staff crafting appropriate stakeholder messaging.

The psychological impact of cyber incidents requires specific attention. Incident fatigue and stress can compromise decision-making when clear judgment is most needed. Your organisation needs structured rest periods, rotation schedules and psychological support systems to sustain team effectiveness during extended recovery operations. Financial institutions often implement dedicated wellness programs and mandatory breaks during incident response to maintain cognitive clarity because burnt-out responders make poor decisions when the stakes are highest.

Measuring and Improving Recovery

Defining Recovery Metrics

You'll need objective metrics to measure both your recovery readiness and actual performance during incidents. Start by establishing your baseline performance, then identify specific areas for improvement and systematically track your progress over time.

Capability metrics help gauge how prepared an organisation is to recover from disruptions. These include the proportion of systems with documented recovery procedures, how often recovery tests are conducted and the rate at which employees complete recovery training.

Performance metrics evaluate how effectively you recover during incidents or exercises. You'll measure time to restore critical services compared to RTOs, data loss compared to RPOs and accuracy of recovery time estimates provided to stakeholders.

More sophisticated recovery programs incorporate stakeholder-focused metrics that evaluate recovery from the perspective of those impacted by the incident. Customer satisfaction scores following recovery, employee feedback on communication clarity, partner assessments of coordination effectiveness and executive confidence in recovery capabilities provide valuable insights beyond technical measurements.

Financial services organisations track customer retention rates post-incident as a key indicator of successful recovery. They recognise that technical restoration alone doesn't guarantee business continuity without maintaining stakeholder trust.

Conducting Post-Incident Reviews

The After-Action Review (AAR) represents one of recovery's most valuable contributions to organisational resilience. This structured analysis examines what happened, why it happened, how teams responded and what should change.

Effective AARs require psychological safety—participants must feel comfortable discussing failures without fear of blame—and methodical analysis that distinguishes between root causes and symptoms.

The AAR process follows a progressive structure that maximises learning value. Begin with immediate operational debriefs within days to capture technical details while memories remain fresh. Conduct formal reviews weeks later with broader participation to examine systemic issues. Finally, implement long-term follow-ups months after the incident to verify the implementation of improvements and assess effectiveness.

Healthcare organisations conduct multi-phase AARs following data breaches, using insights from initial technical debriefs to inform broader governance reviews. They track implementation of security enhancements over 6-12 month periods (because real change takes time to verify).

Continuous Improvement and Lessons Learned

Recovery generates insights that benefit all security functions through systematic improvement processes. Detection capabilities improve through the identification of monitoring gaps. Response playbooks incorporate streamlined procedures based on recovery timeline analysis. Protection measures address vulnerable configurations identified during restoration.

Effective improvement processes tend to follow a familiar progression: starting with documented observations from recovery activities, moving to an analysis of underlying causes rather than surface symptoms, followed by the development of targeted recommendations with clear ownership. Progress is then tracked over time, and the effectiveness of each improvement is verified through subsequent testing.

This disciplined approach transforms recovery from a one-time restoration activity into a continuous improvement engine that systematically strengthens your organisation's entire security program. Australian telecommunications companies implement quarterly reviews of lessons learned from recovery exercises, using these insights to refine network resilience strategies and enhance service continuity planning.

Recovery Challenges and Practical Considerations

Technical Complexity and System Interdependence

Modern technology environments present fundamental recovery challenges that'll test even experienced teams. Think of today's systems like a spider's web; touch one strand, and vibrations ripple throughout the structure. Cloud-native applications spanning multiple service providers create recovery dependencies on third-party systems you can't control. Microservice architectures with hundreds of interconnected components generate complex recovery sequencing requirements that single-system recovery plans can't address.

These complexities require sophisticated recovery approaches. Service dependency mapping tools create accurate visualisations of recovery sequencing requirements. API-driven automation accelerates and standardises recovery processes while reducing human error. Infrastructure-as-code approaches ensure recovered environments match pre-incident configurations. Some organisations employ service mesh technologies to manage complex interdependencies between various system components, enabling more precise recovery prioritisation during outages.

Recovery Process in Action

A destructive ransomware attack instantly transforms your organisation's focus from routine operations to crisis management. When system crashes and boot failures cascade across your network, security analysts must swiftly distinguish between conventional ransomware and more destructive variants. This identification triggers immediate incident reporting through established channels, mobilising the IRT to establish a command centre isolated from compromised networks.

First responders execute urgent containment measures by disconnecting infected systems without powering them down, preserving critical forensic evidence. Concurrently, network administrators implement emergency segmentation protocols to halt lateral movement. The speed and effectiveness of these initial containment actions significantly influence your overall recovery timeline.

After stabilising containment, eradication commences with detailed forensic analysis to determine how the ransomware infiltrated and spread. Digital forensics specialists analyse system logs, memory dumps and network traffic captures to locate patient zero and map the malware's movement patterns. This intelligence then directs the systematic removal of malware components from affected systems.

The restoration phase uses offline backups stored in immutable or air-gapped environments, safeguards that protect backups from encryption during an attack. Throughout recovery, business continuity plans function alongside technical restoration efforts, with critical services shifting to alternate processing sites or cloud-based failover environments while teams rebuild primary systems. Australian government agencies, for example, employ dedicated cyber IRTs who conduct ransomware tabletop exercises, providing them with the capability to reduce containment times during actual incidents through practised coordination.

Balancing Recovery with Legal and Forensic Requirements

Recovery efforts operate within legal constraints that sometimes conflict with business pressure for rapid restoration. Much like trying to repair a car while the police still need it for evidence, both can't occur simultaneously. Evidence preservation requirements may prevent immediate reimaging of compromised systems needed for service restoration. Legal teams may restrict your recovery communications to avoid statements that could create liability or complicate insurance claims. Privacy regulations might limit information sharing that would otherwise accelerate coordinated recovery across organisations experiencing similar attacks.

Forensic requirements present similar challenges, particularly in cases involving sophisticated threat actors or potential criminal activity. Forensic investigation requires preserving affected systems in their compromised state, directly conflicting with the recovery needs to restore clean, functional environments. This tension often necessitates compromise approaches: creating forensic images before restoration, preserving logs while rebuilding systems or maintaining representative compromised systems while recovering others. Australian financial services firms typically maintain dedicated forensic environments where compromised system images can be analysed without delaying restoration of critical banking services, balancing regulatory obligations for both service continuity and proper incident investigation.

Summary

Recover, the final principle in the cybersecurity lifecycle, encompasses the strategic capabilities needed to rebuild systems and restore operations following security incidents. This chapter explored how recovery transforms crisis management into operational restoration, creating continuous improvement cycles that enhance organisational resilience.

Recovery extends beyond technical restoration, including rebuilding stakeholder trust, addressing vulnerabilities and strengthening security posture. Unlike disaster recovery, cybersecurity recovery specifically addresses malicious activities, requiring you to consider evidence preservation, attacker persistence and potential legal ramifications. Recovery timeline varies significantly based on incident severity, your preparedness and available resources, emphasising the need for realistic recovery objectives for critical systems.

Two key metrics guide your recovery planning: RTO, which specifies maximum acceptable downtime, and RPO, which defines acceptable data loss measured in time. These metrics establish stakeholder expectations, guide investment decisions and provide benchmarks for evaluating recovery performance. BIA forms the foundation for your recovery planning by identifying critical functions and evaluating potential consequences across financial, operational and reputational dimensions.

Data protection strategies should include the 3-2-1 Backup Strategy, maintaining three copies on two different media with one copy stored off-site. Immutable and air-gapped backups serve as essential defences against ransomware variants targeting backup infrastructure. Recovery validation testing verifies both backup integrity and restoration procedures through regular automated checks and complete recovery simulations.

When selecting recovery sites, you can choose hot sites with real-time replication for near-immediate failover, warm sites requiring only data restoration or cold sites offering economical solutions when longer recovery timeframes are acceptable. Cloud-based recovery environments provide compelling alternatives through flexibility and scalability, while DRaaS offers automated failover capabilities that minimise manual intervention.

Effective communication spans pre-incident planning, active incident management and post-incident trust rebuilding. The human dimension represents the most critical aspect of recovery, requiring you to foster a culture of preparedness, implement resilience-focused training and attend to psychological impacts. Measuring recovery effectiveness involves capability metrics assessing readiness, performance metrics evaluating actual recovery and stakeholder-focused metrics examining recovery from those impacted.

The AAR significantly strengthens organisational resilience by examining what happened, why, how teams responded and what should change. Recovery generates insights benefiting all security functions through systematic improvement processes that document observations, analyse root causes, develop targeted recommendations, track implementation and validate effectiveness.

Modern technology environments present recovery challenges through architectural complexity and tight interdependencies. Recovery efforts must balance legal constraints and forensic requirements that sometimes conflict with business pressure for rapid restoration. By mastering the Recover principle, you'll transform inevitable security incidents from potential disasters into opportunities for advancement and improved security posture.

Revision Questions

1. What is the Recover principle in cybersecurity, and how does it differ from disaster recovery?

2. What are RTOs and RPOs, and why are they important for recovery planning?

3. What is a Business Impact Analysis (BIA), and what role does it play in recovery planning?

4. Explain the 3-2-1 Backup Strategy and why it's effective for cybersecurity recovery?

5. What are the differences between hot, warm and cold recovery sites?

6. How do cloud-based recovery environments compare to traditional on-premises recovery approaches?

7. Why is communication critical during recovery, and what phases should it address?

8. What is an After-Action Review (AAR), and how does it contribute to organisational resilience?

9. What metrics can be used to evaluate recovery effectiveness?

10. What challenges do modern technology environments present for recovery efforts?

Cybersecurity Hacks

Hack 1: The "Recovery Time Machine" Validation Method

Establish a quarterly "time machine" exercise where you randomly select a date from the past 3 months and attempt to restore systems to that exact point in time using only your existing backup and recovery procedures. This exercise tests both your backup integrity and recovery processes without the pressure of an actual incident. Set a timer matching your established RTO and work through the complete restoration process, documenting any issues, delays, or missing steps encountered.

Unlike tabletop exercises, this method provides hands-on validation of your actual recovery capabilities using real backup data and procedures. The random date selection prevents teams from preparing for specific scenarios, ensuring genuine testing of recovery readiness. Additionally, this exercise helps staff maintain their recovery skills through regular practice, building confidence and competence that proves invaluable during actual incidents.

Hack 2: The 'Tabletop RPG' Approach to Recovery Planning

Transform conventional cybersecurity recovery exercises into engaging role-playing scenarios to improve team preparedness and response coordination. This innovative approach involves creating detailed responsibility profiles for each IRT member, developing scenario cards describing various cyber incidents with realistic complications and introducing controlled unpredictability through randomisation techniques. A designated facilitator from management guides the exercise while team members navigate the simulated crisis, earning recognition for effective decision-making and creative problem-solving. This method makes recovery training more memorable and reveals potential procedural weaknesses before they impact actual incidents. It also strengthens cross-functional understanding as team members experience challenges from different perspectives, ultimately improving coordination during real-world recovery operations when roles and responsibilities must function seamlessly under pressure.

Part III

Human Factors

Chapter 16: Rules of Engagement

– Legal and Ethical Boundaries in Cybersecurity –

Objectives

By the end of this chapter, you will be able to:

- Explain the relationship between cybersecurity, legal frameworks and ethical considerations
- Identify key data privacy and cybersecurity legislation in Australia and internationally
- Understand legal requirements for data breach reporting and their impact on cybersecurity practices
- Apply core ethical principles of confidentiality, privacy and transparency to cybersecurity scenarios
- Identify elements of a professional legal and ethical toolkit for making defensible cybersecurity decisions.

In February 2021, a hacker breached the water treatment facility in Oldsmar, Florida, attempting to poison the water supply by drastically increasing sodium hydroxide levels. Staff intervened before harm occurred, but the incident exposed serious vulnerabilities in critical infrastructure and highlighted why legal and ethical frameworks aren't just bureaucratic red tape—they're essential safeguards for cybersecurity.[112]

The intersection of cybersecurity with legal and ethical considerations creates a complex and dynamic landscape for security professionals. These experts must navigate a world where technology often outpaces regulation, demanding decisions beyond what's technically possible. While tools can plug security gaps, legal boundaries and ethical implications deserve equal weight in a digitally linked society.

Previous chapters examined core cybersecurity principles aligned with the National Institute of Standards and Technology Cybersecurity Framework (NIST CSF) and the Australian Signals Directorate-Australian Cyber Security Centre (Information Security Manual (ASD-ACSC ISM). These principles, however, must be implemented with careful attention to legal and ethical constraints. Such constraints establish the operational boundaries that ensure security measures protect systems without infringing upon individual rights or violating ethical standards.

Legal frameworks in cybersecurity include diverse laws and regulations dictating how organisations must safeguard data, address breaches and maintain system resilience. These regulatory structures establish standardised approaches with explicit guidelines and defined penalties for non-compliance. They form baseline operational requirements, shielding organisations from liability while clarifying legitimate security boundaries.

Security professionals encounter complex ethical questions daily: Is system penetration justifiable for defensive purposes? When should privacy protections take precedence over security measures? These questions exemplify the tension between technological capability and societal acceptability. Emerging technologies (AI-powered security tools, automated threat detection, offensive security techniques) intensify this tension by introducing capabilities that current legal frameworks haven't anticipated. Consequently, you'll often find yourself in uncharted territory.

Ethical principles frequently extend beyond explicit legal requirements, notably when regulations lag behind innovation. The foundational trust in digital systems requires security professionals to make principled decisions without explicit legal direction.

Understanding the relationship between cybersecurity, law and ethics establishes the groundwork for responsible security practices. These practices effectively protect systems while honouring societal values and individual rights, because ultimately, we're not just protecting data and systems, we're protecting people.

Legal Foundations of Cybersecurity

The Evolution of Cybersecurity Law

At a BSides conference in Adelaide, where I once volunteered, I discovered a unique Capture The Flag (CTF) challenge designed for cybersecurity Governance, Risk and Compliance (GRC) professionals. Participants had to match descriptions of laws and regulations to their titles, which was deceptively complex even for experienced practitioners. This exercise revealed an essential truth: cybersecurity professionals need a sound understanding of their field's legal frameworks.

The cybersecurity legal landscape has evolved dramatically from basic computer crime statutes to intricate regulatory systems. Foundational legislation such as *The Computer Fraud and Abuse Act (CFAA)* (1986) in the US and *The Computer Misuse Act* (1990) in the UK established core principles criminalising unauthorised access. Key incidents accelerated legal development, such as the Morris Worm, which in 1988 highlighted systemic vulnerabilities. Estonia's 2007 cyber-attacks catalysed international cooperation frameworks. The Sony data breach (2011) and Cambridge Analytica scandal (2018) transformed privacy regulations, leading to rigorous notification requirements and comprehensive legislation like the GDPR.

Understanding this legal evolution should be a priority in your profession development. When you grasp these frameworks, you can effectively navigate complex regulatory requirements and ensure compliance with data protection laws and industry-specific mandates. This knowledge enables you to implement precise security measures that address threats while satisfying legal obligations. It enhances your ability to communicate risks clearly to organisational leadership and supports the development of effective incident response procedures. Your legal awareness becomes instrumental in creating cybersecurity policies that align with current standards and help your organisation avoid significant penalties.

This progression signifies cybersecurity's evolution from a specialised technical concern to an international policy priority that requires increasingly sophisticated and coordinated legal responses.

Data Privacy Legal Frameworks

The Privacy Act 1988 (AU)

The Privacy Act 1988 forms Australia's primary privacy protection framework, establishing clear guidelines for handling personal information by government agencies and private sector organisations, including healthcare providers. At its core are the 13 Australian Privacy Principles (APPs), which establish standards for collecting, using, storing and disclosing personal information. These principles maintain technology-neutrality, allowing organisations to adapt practices to their specific operational needs. The APPs address numerous privacy concerns, from transparent management of personal information to the collection of solicited and unsolicited data and the appropriate use and disclosure of personal information.[113]

The Privacy Act contains several critical components, notably the Notifiable Data Breaches (NDB) scheme, implemented in February 2018. This scheme requires organisations covered by the Act to notify affected individuals and the Office of the Australian Information Commissioner (OAIC) when a data breach occurs. Additionally, the Act empowers the OAIC with enforcement capabilities, including conducting privacy assessments, accepting enforceable undertakings and pursuing civil penalties for serious or repeated privacy violations. These penalties can be significant, encouraging organisations to implement comprehensive data protection and security measures. The Act undergoes regular review and updates to address emerging technologies and ever-growing privacy considerations.[114]

The General Data Protection Regulation (GDPR) (EU)

The GDPR stands as one of the most influential data privacy laws affecting organisations worldwide. Implemented by the European Union in 2018, the GDPR protects EU citizens' data by enforcing strict requirements on data processing activities, regardless of the organisation's location. The regulation broadly defines personal data as any information relating to an identifiable person, including names, email addresses and IP addresses.

The GDPR provides individuals with specific rights: accessing their data, correcting inaccuracies, requesting deletion (the 'right to be forgotten') and transferring data between service providers (data portability). Organisations must secure explicit consent before collecting and processing personal data and must implement effective data security measures. Failing to comply with the GDPR can trigger substantial penalties, up to €20 million or 4% of global annual turnover, whichever amount is higher.[115]

California Consumer Privacy Act (CCPA) (US)

The CCPA, enacted in 2018, marks a significant advancement in data privacy rights within the United States. The CCPA grants California residents the right to know what personal data companies collect about them, why companies use it and which third parties receive it. This legislation also enables consumers to access their data, request its deletion and opt out of its sale.

Companies subject to the CCPA must provide transparent and accessible privacy notices and implement systems to respond to consumer requests within specified timeframes. The legislation enforces compliance through penalties reaching $7,500 for each intentional violation.[116]

The Health Insurance Portability and Accountability Act (HIPAA) (US)

HIPAA, passed in 1996, establishes the standard for protecting sensitive patient data in the United States. HIPAA requires healthcare providers, health plans and other entities handling health information to implement safeguards that ensure the confidentiality, integrity and availability of electronic protected health information (ePHI).

HIPAA contains provisions including the Privacy Rule, which establishes national standards for protecting health information, and the Security Rule, which sets requirements for securing ePHI. HIPAA requires organisations to conduct risk assessments, implement access controls and train employees in compliance procedures. Violations can result in substantial fines, ranging from $100 to $50,000 per violation.[117]

Cybersecurity-Specific Legislation

Let's explore the legislative landscape shaping our approach to cybersecurity today. Think of these laws as the 'rules of the road' for our digital highways, designed to keep everyone safe while enabling innovation.

The Cyber Security Act (CSA) 2024 (AU)

CSA 2024 establishes Australia's comprehensive legislative framework addressing cyber threats across the digital landscape. This isn't just a regulatory burden; it's designed to strengthen national cyber resilience by imposing mandatory security obligations on businesses, digital service providers and critical technology companies.

The Act creates a unified national cyber incident reporting scheme and sets minimum security standards for organisations. It implements sector-specific regulatory approaches based on risk profiles. The legislation significantly expands the ASD's authority during major cyber-attacks and establishes certification programs for cybersecurity products and services. Non-compliance triggers substantial financial penalties ranging from thousands to millions depending on violation severity.[118]

The Security of Critical Infrastructure Act 2018 (SOCI Act) (AU)

The SOCI Act provides essential cybersecurity legislation for Australia's critical infrastructure protection. Consider it the digital equivalent of protecting our power plants and hospitals—that's exactly what it does.

The Act places specific obligations on owners and operators across vital sectors, including energy, communications, transport and healthcare. These requirements include identifying critical infrastructure assets, implementing robust risk management programs and fulfilling mandatory incident reporting for cyber events. The legislation grants the Australian Government enhanced intervention powers for serious security threats, including issuing risk mitigation directions or intervening directly in severe cases. *The SOCI Act* enforces compliance through significant penalties.[119]

The Telecommunications Act 1997 (AU)

The Telecommunications Act 1997 contains specific provisions safeguarding network security and integrity. It requires carriers and carriage service providers to protect communication confidentiality, maintain network integrity and implement measures against unauthorised access.

The Act mandates that industry participants cooperate with law enforcement and national security agencies when responding to cyber threats or security incidents. It also requires reporting significant risks or breaches that might compromise national security or critical infrastructure. Companies that do not comply with these security provisions face substantial fines and regulatory actions.[120]

The Criminal Code Act 1995 (AU)

The Criminal Code Act 1995 defines and regulates criminal offences related to cybercrime and cybersecurity in Australia. Part 10.7 specifically addresses computer offences, encompassing unlawful activities such as unauthorised access to data, modification of information and impairment of computer systems.

These provisions target cyber-attacks, hacking and malicious activities threatening network integrity and information security. The Act specifies severe penalties for cyber-related offences, with sanctions including imprisonment and substantial fines based on crime severity. The legislation also provides a framework for international collaboration in addressing cross-border cybercrime challenges.[121]

The Computer Fraud and Abuse Act (CFAA) (US)

The CFAA, enacted in 1986, is the primary US legislation against computer-related crimes. Originally focused on protecting government and financial systems, the CFAA now covers broader unauthorised computer activities—it's had quite the career expansion over nearly four decades.

The Act criminalises accessing computers without authorisation or exceeding authorised access, particularly when involving fraud, damage or data theft. It addresses various cyber threats, including hacking, malware distribution and cyber extortion. CFAA violations can result in both civil and criminal penalties, including imprisonment and significant financial penalties.[122]

The Federal Information Security Modernization Act (FISMA) (US)

FISMA, updated in 2014 from its 2002 predecessor, strengthens the cybersecurity posture of US federal agencies. FISMA requires agencies to implement information security programs that protect sensitive data and reduce system vulnerabilities.

The Act necessitates developing security policies and procedures that emphasise risk management and continuous monitoring. It assigns oversight responsibility to the Office of Management and Budget (OMB), while the Department of Homeland Security (DHS) provides operational cybersecurity assistance to federal agencies. Non-compliance with FISMA typically results in audit findings, potential budget impacts and increased oversight scrutiny.[123]

International Law and Cooperation

The Global Cybersecurity Legal Landscape

The borderless nature of cyberspace challenges traditional legal frameworks founded on geographical jurisdictions. Daily international exchanges via social media, emails and digital platforms occur alongside cybercriminals who exploit jurisdictional gaps, launching attacks from countries with minimal enforcement capabilities while targeting victims thousands of kilometres away. This disconnect between cyber threats and conventional legal boundaries necessitates innovative approaches to international cooperation and harmonised legislation.

Cross-border cybercrime investigations confront significant obstacles: differing legal definitions, investigative procedures and evidence standards. A single attack chain typically spans multiple countries (e.g., command servers in Russia, compromised infrastructure in Brazil and financial transactions through Caribbean nations), requiring coordinated responses across jurisdictions with varying capabilities and priorities. This intricate web allows sophisticated threat actors to operate with relative impunity, evading prosecution through jurisdictional complexity.

The Budapest Convention on Cybercrime, adopted by the Council of Europe in 2001 and ratified by Australia, is the most significant international response to these challenges. This landmark treaty establishes standard definitions of cybercrime offences, harmonises national laws and creates mechanisms for international cooperation. Signatories commit to criminalising specific digital offences, implementing procedural powers for electronic evidence collection and providing mutual legal assistance. The convention also balances enforcement with protecting human rights and fundamental freedoms.[124] However, the notable absence of major cyber powers like Russia and China creates critical gaps in its global effectiveness.

Extradition presents a persistent challenge in cybersecurity enforcement. Many nations function as 'safe havens' by refusing extradition for cyber offences or implementing restrictions that effectively shield domestic hackers targeting foreign entities. These havens serve as operational bases for both cybercriminal organisations and state-sponsored threat actors, undermining international enforcement efforts.

Cultural and jurisdictional differences add layers of complexity to global cybersecurity governance. Western democracies typically value individual privacy and corporate responsibility, while authoritarian regimes often prioritise national security and information control. These divergent approaches create conflicting requirements for data sovereignty, surveillance authorities and encryption policies that you will encounter in international cybersecurity practice.

Despite these challenges, progress advances through bilateral agreements, regional frameworks and international organisations. As a cybersecurity practitioner, you must learn to navigate this complex legal landscape while balancing security objectives against varying compliance requirements across global operations.

When Security Goes Wrong: Breach Response Law

Legal Requirements for Data Breach Reporting

Data breach notification laws require organisations to promptly inform affected individuals and regulatory authorities when a data breach occurs. These laws promote transparency, allowing individuals to take protective measures while holding organisations accountable.

In Australia, the Notifiable Data Breaches (NDB) scheme mandates organisations report breaches to the OAIC and affected individuals when breaches could result in serious harm. This notification must occur as soon as practicable after the organisation becomes aware of the breach. Additionally, organisations must provide specific recommendations to affected individuals on protective steps they can take. This scheme applies to various entities, including Australian government agencies, businesses and not-for-profit organisations with annual turnover exceeding $3 million, as well as credit reporting bodies and health service providers regardless of size.

The GDPR takes a similar approach to the EU, requiring organisations to report data breaches to the relevant supervisory authority within 72 hours of discovery unless the breach is unlikely to risk individuals' rights and freedoms. When you examine the GDPR requirements more closely, you'll notice that organisations must notify affected individuals without undue delay if the breach poses a high risk to their rights and freedoms.[125]

The regulatory framework for data protection in the US presents significant complexity, with data breach notification requirements differing across states. *The California Data Breach Notification* Law, frequently referenced as a national benchmark, mandates that organisations inform affected individuals "in the most expedient time possible and without unreasonable delay" following unauthorised access to personal information. This California standard has influenced their national discussions, yet each state has developed distinct notification timelines, thresholds for reporting and enforcement mechanisms. Non-compliance with these state-specific regulations carries substantial consequences, including civil penalties reaching millions of dollars and exposure to class action lawsuits.

Impact on Cybersecurity Practices

These notification requirements have profoundly influenced cybersecurity practices by motivating organisations to implement stronger security measures and enhance their incident response capabilities. The obligation to disclose breaches promptly has led organisations to invest in sophisticated technologies and processes that support the efficient detection, mitigation and reporting of breaches.

Furthermore, awareness of potential public scrutiny, reputational damage, and legal and financial repercussions has compelled organisations to prioritise data protection and adopt a proactive approach to cybersecurity. As a professional in this field, you'll observe that organisations increasingly conduct regular security assessments, implement effective employee training programs, strengthen data encryption protocols, and develop and regularly test IRPs to ensure compliance and protect sensitive information. While legal compliance provides a key element for cybersecurity practices, the rapidly evolving digital landscape demands that professionals also navigate complex ethical considerations that extend beyond regulatory requirements.

Ethical Dimensions of Cybersecurity

Core Ethical Principles

Confidentiality

Building on our Chapter 6 discussion, confidentiality is the foremost ethical obligation for cybersecurity practitioners. You're generally entrusted with everyone's sensitive information. This principle requires protecting sensitive data from unauthorised access and ensuring information remains available only to properly authorised individuals. Without this foundation, trust crumbles like a house of cards.

You'll routinely handle extensive volumes of sensitive information such as personal details, financial records and proprietary business data. Your ethical responsibility extends to shielding this information from breaches and unauthorised disclosures. However, maintaining confidentiality presents significant challenges, particularly when it conflicts with other ethical imperatives or legal requirements. When you uncover illegal activities or critical vulnerabilities requiring reporting, balancing confidentiality obligations with disclosure responsibilities creates a complex ethical minefield.

The Volkswagen emissions scandal of 2015 exemplifies such an ethical quandary in cybersecurity. A Volkswagen engineer, James Liang, participated in developing software designed to circumvent emissions testing procedures. This situation created an acute ethical conflict: Liang's awareness of Volkswagen's deliberate deception contradicted his confidentiality obligations as an employee while simultaneously raising significant public interest concerns regarding environmental harm and corporate fraud. Liang eventually chose to cooperate with authorities, exposing a widespread corporate deception. This decision, however, meant breaching his professional obligations to his employer and resulted in personal legal consequences.[126]

Privacy

Privacy constitutes another critical ethical consideration in cybersecurity—it's where the rubber meets the road. While data collection, storage and analysis form the core of effective cybersecurity efforts, these activities can significantly impact individual privacy rights. Typical cybersecurity measures involve monitoring network traffic, analysing user behaviour patterns and maintaining comprehensive data logs, all of which may inadvertently infringe on users' privacy.

Consider your organisation's policy regarding emails exchanged on their infrastructure: they likely retain the right to examine your work communications. As a cybersecurity practitioner, you may need to review other employees' emails, including personal messages occasionally appearing in work accounts, potentially compromising their privacy.

The ethical implications surrounding privacy in cybersecurity centre on achieving an appropriate balance between security requirements and respect for individual privacy rights. While intrusive monitoring techniques can enhance threat detection capabilities, they may simultaneously lead to excessive surveillance and privacy erosion. This balancing act grows increasingly challenging as data generation and collection continue at unprecedented rates.

The Cambridge Analytica incident provides a compelling illustration of the consequences of privacy breaches. This company harvested personal data from millions of Facebook users without obtaining proper consent and then employed this information for targeted political advertising.[127] Such cases highlight the importance of ethical data handling practices and the necessity for privacy protections embedded within legal frameworks.

Transparency and Disclosure

Conventional organisational operations prioritise information management, business interests and reputations. Transparency and disclosure principles typically contradict this instinct—they're akin to asking companies to voluntarily expose their dirty laundry. While this perspective has merit, organisations simultaneously bear an ethical responsibility to disclose security breaches to users and stakeholders promptly and honestly.

This ethical obligation focuses primarily on customer data and potential privacy losses that could lead to adverse customer outcomes or compromise valuable trade secrets. Failing to disclose breaches involving such data might significantly impact an organisation's financial position or even threaten its market viability. Applying transparency and disclosure principles effectively helps mitigate potential damage while fostering trust and accountability.

Implementing transparency and disclosure also presents challenges due to potential negative organisational repercussions. Companies may hesitate to reveal security incidents, fearing legal consequences or reputational damage. Despite these concerns, ethical considerations demand that organisations prioritise their users' safety and well-being above immediate institutional interests.

The 2017 Equifax breach demonstrates the consequences of inadequate transparency practices. This massive security failure exposed the personal information of over 147 million individuals. Equifax's delayed and insufficient breach disclosure generated widespread criticism and resulted in substantial legal and financial penalties.[128] This case effectively illustrates the ethical importance of timely and transparent security incident disclosure.

Professional Ethics in Practice

Professional ethics in cybersecurity draws clear boundaries between legitimate security work and criminal activity. Many students start their cybersecurity careers imagining themselves as elite hackers, perhaps fantasising about gaining wealth through system breaches. Reality proves starkly different. Practitioners either face legal consequences or avoid the 'dark side' altogether due to the personal toll and extraordinary effort required (Kevin Mitnick vividly illustrates this in his book *The Art of Invisibility*). Consequently, many redirect these interests toward ethical hacking.

Ethical or 'white hat' hacking operates within specific parameters: explicit authorisation, defined scope and formal agreements. This permission-based approach converts potentially illegal activities into valuable security services that enhance organisational defences. When you work as an ethical hacker, you'll gain legitimate authority to examine systems for vulnerabilities, enabling organisations to identify and fix weaknesses before malicious actors exploit them.

An ethical practitioner follows essential principles of confidentiality, integrity and minimal system disruption. Effective penetration testing requires detailed planning with comprehensive rules of engagement that define testing periods, system exclusions and permissible techniques. This methodical approach prevents operational disruptions (critical for environments with Industrial Control Systems) while successfully identifying security vulnerabilities.

Professional organisations, including the International Information System Security Certification Consortium (ISC2) and Information Systems Audit and Control Association (ISACA), develop and maintain comprehensive ethical frameworks through certification requirements and conduct codes. These guidelines prioritise responsible disclosure protocols, requiring ethical hackers to report discoveries to affected organisations and provide sufficient remediation time before considering public disclosure.

Corporate cybersecurity ethics transcends basic compliance, requiring meaningful security investments. Ethical organisations view cybersecurity as a collective benefit, contributing to digital safety through transparent vulnerability disclosure initiatives and threat intelligence collaboration.

The practice of ethical cybersecurity ultimately demands balancing various priorities: client requirements versus public protection, thorough security testing versus system stability and organisational privacy versus necessary transparency. By adhering to established ethical principles, cybersecurity professionals elevate technical expertise into a discipline that serves security objectives and broader community interests.

Strategic Briefing: The 2020 Toll Group Cyber Attack

In early 2020, Toll Group, one of Australia's largest logistics companies, experienced a significant ransomware attack that disrupted operations across its transportation and logistics network. The Mailto ransomware attack forced the company to shut down numerous IT systems to contain the malicious software's spread.[129] This incident illuminated the growing ransomware threat to critical infrastructure and supply chain operations, especially for organisations managing sensitive customer data and complex logistical systems.

The attack revealed the severe operational and reputational consequences of cybersecurity breaches in the logistics sector, demonstrating the necessity of comprehensive security measures, effective incident response protocols and strategic decision-making when confronting ransom demands.

Impact and Scope

The cyberattack caused widespread disruption to Toll Group's operations and services. Key impacts included customer-facing delivery tracking systems becoming inoperable, significantly impaired order processing systems, offline freight booking systems, disabled supply chain management platforms and affected communications networks across multiple sites.

These disruptions led to substantial operational delays and service interruptions, affecting Toll's global customer base and disrupting critical supply chains dependent on the company's logistics services. The incident potentially exposed sensitive customer and operational data, though Toll Group actively assessed and mitigated these risks throughout their response efforts.

Attack Methodology

The attackers employed Mailto ransomware (also known as Netwalker), a sophisticated malware variant that encrypted files and demanded payment for their release. While Toll Group did not publicly disclose the initial access vector, security experts suggest the attack likely originated through phishing emails, exploitation of unpatched vulnerabilities or compromised credentials.

After gaining access to Toll's network, the ransomware propagated through the company's systems, encrypting critical files and making them inaccessible. The attackers demanded cryptocurrency payment in exchange for decryption keys, threatening to publish exfiltrated data if Toll refused to pay the ransom.

Response and Mitigation

Upon detecting the breach, Toll Group implemented a comprehensive incident response strategy. Their initial critical step involved containing the incident by disconnecting affected systems and isolating network segments to prevent the ransomware from spreading further.

The company strategically decided against paying the ransom, instead focusing on restoring systems from secure backups. They enlisted specialised cybersecurity firms and notified relevant authorities, including law enforcement agencies and the Australian Cyber Security Centre (ACSC), to assist with the investigation and recovery.

Throughout the systems outage, Toll maintained essential operations using manual processes while providing regular customer updates about service disruptions and recovery progress. Their recovery process involved methodical restoration of critical systems, with thorough security verification before bringing each system back online.

Lessons Learned

The Toll Group ransomware attack revealed valuable lessons for logistics and enterprise organisations regarding cybersecurity resilience, incident response and strategic decision-making. The breach demonstrated the importance of implementing appropriate security controls, including regular vulnerability assessments and timely patch management. Network segmentation proved vital for limiting ransomware spread, while secure backup systems emerged as priorities for recovery without ransom payment.

Effective incident response planning significantly influences recovery success. Organisations must maintain regularly tested backup systems isolated from the main network to prevent compromise during attacks. This preparation should include documented manual procedures for maintaining operations during system outages and detailed incident response protocols to guide breach management and stakeholder communication.

The attack emphasised the significance of strategic decision-making regarding ransom payments. Toll's refusal to pay reflected their commitment to ethical principles and adherence to law enforcement recommendations, while necessitating reliable recovery capabilities to minimise operational impact. Organisations should establish clear policies on ransom payments and ensure they possess the technical resilience to recover independently without funding cybercriminals.

The Cybersecurity Professional's Legal and Ethical Toolkit

Having processed all of that legal and ethical information, what does putting it into practice look like? A legal and ethical toolkit will equip you to make defensible decisions while protecting organisational assets and stakeholder interests.

Documentation and evidence preservation form the foundation of this toolkit. Meticulous chain of custody protocols ensure digital evidence remains admissible in legal proceedings (just like police evidence), while standardised documentation frameworks capture incident details without interpretation bias. You'll need long-term storage solutions that maintain evidence integrity throughout required retention periods, particularly when confronting cross-jurisdictional situations demanding compliance with multiple legal standards.

Effective collaboration with legal counsel transforms potential liability into a strategic advantage. Establish clear communication channels between technical and legal teams to prevent critical misunderstandings during crises. Developing a shared vocabulary bridges the terminology gap, allowing counsel to translate technical concepts into legal frameworks. Pre-incident legal reviews identify regulatory obligations, while joint tabletop exercises develop the collaborative expertise needed during actual incidents.

Ethical decision-making frameworks offer structured approaches to complex situations where legal guidance may prove insufficient. You can apply established frameworks such as utilitarianism or deontology to cybersecurity scenarios, enabling you to balance organisational interests against broader societal responsibilities. For instance, when discovering a zero-day vulnerability in your organisation's widely-used software, a utilitarian approach might prioritise public disclosure to protect the greatest number of users, while a deontological perspective could emphasise the duty to shareholders and controlled disclosure through proper channels. This approach becomes particularly valuable when resolving ethical conflicts where legal compliance differs from ethical best practices, or when addressing variations in international ethical standards.

Maintaining current legal knowledge constitutes an ongoing professional obligation; let's face it, the legal landscape changes almost as quickly as technology itself. Personalised monitoring systems tracking relevant jurisdictions provide early awareness of regulatory developments. Active participation in industry working groups focused on emerging regulations delivers insight into implementation expectations, while specialised legal resources contribute domain-specific expertise that general counsel may lack.

The practical centrepiece of this toolkit is a legal and ethical decision checklist integrating documentation templates with justification sections, stakeholder impact assessments, compliance verification steps and post-incident review processes. There are several templates and frameworks that can support the legal and ethical decision checklist concept including:

- *NIST Incident Response Framework (SP 800-61r2)*—Provides a comprehensive incident response framework with structured templates and checklists. See https://csrc.nist.gov/pubs/sp/800/61/r3/final

- *ACSC Cyber Incident Response Plan Template*—A comprehensive template providing structured guidance for cyber incident response, including detailed procedures, role definitions, communication protocols and multiple supporting templates. See https://www.cyber.gov.au/sites/default/files/2023-03/ACSC%20Cyber%20Incident%20Response%20Plan%20Guidance_A4.pdf

- *CIS Control 17 Incident Response Policy Template*—A policy template designed to supplement *CIS Controls v8*, specifically for Control 17: Incident Response Management. Provides comprehensive guidance for establishing incident response capabilities. See https://www.cisecurity.org/insights/white-papers/incident-response-policy-template-for-cis-control-17

This systematic approach transforms abstract ethical principles into actionable guidance during high-pressure security events, enabling you to respond confidently within legal and ethical boundaries by building upon existing frameworks that provide the structural foundation to integrate legal and ethical decision-making components, compliance verification steps, and justification documentation into a unified toolkit.

Summary

Chapter 16 examined cybersecurity's intersection with legal and ethical frameworks, establishing operational boundaries that ensure security measures protect systems without violating individual rights or ethical standards. This complex landscape requires you to make judgments beyond technical possibilities to consider legal permissibility and ethical implications.

Cybersecurity's legal foundations evolved from basic computer crime statutes to sophisticated regulatory systems reflecting technological advancement and responding to significant incidents. Key frameworks include Australia's *Privacy Act 1988* with its 13 APPs and NDB scheme, the EU's GDPR with strict requirements and substantial penalties, and the US landscape featuring CCPA and sector-specific legislation like HIPAA.

Australia's cybersecurity legislation includes the recent *Cyber Security Act 2024*, establishing comprehensive national cyber resilience measures. *The Security of Critical Infrastructure (SOCI) Act 2018* protects essential services, while provisions within *The Telecommunications Act 1997* and *The Criminal Code Act 1995* provide additional coverage. These frameworks are complemented by breach response laws mandating notification to affected individuals and authorities, profoundly influencing organisational security practices.

Cyberspace's borderless nature challenges traditional jurisdictional frameworks, requiring international cooperation through mechanisms like the *Budapest Convention*. Despite progress, obstacles remain in cross-border investigations, extradition challenges and cultural differences in cybersecurity governance approaches. Hopefully, international cooperation will improve as cyber threats evolve.

Cybersecurity ethics centre on three core principles: confidentiality, protecting sensitive information from unauthorised access; privacy, balancing security requirements against individual rights; and transparency, encouraging prompt, honest disclosure of security incidents. Professional ethics distinguishes legitimate security work from criminal activity through ethical hacking practices with explicit authorisation, defined scope and formal agreements.

The 2020 Toll Group ransomware attack illustrated these principles, demonstrating cybersecurity breaches' operational consequences and highlighting the importance of comprehensive security measures, effective incident response and strategic decision-making regarding ransom demands (a decision no organisation wants to face).

You'll need a comprehensive legal and ethical toolkit, including documentation protocols, collaboration with legal counsel, ethical decision-making frameworks, ongoing legal education and structured checklists for navigating complex scenarios. This integrated approach transforms abstract principles into actionable guidance, enabling confident responses within appropriate boundaries during high-pressure security events.

Revision Questions

1. Why is understanding the relationship between cybersecurity, legal frameworks and ethical considerations important?
2. What are the three core ethical principles in cybersecurity discussed in the chapter?
3. What does Australia's Notifiable Data Breaches (NDB) scheme require?
4. How has breach notification law impacted cybersecurity practices?
5. What distinguishes ethical hacking from criminal hacking activities?
6. What key Australian legislation governs cybersecurity practices?
7. What is the *Budapest Convention on Cybercrime* and why is it significant?
8. What were the key components of Toll Group's response to their 2020 ransomware attack?
9. What elements should be included in a cybersecurity professional's legal and ethical toolkit?
10. What challenges exist in international cybersecurity law enforcement?

Cybersecurity Hacks

Hack 1: Legal Framework Mind Map

Visualise the complex legal landscape by creating a mind map that organises cybersecurity legislation by geographical jurisdiction. Begin with primary branches representing key regions (Australia, European Union, United States), then extend secondary branches for specific laws within each region. For Australian legislation, include *The Privacy Act 1988*, *The Cyber Security Act 2024* and *The Security of Critical Infrastructure (SOCI) Act 2018*. Under each law, note essential elements including scope, compliance requirements and potential penalties. This visual tool transforms abstract legal concepts into a practical reference that assists you to identify applicable regulations when facing security challenges across different jurisdictions.

Hack 2: Ethical Dilemma Role-Play

Strengthen your ethical decision-making by developing scenario cards based on real-world ethical tensions discussed in the chapter. Create situations that place confidentiality against disclosure requirements or privacy against security needs, such as discovering a critical vulnerability in a client's system that potentially exposes sensitive data. Assume different stakeholder roles (security professional, executive, regulator, affected user) with colleagues or study partners and debate appropriate responses. This exercise develops your ability to navigate complex ethical terrain while appreciating multiple perspectives that influence security decisions in professional practice.

Chapter 17: Metrics That Matter

– Tracking Cybersecurity Success –

Objectives

By the end of this chapter, you will be able to:

- Differentiate between cybersecurity metrics and Key Performance Indicators (KPIs) and explain their roles in security assessment
- Identify and categorise core cybersecurity metrics across security posture, operations and business impact domains
- Select appropriate KPIs aligned with security goals and risk profiles
- Implement effective data collection and analysis strategies for measuring cybersecurity effectiveness
- Evaluate the balance between security controls and operational usability using a risk-based approach.

Defining cybersecurity effectiveness is like trying to hit a dartboard in a windstorm. The dynamic nature of cyber threats and constantly developing technologies make this particularly challenging. As we've explored throughout this book, cybersecurity encompasses protecting data, systems and organisational reputations, so it's a multifaceted beast that needs assessment through multiple lenses. This complexity demands a structured measurement approach, with metrics and Key Performance Indicators (KPIs) as your essential evaluation toolkit.

Effective cybersecurity must ensure data confidentiality, integrity and availability (see Chapters 6, 7 and 8), while safeguarding the organisation's reputation and operational continuity. A security breach can trigger far-reaching consequences, from financial losses to damaged trust and legal ramifications. This scope requires measurement approaches that extend beyond surface-level assessments to capture the full spectrum of security considerations.

The pitfalls of subjective assessment become evident when examining real-world practices. Once, when I managed a system undergoing security assessment, I meticulously prepared performance indicator data to demonstrate security control effectiveness. Remarkably, the assessor disregarded this evidence, basing his evaluation solely on intuition and establishing what he termed a "gentleman's agreement". This approach directly contradicts professional security standards, and (hopefully) no one uses this approach anymore!

Intuition and anecdotal evidence are inherently compromised by cognitive biases and lack the analytical depth necessary for accurate security assessment. It's much like diagnosing an illness based on how the patient looks rather than running tests. In contrast, data-driven decision-making, supported by well-defined metrics and KPIs, provides the foundation for accurately understanding your cybersecurity status and identifying specific areas for improvement.

Metrics are quantifiable measures that track specific cybersecurity processes, generating valuable raw data about performance trends. However, KPIs operate more strategically by aligning with an organisation's broader goals, indicating meaningful progress toward these targets. It's critical to understand that while all KPIs are metrics, not all metrics qualify as KPIs. For instance, the 'number of vulnerabilities identified' represents a basic metric or statistic, whereas the 'percentage reduction in vulnerabilities over six months' constitutes a KPI that actively drives remediation priorities and resource allocation decisions.

The value of metrics and KPIs extends significantly beyond internal assessments to fulfil requirements in external reporting and regulatory compliance. They provide tangible evidence for stakeholders, auditors and regulatory bodies, ensuring the transparency and accountability increasingly demanded in today's digital landscape. Australian organisations must particularly consider frameworks like the ASD ISM and *The Privacy Act 1988* requirements when developing their measurement approaches.

Developing an effective cybersecurity measurement framework requires a comprehensive approach acknowledging the field's complex nature. Through thoughtful implementation of appropriate metrics and KPIs, you'll establish data-driven decision-making processes that support continuous improvement in your organisation's security posture and resilience.

Core Cybersecurity Metrics

Core cybersecurity metrics reveal an organisation's cybersecurity readiness across three critical areas that every security leader must monitor:

- Security Posture Metrics—Evaluate preparedness against threats by focusing on system vulnerabilities and risks to assess your defensive capabilities

- Operations Metrics—Assess the efficiency of incident response and monitoring processes to identify improvement opportunities in your security operations

- Impact Metrics—Connect cybersecurity performance to business outcomes, ensuring complete understanding of financial and operational risks faced by the organisation.

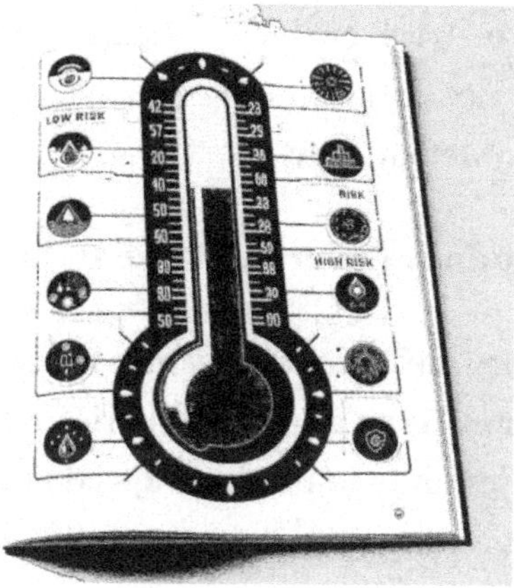

Security Posture Metrics

Vulnerability Management

Number of Vulnerabilities Identified, Prioritised and Remediated

Chapter 3 established the foundation for vulnerability management, describing its types and potential impact. You'll need specific metrics to effectively manage vulnerabilities and assess their identification, prioritisation and remediation efforts. These metrics offer valuable insights into current security weaknesses.

The number of vulnerabilities identified demonstrates the effectiveness of scanning and assessment processes. In contrast, the rate of addressed high-priority vulnerabilities shows an organisation's ability to focus on critical issues posing the highest risk. The time-to-remediation metric indicates how quickly an organisation can mitigate identified risks.

Tracking these metrics enables a holistic approach to vulnerability management. You're not just counting problems but also measuring your ability to solve them systematically.

Patching Frequency (Average Time to Patch Vulnerabilities)

Patching frequency measures the average time to apply fixes for identified vulnerabilities and serves as an elemental metric in vulnerability management. This measure quantifies an organisation's speed and efficiency in responding to known vulnerabilities and reflects its commitment to maintaining current systems and software.

Implementing a shorter patching cadence reduces the window of opportunity for attackers to exploit vulnerabilities. This metric also relates directly to compliance requirements, as many regulatory frameworks establish specific timeframes for patching. The ISM specifies a security control that many organisations must follow:

Control: ISM-1877; Revision: 0; Updated: Sep-23; Applicability: All; Essential Eight: ML1, ML2, ML3

Patches, updates or other vendor mitigations for vulnerabilities in operating systems of internet-facing servers and internet-facing network devices are applied within 48 hours of release when vulnerabilities are assessed as critical by vendors or when working exploits exist[130].

Network Security

Intrusion Detection/Prevention System (IDS/IPS) Events and Alerts

Monitoring events and alerts generated by IDS/IPS systems helps organisations understand the frequency and nature of attempted intrusions. High alert volumes may indicate targeted attacks or suggest the need to fine-tune systems to reduce false positives. Security teams analyse these events to identify patterns and trends in attack vectors, enabling more effective preparation and response strategies.

Unauthorised Access Attempts

Analysis of unauthorised access attempts provides a crucial metric for evaluating access control effectiveness. This metric serves dual purposes: it evaluates existing access control mechanisms and identifies potential security breaches in their early stages. By tracking the frequency, source and patterns of these attempts, security teams gain insights into potential threat profiles. A sudden spike in unauthorised access attempts from a specific geographic region or targeting a particular system might indicate a coordinated attack. In contrast, consistent low-level attempts could suggest ongoing reconnaissance by potential attackers. Security teams should establish normal activity baselines and implement alerts for anomalies.

Network Traffic Analysis (NTA)

NTA for anomalies serves as a fundamental method for detecting suspicious cybersecurity activities. This technique has been central to network security since its earliest days, helping identify behaviours that might indicate cyber-attacks. Anomalous traffic patterns can reveal malware infections, data exfiltration attempts or other malicious activities. Through continuous monitoring and analysis of network traffic, organisations enable early detection and mitigation of security incidents, thereby protecting data integrity and confidentiality.

Endpoint Security

Malware Detection and Prevention Rates

Malware detection and prevention rates indicate endpoint security effectiveness. High detection rates combined with low infection rates typically suggest effective endpoint security practices. However, high detection rates might also point to other concerns, such as users connecting infected devices or accessing suspicious websites.

Security teams must also consider potential false positives when interpreting these metrics. These indicators measure the effectiveness of antivirus and anti-malware solutions and guide endpoint protection enhancements.

Application Control Effectiveness

Application control forms one of the ASD's Essential Eight strategies, protecting organisations against various cyber threats. It encompasses measures controlling which applications can run on endpoints.

Metrics in this area measure how successfully these practices prevent unauthorised applications from installing and executing. Effective application control policies aid in the reduction of malware infection risks and unauthorised software usage, commonly known as shadow IT.

Endpoint Detection and Response (EDR) Findings

EDR findings comprise information collected by EDR solutions monitoring endpoint activities in real-time. They include details on detected threats, suspicious behaviours and security incidents on endpoint devices.

EDR solutions analyse this data to identify potential breaches or malware that may have bypassed traditional security measures. Metrics associated with EDR findings provide valuable information about endpoint security posture. By analysing these metrics, your security teams can assess threat detection capabilities and enhance cybersecurity strategy.

Access Management

User Access Reviews and Privilege Escalation Monitoring

Regular user access reviews and monitoring for privilege escalation attempts are essential for maintaining secure access controls. Metrics in this area include access review frequency, detected privilege escalations (which should always total zero for unauthorised privilege escalations) and corrective actions taken. These metrics ensure appropriate access rights and promptly address any unauthorised privilege escalations.

Password Strength and Multi-Factor Authentication (MFA) Adoption Rates

Effective access management depends on strong passwords and MFA implementation. Although password strength quantification presents challenges, organisations can track password policy compliance and password manager usage. MFA adoption rates offer a clearer metric by measuring the percentage of users or accounts protected by this additional security layer. Strong password practices combined with high MFA adoption significantly decrease unauthorised access risks. For organisations that have yet to implement MFA, prioritising its deployment remains key.

Security Operations (SECOPS) Metrics

Incident Response

Mean Time to Detect (MTTD), Mean Time to Respond (MTTR), Mean Time to Contain (MTTC)

These three metrics are critical indicators of an organisation's incident response capabilities. MTTD measures the average time to detect a security incident. MTTR has multiple meanings, including Mean Time to Respond (measuring average time from detection to response initiation) and Mean Time to Repair/Resolve/Recover (measuring average time from detection to full incident resolution). MTTC measures the average time from detection to incident containment. Organisations typically measure these metrics in hours or days. Shorter times generally indicate more efficient incident response processes, though complex incidents may require extended resolution periods. Consistent tracking and gradual reduction of these times demonstrate a maturing and coordinated incident response capability.

Number and Severity of Security Incidents

Tracking security incident numbers and severity provides insights into the threat environment and an organisation's incident handling ability. A higher number of low-severity incidents might indicate effective detection capabilities, while fewer high-severity incidents suggest successful prevention measures. These metrics help identify trends and areas for improvement in incident management.

Security Awareness and Training

Employee Participation Rates in Security Awareness Training Programs

Participation rates in security awareness training programs reflect an organisation's commitment to building a security-conscious culture. High participation indicates employees are engaged and informed about cybersecurity best practices. These metrics assist you to assess training program effectiveness and identify areas for improvement.

Phishing Simulation Click-Through Rates to Assess Employee Awareness

Phishing simulations provide valuable tools for assessing employee awareness and readiness to handle phishing attacks, among the most common threats organisations face. Click-through rate metrics in these simulations offer insights into security awareness training effectiveness. Lower click-through rates indicate heightened employee vigilance and awareness, strengthening the overall security posture.

Security Knowledge Retention Measured Through Post-Training Assessment

Knowledge retention measurements through post-training assessments help evaluate security awareness program effectiveness. These metrics indicate how well employees retain and apply critical security information daily. High retention rates demonstrate successful training programs.

Impact Metrics

Cost of Security Incidents

Estimating the Financial Impact of Security Incidents

Security measures carry financial costs, but the consequences of security incidents can far exceed initial investments. Quantifying the financial impact of incidents such as data breaches and downtime is essential for understanding cybersecurity business risks. These metrics help estimate direct and indirect costs, including revenue losses, legal fees, regulatory fines and reputational damage. Understanding these financial implications allows organisations to justify security measure investments and demonstrate cybersecurity program value.

Demonstrating the Potential Return on Investment (ROI) for Improved Security Measures

Beyond estimating incident financial impact, metrics calculating security investment ROI provide compelling business cases for enhanced cybersecurity measures. You can demonstrate the financial benefits of a security strategy by comparing security incident costs with preventive measure investments. This metric supports data-driven decision-making and helps secure funding for security initiatives. Common calculation methods include Cost-Benefit Analysis (CBA) and ROI.

Tactical Briefing: Example ROI Calculation for Security Investment

Consider a mid-sized financial services company implementing an advanced SIEM system with automated threat response.

Investment Analysis:

Year 1 Costs:

- SIEM licensing and implementation: $170,000
- Training and maintenance: $41,000
- Total Investment: $211,000

Annual Benefits:

Incident Prevention: Historical average: 12 incidents/year at $45,000 each. SIEM prevents 70% (8.4 incidents). Savings: $378,000

Efficiency Gains: Reduced MTTD from 48 to 2 hours, less overtime. Savings: $39,800

Compliance: Avoided fines and reduced audit prep. Savings: $60,000

Total Annual Benefits: $477,800

ROI Results:

First Year ROI = (Benefits - Costs) / Costs × 100

= ($477,800 - $211,000) / $211,000 × 100 = 126%

Three-Year ROI (including ongoing maintenance):

= ($1,422,900 - $259,000) / $259,000 × 100 = 449%

Break-even: 5.3 months

This example demonstrates how quantifying incident prevention, operational efficiency and compliance benefits can build a compelling business case. The methodology, identifying direct costs (incident response) and indirect benefits (productivity gains), applies across various security investments.

System Uptime and Availability

Monitoring uptime and downtime rates effectively assesses how security events impact business operations. High uptime indicates your security measures prevent disruptions, while low downtime demonstrates quick recovery ability. These metrics provide insights into disaster recovery and business continuity effectiveness.

Furthermore, tracking these metrics highlights your cybersecurity program's value. By showing how security measures maintain uninterrupted operations, you can emphasise investment importance and align security objectives with business goals. This ensures security initiatives support organisational resilience and business success.

KPIs for Security Management

Selecting the Right KPIs

Selecting appropriate KPIs begins with aligning them to your organisation's security goals and risk profile. This alignment ensures your metrics directly support strategic objectives rather than becoming vanity numbers on a dashboard. For instance, if you aim to reduce security incidents, relevant KPIs might track incident numbers and vulnerability remediation times.

Limiting yourself to a manageable set of KPIs prevents information overload. When organisations track too many metrics, they often struggle to extract meaningful insights (like trying to watch 20 TV channels simultaneously). Instead, identify a few critical indicators that address high-priority security risks and significantly impact your overall security posture. This focused approach enables more effective monitoring and response to cybersecurity risks.

Effective KPIs must be measurable, actionable and relevant to stakeholders. Measurable KPIs provide quantifiable data that can be tracked consistently over time. An example would be the 'percentage of critical vulnerabilities patched within 48 hours', a metric that is both measurable and actionable while remaining relevant to IT teams and executive stakeholders. When KPIs drive specific security improvements and provide useful information for decision-making across various organisational levels, they become powerful tools for enhancing your security program.

KPIs for Continuous Improvement

Establishing baseline metrics forms the foundation of using KPIs for continuous improvement. To create these baselines, initial data for each indicator should be collected over a defined period, such as several months. By tracking trends at regular intervals, weekly, monthly or quarterly, you can identify areas needing improvement and measure the effectiveness of your security measures.

Once baselines are established, set realistic improvement targets considering industry benchmarks and your organisational context. This cycle of measuring, analysing and improving should be ongoing, with periodic reviews to ensure KPIs remain relevant. For example, if your baseline shows an average of 72 hours to patch critical vulnerabilities, you might aim to reduce this to 48 hours over the next quarter, with weekly progress tracking.

Benchmarking your KPIs against industry standards allows you to compare performance with peers and identify best practices. This comparative analysis assists you to set achievable targets and recognise growth opportunities. For instance, you might benchmark incident response times against industry averages to ensure competitiveness. However, alignment with your specific organisational goals and risks takes precedence over peer comparisons; each organisation faces unique cybersecurity risks and makes different investment decisions.

Using KPIs to communicate security posture and progress to stakeholders is essential for maintaining transparency, meeting compliance requirements and building trust. It also directly supports business growth. Regular reporting keeps stakeholders informed about the effectiveness of cybersecurity initiatives and enables data-driven decisions. This communication proves crucial for securing ongoing support and funding for security programs.

Challenges and Considerations

Data Collection and Analysis

One of the primary challenges in cybersecurity involves collecting and analysing data efficiently. The sheer volume of security data generated daily, from firewalls, intrusion detection systems, endpoint devices and other sources, can be overwhelming.

This quantity often leads to data overload, making identifying critical security events and meaningful trends difficult. You'll find yourself drowning in alerts while real threats slip through unnoticed.

Data quality and integrity serve as the foundation for effective analysis. Poor-quality data frequently results in false positives, missed threats and inaccurate trend analysis. You need strong data validation processes, consistent data formats, regular audits of collection mechanisms and protection from tampering or corruption.

Organisations increasingly adopt advanced analytics and Machine Learning (ML) techniques to manage vast amounts of security data. These include behavioural analytics to detect anomalies, predictive analytics to forecast potential threats and ML algorithms to identify patterns indicative of sophisticated attacks. The need for real-time analysis adds another layer of complexity, requiring stream processing technologies, automated response systems and continuous monitoring.

Data collection and analysis represent multifaceted and rapidly evolving challenges. Protecting your organisation from constantly changing threats requires a combination of skilled personnel, effective policies and practices, and advanced technologies.

Balancing Security and Usability

Gene Spafford famously noted: "The only truly secure system is one that is powered off, cast in a block of concrete and sealed in a lead-lined room with armed guards". This quote highlights the tension between security and usability. A system rendered unusable by excessive security controls serves no practical purpose.

Overly stringent security measures impede productivity, causing user frustration and potential circumvention of security policies. For example, complex password requirements and frequent password changes might enhance security but also burden users. This burden can lead to insecure practices, such as writing down passwords on sticky notes and hiding them under keyboards.

A risk-based approach supports your ability to strike the right balance between security and usability. This approach involves assessing risks across all assets, systems and processes, then implementing security measures proportionate to the identified risk level. Critical systems may require MFA and strict access controls, while less sensitive systems might employ more user-friendly security measures. The goal is to align security controls with your organisation's risk tolerance without compromising usability.

Metrics measuring risk tolerance can help determine whether an appropriate balance has been achieved. Two key metrics include:

1. Key Risk Indicators (KRIs):

- These quantifiable measurements track specific risks.
- KRIs typically have defined thresholds aligned with an organisation's risk tolerance.

2. Residual Risk:

- This metric measures the risk level remaining after implementing security controls.

- Organisations often establish acceptable residual risk ranges reflecting their overall risk tolerance.

Security Culture and Mindset

Effective cybersecurity requires employees at all levels to understand the importance of security and their role in maintaining it. Organisations must foster a security-conscious culture through ongoing education and awareness programs. These initiatives keep security a constant priority and equip employees with the knowledge and skills to recognise, respond to and report potential threats. Emphasising security in daily operations creates a positive security culture and strengthens cyber defence resilience.

Developing a positive security culture requires committed leadership. Senior management must demonstrate this commitment by actively prioritising security in organisational policies, decisions and communications. This approach creates an environment where security becomes valued and integrated into daily operations rather than viewed as an obstacle or afterthought. Leaders can regularly address security topics in team meetings, acknowledge employees who demonstrate good security practices and ensure security considerations form part of all project planning processes.

Assessing security awareness program effectiveness requires metrics focused on behavioural change. These provide insights into your organisation's security culture:

1. Participation rates in security training programs

- High rates indicate employee engagement with security initiatives.

2. Results of phishing simulations

- Improved performance over time demonstrates increasing security awareness.

3. Frequency of reporting suspicious activities

- An increase in reports suggests heightened vigilance among employees.

When analysed collectively, these metrics offer valuable insights into your organisation's security posture. High participation in training programs and improved outcomes in phishing simulations indicate an engaged workforce. Meanwhile, frequent and accurate reporting of suspicious activities suggests employees actively apply their security knowledge in everyday operations. These metrics should be tracked over time to identify trends and improvement areas—congratulations if you're seeing positive trends across all three.

Summary

Measuring cybersecurity effectiveness presents a complex challenge akin to hit a moving target in a windstorm. This chapter examined the essential metrics and KPIs organisations need to evaluate their security posture, operational efficiency and impact within an increasingly sophisticated threat landscape.

Effective cybersecurity measurement requires a structured, data-driven approach rather than subjective assessments. Metrics provide quantifiable measures tracking specific security processes, while KPIs function strategically to indicate meaningful progress toward organisational goals. These complementary tools satisfy requirements for both internal assessment and external reporting to stakeholders and regulatory bodies.

Core cybersecurity metrics encompass three critical categories. Security posture metrics assess preparedness through vulnerability management, network security, endpoint protection and access management, including vulnerability remediation rates, patching frequency, intrusion detection alerts, unauthorised access attempts and malware prevention effectiveness. Operational metrics evaluate incident response capabilities via measurements such as Mean Time to Detect (MTTD) and Mean Time to Respond (MTTR), alongside security awareness indicators like phishing simulation results and knowledge retention. Impact metrics link security performance to organisational outcomes through financial impact estimations, security investment ROI calculations and system availability tracking.

Selecting appropriate KPIs necessitates alignment with organisational security goals and risk profiles while maintaining a focused set of indicators. Effective KPIs must be measurable, actionable and relevant to drive continuous improvement. This involves establishing baseline metrics, defining realistic targets, monitoring trends and benchmarking against industry standards while communicating progress transparently to stakeholders.

This chapter highlighted several significant challenges in cybersecurity measurement. Data collection and analysis encounter obstacles from overwhelming information volume and quality concerns, necessitating advanced analytics and ML techniques. Balancing security with usability requires a risk-based approach that implements proportionate security measures without hindering operational efficiency. Cultivating a strong security culture depends on ongoing education, visible leadership commitment and metrics focused on behavioural change to evaluate program effectiveness.

Through these metrics and KPIs, organisations can establish data-driven decision-making processes that foster continuous security improvement. This structured measurement approach provides the foundation for accurately understanding cybersecurity status and identifying specific enhancement areas, ultimately strengthening overall organisational resilience against evolving cyber threats. You'll find that consistent measurement transforms security from guesswork into strategic advantage.

Revision Questions

1. What is the difference between metrics and KPIs in cybersecurity?
2. Why is data-driven decision-making important in cybersecurity assessment?
3. What are the three critical areas covered by core cybersecurity metrics?
4. Name three key security posture metrics discussed in the chapter.
5. What do the acronyms MTTD, MTTR and MTTC represent, and why are they important?
6. How can organisations measure the financial impact of security incidents?
7. What criteria should be considered when selecting appropriate KPIs?
8. What is the process for using KPIs to drive continuous improvement?
9. What challenges do organisations face in cybersecurity data collection and analysis?
10. How can organisations balance security and usability effectively?

Cybersecurity Hacks

Hack 1: The "Rule of 5" Metric Selection Framework

Start your cybersecurity measurement journey by implementing the "Rule of 5" - select exactly five core metrics that cover the security fundamentals. Choose one metric from each category:

- Vulnerability Management (e.g., percentage of critical vulnerabilities patched within 48 hours)
- Incident Response (e.g., Mean Time to Detect)
- User Awareness (e.g., phishing simulation click-through rates)
- Access Control (e.g., MFA adoption rate)
- Business Impact (e.g., security incident downtime hours)

This approach prevents the common mistake of trying to measure everything at once, which leads to analysis paralysis. By focusing on just five key indicators, you can establish a solid measurement foundation, develop expertise in data collection and demonstrate early wins to stakeholders. Once you've mastered these five metrics and can consistently collect and act on the data, you can gradually expand your measurement program.

Hack 2: The "Baseline Snapshot" Quick Start

Create an immediate cybersecurity baseline by conducting a snapshot assessment using free or low-cost tools already available in your environment. Spend one week collecting basic data: run a vulnerability scan using tools like Nessus Essentials or OpenVAS, check your firewall logs for the number of blocked attempts and document your current patch deployment times. Even with limited resources, this exercise consolidates your understanding of the measurement concepts and provides real data about your organisation's current state. This baseline becomes your "before" picture for measuring improvement and helps you identify which areas need immediate attention. The key insight is that imperfect data collected consistently beats perfect data that never gets gathered.

Chapter 18: Cybersecurity Begins with You

– Building a Human Firewall –

Objectives

By the end of this chapter, you will be able to:

- Understand the concept of the 'human firewall' and its importance in cybersecurity defence
- Identify key components of an effective security awareness training program
- Recognise essential security topics including phishing awareness, password security and data handling protocols
- Implement various measurement techniques to evaluate the effectiveness of security training
- Identify targeted security training approaches based on audience roles, access levels and learning preferences.

The term 'human firewall' highlights an all-important truth: people, not just technology, are your first line of defence against cyber threats. Think of it like a neighbourhood watch program for your digital infrastructure. When employees are trained to recognise and respond to security risks, they transform from potential vulnerabilities into active defenders of an organisation's digital assets. This concept is foundational in cybersecurity strategy, acknowledging that even the most sophisticated technical safeguards can't compensate for untrained personnel.

Security awareness training plays an important role in this transformation. It doesn't just teach technical skills; it fosters a security-conscious mindset at every level of an organisation. Through structured education, employees learn to identify phishing attempts, recognise social engineering tactics and handle sensitive data responsibly. These competencies enable informed decisions during everyday technology interactions, significantly reducing the likelihood of cyber incidents. Effective programs blend theoretical knowledge with practical scenarios that simulate real-world threats, creating muscle memory for security responses.

The impact of well-implemented awareness training is substantial and measurable. A recent Proofpoint study found that organisations that increased investment in phishing training by 52% saw successful spear-phishing attacks drop dramatically from 88% in 2022 to 56% in 2023.[131] This significant reduction illustrates how systematic training directly strengthens organisational security posture. Moreover, trained employees become active participants in SECOPS, promptly reporting suspicious activities and giving security teams early warning of potential threats before they escalate into major incidents.

Beyond threat prevention, effective security awareness delivers multiple organisational benefits. When personnel understand security principles, they consistently implement best practices like strong authentication, verifying communication sources and following data protection protocols. This behaviour establishes a workplace culture where security becomes ingrained, second nature, rather than an afterthought. The resulting security-conscious environment substantially improves resilience against evolving threats while supporting regulatory compliance efforts.

The consequences of neglecting this critical aspect of cybersecurity can be severe. Employee-related security failures frequently lead to data breaches, ransomware infections and intellectual property theft. These incidents carry heavy financial penalties, damage corporate reputation and erode stakeholder trust, impacts far exceeding the cost of prevention through training. By investing in human firewall development, organisations protect not only their information assets but also their market position and customer relationships.

This chapter examines effective security awareness training approaches, providing practical guidance for developing human firewalls. You'll learn how to design, implement and measure awareness programs that transform personnel from security liabilities into vigilant defenders. By the end of the chapter, you'll understand the essential components of programs that create lasting security behaviours rather than merely transferring information. The skills presented here apply equally to corporate environments and personal security practices, making them valuable regardless of your goals.

Developing an Effective Training Program

Creating a successful security awareness training program requires more than distributing information; it requires a strategy that captivates employees' attention and embeds cybersecurity best practices into their daily workflows. A thoughtfully designed program transforms your workforce from potential weak spots into a proactive, intelligent defence against evolving cyber threats.

Conducting a Comprehensive Needs Assessment

An effective security awareness program starts with a detailed needs assessment to identify your organisation's vulnerabilities. This process extends beyond basic risk identification; it reveals knowledge deficiencies, highlights variations in departmental threat exposure and customises training according to employees' particular roles.

When conducting your assessment, employ several proven techniques for optimal results. Anonymous security questionnaires can measure employees' existing knowledge of topics such as data protection protocols and incident reporting procedures. Controlled phishing exercises deliver practical insights by evaluating how staff respond to convincing but fabricated emails, thereby identifying high-risk groups. Further, examining previous security breaches can uncover recurring patterns and policy non-compliance, providing valuable insights for enhancing your organisation's security measures.

Developing Compelling Content

Content development should focus on clear, engaging and current resources covering security topics ranging from sophisticated phishing identification to effective credential management. When creating materials, employ diverse formats including interactive e-learning modules, video demonstrations, infographics and case studies, to maintain interest and enhance knowledge retention across different learning preferences.

Incorporate storytelling techniques that transform abstract security concepts into memorable scenarios. For example, trace the timeline of a data breach from initial compromise through resolution, highlighting critical decision points where employee actions determined outcomes. When you integrate practical exercises and scenario-based challenges, cybersecurity principles become intuitive behavioural patterns rather than merely theoretical concepts that employees struggle to implement.

Implementing Effective Delivery Methods

Your delivery methods significantly impact program effectiveness and knowledge retention. Successful implementation balances structured training sessions with accessible on-demand resources and contextual, just-in-time (JIT) learning opportunities.

For maximum impact, combine instructor-led workshops for complex topics with self-paced digital modules for foundational concepts, creating a blended learning environment. Deliver concise, focused training units that address specific security behaviours during natural workflow pauses, allowing for microlearning that doesn't overwhelm participants. Implement simulation platforms where employees can safely encounter and respond to simulated threats, building practical experience without risk.

These varied delivery methods create a sustainable security culture by accommodating diverse work styles, locations and learning preferences while maintaining consistent security messaging throughout your organisation.

Measuring Effectiveness and Continuous Improvement

An effective training program remains dynamic through careful measurement and systematic refinement. Establish success metrics, obtain participant feedback and conduct regular assessments to identify improvement opportunities. Track security behaviours such as phishing report rates, password manager adoption and policy compliance scores to gauge real-world impact.

Implement knowledge retention assessments regularly following training to measure long-term retention and identify concepts requiring reinforcement. Analyse the relationship between training completion and security incident involvement to determine which modules most effectively reduce specific incident types and where additional focus may be needed.

Systematically updating content and adjusting strategies based on emerging security challenges and measured outcomes ensures your organisation's employees remain informed, vigilant and adequately prepared to defend against sophisticated threat actors.

Key Topics in Security Training

Just as network firewalls require diverse settings to be effective, a security training program must cover various topics to configure the human firewall. Organisations can empower their employees by focusing on critical areas such as phishing awareness, password security and data handling protocols. This empowerment enables staff to recognise and mitigate a wide spectrum of threats, much like a well-configured network firewall.

Phishing and Social Engineering

Phishing and social engineering remain at the forefront of security training curricula. These sophisticated deception techniques continue to evolve, targeting human psychology to bypass technical defences. Equipping employees with the skills to identify and report suspicious communications transforms them from potential vulnerabilities into sentinels capable of preventing malicious attempts at data breaches or system compromise.

Password Security and Authentication

Password security and authentication are essential focal points of effective cybersecurity training. As gatekeepers of sensitive information, employees must understand the importance of creating and maintaining strong, unique passwords for each account. This includes elements we covered in Chapter 1, such as using a combination of uppercase and lowercase letters, numbers and symbols while avoiding easily guessable information.

Staff should also be trained in properly using MFA, which adds an extra layer of security by requiring two or more verification methods to access an account, as discussed earlier. These practices enable employees to protect their individual accounts and strengthen the entire organisational network against unauthorised access and data breaches.

Mobile Device Security

Technology evolution has transformed our work habits. Some professionals started in IT when computers were immobile behemoths, and mobile phones were enormous, much like a brick. Today, a mobile office fits in our pocket; smartphones and laptops are now lighter than books.

This shift to portable, always-connected devices has revolutionised how employees work, necessitating comprehensive mobile device security training. Such training empowers employees to secure devices, recognise public Wi-Fi risks and understand the importance of updating software. By equipping staff with this knowledge, organisations effectively extend their security perimeter beyond office walls, ensuring strong protection in our era of remote work and Bring Your Own Device (BYOD) policies.

Cloud Security

Cloud security training shares the same foundational principles as other employee security training but addresses the unique challenges of cloud-based services. Like all security education, it must be tailored to employees' roles, with security teams requiring deeper technical knowledge of the shared responsibility model that delineates provider and user security obligations (see Chapter 8).

Cloud security training follows familiar security frameworks, covering data classification (identifying and categorising sensitive information), access control (implementing strong authentication and authorisation), encryption (protecting data in transit and at rest), compliance (understanding regulatory requirements for cloud-stored data) and incident response (preparing for potential security breaches). This approach ensures staff can leverage cloud computing for business efficiency while maintaining the strong security posture expected across all organisational systems.

Insider Threats and Data Protection

Another vital security training component is addressing insider threats and data protection. Training on insider threats helps foster a culture of security consciousness. When employees understand their role in maintaining security, they become valuable assets in the overall cybersecurity strategy. Additionally, this training enables employees to recognise the value of the data they handle, understand proper data handling procedures and appreciate their responsibility in preventing intentional and unintentional data leaks.

Artificial Intelligence (AI) and Machine Learning (ML)

AI and ML present both opportunities and challenges in the security landscape, making them essential topics in security training. Employees need to understand how AI-powered threats such as deepfakes and advanced phishing can bypass traditional security measures by mimicking trusted sources with unprecedented accuracy.

Conversely, staff should know how their organisation employs AI for security enhancement, including anomaly detection and threat intelligence. Training should emphasise critical evaluation of AI-generated content, awareness of potential AI biases that might create security blind spots, the ethical considerations surrounding AI in security contexts and inserting sensitive organisational information into external AI systems. By developing this understanding, employees become better equipped to navigate an increasingly AI-influenced security environment while maintaining appropriate trust and scepticism when interacting with digital systems.

Building Your Human Firewall

These key topics form the core of effective cybersecurity training programs, serving the dual purpose of educating staff and safeguarding organisational assets. Focusing on these areas enables organisations to equip their workforce with a comprehensive understanding of cybersecurity challenges. This knowledge empowers employees to make informed decisions about digital security, recognise potential threats across platforms, take proactive measures to protect themselves and the organisation and adapt to new security challenges as they emerge. A well-trained workforce is an effectively configured human firewall, significantly enhancing organisational security posture.

Measuring Training Effectiveness

Let's be honest, you've invested serious time, money and resources into cybersecurity training. But how do you know it's working? What happens if those resources get diverted elsewhere in your organisation? Is your training delivering real value, or are you just ticking boxes?

While some still question whether cybersecurity training is necessary (hopefully that's becoming rarer), mounting evidence shows no organisation escapes cyber risks. Size doesn't matter—threats find everyone. Without proper training, your organisation's ability to adapt and thrive gets compromised. That's why assessing your security awareness programs isn't optional but essential for demonstrating value and refining effectiveness.

To tell the complete story, you'll need both numbers and narratives. Quantitative metrics give you hard data, while qualitative insights reveal the human element. Involving employees in measurement provides valuable perspectives and reinforces the security culture you're building. This approach doesn't just gauge success; it demonstrates the importance of training throughout your organisation.

Phishing Simulation Results

Phishing simulation results are a powerful indicator of training efficacy. By periodically and ethically deploying simulated phishing campaigns, you can track improvements in employee vigilance over time. Declining click rates on suspicious links and increasing rates of reported phishing attempts demonstrate heightened awareness and improved decision-making skills among staff. These metrics showcase the direct impact of training and highlight areas where additional focus may be needed.

Security Incident Metrics

Real-world application matters most, and that's where security incident metrics prove invaluable. Monitor incident frequency and severity to assess how well employees translate training into practice. Reduced incidents from human error (accidental data exposures, malware infections) indicate effective knowledge transfer.

More employee-reported security concerns signal something equally important: proactive security consciousness. When staff start spotting potential issues and speaking up, you've fostered the right culture through training.

Knowledge Assessments

Regular quizzes, surveys and practical exercises allow organisations to evaluate employees' understanding of various security topics. These assessments can be tailored to specific roles or departments, ensuring that each employee possesses the knowledge relevant to their responsibilities. Tracking improvement in assessment scores over time provides tangible evidence of the training program's effectiveness in enhancing security literacy across the organisation.

However, don't overload employees with assessments. Security fatigue is real; overwhelmed staff might ignore policies entirely, defeating your purpose.

Behavioural Change Indicators

Numbers tell part of the story, but behavioural shifts reveal deeper impact. Watch for changing employee attitudes towards security practices, increased engagement in security discussions and proactive improvement suggestions. These signal a deepening security culture that extends beyond formal training sessions.

Focus groups and one-on-one interviews uncover valuable perspectives on training relevance and daily work applicability. Sometimes the most useful insights come from casual conversations, not formal surveys.

Making It Count

Leveraging these measurement techniques validates security awareness training effectiveness and demonstrates clear value. More importantly, it cultivates shared responsibility among employees. This data-driven approach assists staff to recognise their progress and contributions to organisational security posture.

When employees see how they're improving and contributing, everyone understands their role in safeguarding the organisation. That's when training transforms from mandatory sessions into a meaningful security culture.

The Human Element

Know Your Audience: Targeted Training

Here's the thing about security training—one size doesn't fit all. Your employees occupy different roles, have varying access levels and learn in completely different ways. If you've been rolling out a generic training module, it's probably been missing the mark for most of your workforce.

Job roles determine unique security challenges. IT staff require appropriate training on technical threats and system vulnerabilities, whereas general employees benefit more from education on phishing awareness and safe browsing practices. Marketing teams might need focused guidance on protecting customer data, while HR personnel must prioritise the secure handling of sensitive employee records. Personalised training ensures content remains relevant and practical for each group.

Remember: When training aligns with daily responsibilities, employees apply what they've learned instead of filing it under 'compliance completed'.

Access levels similarly influence training intensity. Employees managing sensitive data or critical systems aren't just learning awareness; security becomes their core responsibility. Those with elevated privileges need specialised modules and regular reinforcement to prevent breaches.

Don't forget learning preferences either. Some employees retain information better through visual aids and videos, while others learn more effectively through interactive exercises or hands-on activities. A well-rounded security program incorporates multiple formats, including videos, scenario-based learning and practical simulations, ensuring greater engagement and information retention across the organisation.

By aligning training with roles, access levels and learning styles, you'll build a security culture that's both effective and sustainable. The goal isn't just informing employees, it's creating lasting behavioural change.

Engaging Training Methods for Knowledge Retention

Effective security training isn't about information delivery; it's about making knowledge stick. You need interactive, practical approaches that mirror real-world challenges rather than boring employees with passive content (we've all sat through those painful presentations).

Scenario-based learning offers powerful engagement through interactive modules featuring quizzes and authentic scenarios. When employees work through hypothetical security incidents, they apply security knowledge in practical situations, receiving immediate feedback that enhances their decision-making capabilities. This active engagement with material helps transform theoretical concepts into practical skills.

Role-playing exercises build confidence by letting employees practise incident responses. When staff act out security scenarios, they develop the muscle memory to handle actual threats. These exercises enhance preparedness and reinforce appropriate responses under pressure, critical during real incidents when adrenaline kicks in.

Microlearning and short videos divide complex cybersecurity topics into digestible lessons for efficient learning. Brief modules covering subjects like password security or social engineering integrate seamlessly into busy workdays, facilitating continuous learning without significantly disrupting productivity.

Gamification transforms training into engaging challenges through points, badges and leaderboards. This makes learning feel less like mandatory compliance and more like friendly competition. Recognising employees for completing training promotes continuous improvement and vigilance across your organisation.

Integrating these methods equips employees to stay alert and proactive in their security practices.

Sustaining Security Culture

Creating a lasting security culture requires more than initial training; it requires ongoing commitment throughout daily operations. This starts with visible executive sponsorship, which demonstrates leadership's commitment to security principles. When executives actively participate and allocate sufficient resources, employees recognise security as an organisational priority, not just IT's problem.

Strategic communication campaigns maintain awareness through newsletters, intranet resources and targeted emails highlighting emerging threats with practical tips. Security champion networks extend influence by embedding advocates within business units who serve as local resources and apply practices to specific operational contexts.

What is the true measure of a sustainable security culture? Seamless integration into business processes. When security becomes implicit in project planning, product development and daily workflows, it transcends being viewed as a separate obligation. This creates self-sustaining systems where awareness becomes second nature, ensuring collective defence remains effective against evolving threats.

You'll know you've succeeded when employees instinctively consider security implications without prompting—that's when training transforms into culture.

Summary

This chapter explored the 'human firewall' concept—how security awareness training transforms employees from potential vulnerabilities into active defenders of digital infrastructure. We've demonstrated that effective cybersecurity defence requires technological solutions and knowledgeable, vigilant personnel who recognise and respond appropriately to security threats.

Security awareness training fosters security-conscious mindsets throughout organisations, equipping employees with skills to identify phishing attempts, recognise social engineering tactics and handle sensitive data responsibly. Well-structured programs deliver broad benefits, with research showing increased phishing training correlates with significant reductions in successful attacks.

Beyond prevention, training encourages reporting suspicious activity, strengthens regulatory compliance and creates a workplace culture where security becomes second nature. When security awareness permeates daily operations, it transforms from a compliance requirement into an organisational strength.

This chapter outlined key components for developing effective programs, beginning with comprehensive needs assessments to identify unique vulnerabilities and knowledge gaps. Content development focused on creating clear, engaging resources covering the main topics through diverse formats such as interactive modules, videos and real-world case studies.

Successful implementation balances scheduled sessions with on-demand resources, while continuous measurement through clear metrics and participant feedback ensures programs remain relevant and effective. It's about finding the sweet spot between structure and flexibility.

Critical training topics include phishing and social engineering awareness, password security and authentication practices, mobile device security, cloud security frameworks and data protection protocols. These interconnected areas form foundations for a multifaceted understanding of current and emerging cybersecurity challenges, enabling informed decisions across various platforms and work contexts.

This chapter emphasised measuring training effectiveness through quantitative and qualitative metrics. Phishing simulation results, security incident tracking, knowledge assessments and behavioural change indicators provide tangible evidence of program impact while identifying improvement opportunities. This data-informed approach validates security investments and cultivates shared responsibility among employees.

Finally, the chapter emphasised that building effective programs requires targeted approaches based on job roles, access levels and learning preferences. Engaging training methods (interactive modules, simulated phishing attacks, microlearning and gamification) enhance knowledge retention and application.

This approach strengthens organisational security posture while promoting a culture where security consciousness becomes instinctive across all levels. When security awareness becomes part of your organisation's DNA, you've transformed training investment into a sustainable competitive advantage.

Revision Questions

1. What is the 'human firewall' and why is it important in cybersecurity defence?

2. How does security awareness training impact an organisation's vulnerability to phishing attacks?

3. What is the first step in creating an effective security awareness training program?

4. What delivery methods are recommended for effective security awareness training?

5. What are the five key topics that should be covered in comprehensive security training?

6. How can organisations measure the effectiveness of their security awareness training?

7. Why is it important to provide targeted training based on audience characteristics?

8. What engagement methods help ensure knowledge retention in security training?

9. What are the essential content pillars of security awareness training?

10. Why is continuous reinforcement important in security awareness programs?

Cybersecurity Hacks

Hack 1: Security Awareness Micro-Challenges

Implement monthly or quarterly security micro-challenges that require employees to identify phishing emails, evaluate password strength or distinguish between secure and insecure websites. Delivering these brief, engaging activities via email or a dedicated application maintains security awareness without demanding significant time from staff. Consider offering small rewards to successful participants, enhancing motivation while reinforcing the importance of vigilance in everyday digital interactions.

Hack 2: The "Security Buddy" Text Chain

Establish informal security communication channels using existing workplace messaging platforms (Teams, Slack, Signal groups) where employees can quickly share suspicious emails, unusual system behaviours, or security questions without formal reporting procedures. Create department-specific or team-based groups with a simple rule: "When in doubt, shout out." This creates a rapid response network where employees can get quick feedback from colleagues and security-aware team members. The informal nature removes barriers to reporting while building collective security intelligence. Include one security champion in each group who can provide guidance and escalate genuine threats to the security team when necessary.

Chapter 19: From Learner to Leader

— Empowering Your Cybersecurity Journey —

Objectives

By the end of this chapter, you will be able to:

- Identify key components of continuous learning essential for cybersecurity career growth
- Distinguish between technical specialist and management career paths in cybersecurity
- Understand how to create personal learning plans for ongoing professional development
- Develop strategies for effectively communicating security concepts to non-technical stakeholders
- Apply the principles of security advocacy within organisations to drive meaningful change.

Your cybersecurity journey doesn't end with this book—it's just the beginning. Every day, organisations across Australia and globally face sophisticated cyber threats that exploit the smallest vulnerabilities. A career in cybersecurity demands continuous learning, adaptability and a steadfast commitment to protecting digital assets.

Whether starting your career or advancing your expertise, the principles and methodologies covered in these pages provide your foundation. But here's the thing: staying ahead of emerging threats requires you to actively expand your knowledge. The threat landscape evolves daily, and your defences must evolve with it.

This chapter will inspire and equip you for what comes next in your cybersecurity career. You now understand core concepts like confidentiality, integrity, availability and the cybersecurity lifecycle. True success, however, depends on your ability to grow alongside the field. As threats transform, so must your approach to defending against them.

The cybersecurity landscape presents both exciting opportunities and significant challenges. New technologies emerge at breakneck speed. Innovations like AI, quantum computing and Internet of Things (IoT) devices are fundamentally reshaping how we approach security. These advancements drive the need for specialised roles with distinct expertise, from quantum security specialists to protectors of autonomous systems, such as those securing self-driving cars. Your ability to anticipate and adapt in this dynamic environment ultimately defines your effectiveness.

As your career progresses, you'll face critical decisions about specialisation. Will you pursue technical depth or management breadth? How do you effectively communicate security concepts to executives who think malware is something you catch from opening windows? This chapter provides frameworks for making such choices, with strategies for ongoing learning, professional development and security advocacy within organisations.

Beyond technical expertise, your role carries significant responsibility. Your work protects critical infrastructure, safeguards sensitive information and builds trust throughout our digital ecosystem. The skills you develop don't just advance your career; they contribute to the collective security of our interconnected world, including Australia's digital landscape governed by frameworks like the Australian Signals Directorate (ASD) guidelines.

In the following sections, we'll explore the importance of lifelong learning. You'll find practical guidance on staying informed, sharpening your expertise and preparing for emerging challenges. Whether you're beginning your journey or well into your career, this final chapter provides direction for sustained growth and tools to not just navigate but excel in this dynamic field.

The Cybersecurity Professional's Growth Path

The cybersecurity landscape presents a complex yet rewarding career journey with diverse pathways and specialisations. Understanding this general progression framework provides clarity for professionals navigating this dynamic field. Unlike traditional career ladders with predictable rungs, cybersecurity offers a multidimensional matrix of technical and leadership trajectories that professionals can traverse based on their interests, aptitudes and organisational needs. This segmented approach to career development acknowledges the field's breadth while providing structured guidance appropriate to each professional's experience level and aspirations.

Early Career Professionals (0-2 Years Experience)

You're entering an exciting phase filled with opportunities and challenges as you establish your foundation. During these formative years, focus on developing fundamental technical competencies across multiple domains rather than diving into premature specialisation.

Begin by mastering cybersecurity principles, networking principles, operating system security, basic cryptography and security frameworks; these form the bedrock upon which all specialised knowledge builds.

Practical experience proves invaluable at this stage. Seek positions in Security Operations Centres (SOCs), help desk roles with security responsibilities or junior analyst positions where you'll encounter real-world security scenarios under supervision. These roles provide the hands-on experience that transforms theoretical knowledge into practical skills.

Mentorship becomes particularly important at this stage. Identify experienced professionals willing to provide guidance, whether through formal workplace programs or industry associations. These relationships offer contextual knowledge that is difficult to acquire through self-study alone. Complement this guidance by participating in capture-the-flag competitions, building home labs and contributing to open-source security projects.

Document your learning journey, maintain detailed records of incidents handled, technologies mastered and projects completed. This portfolio provides tangible evidence of your growing capabilities when pursuing advancement opportunities. Target entry-level certifications like CompTIA Security+ or EC-Council's Certified Ethical Hacker (CEH) to demonstrate baseline competency to employers while establishing your commitment to the profession's standards.

Mid-Career Specialists (3-7 Years Experience)

Mid-career represents the specialisation phase where professionals develop deeper expertise in specific domains while broadening their understanding of cybersecurity's business dimensions. Having established fundamental skills, you can now make informed decisions about specialisation paths, whether in penetration testing, security architecture, cloud security or incident response.

This specialisation should align with both market demands and personal aptitudes. Pursue areas where you demonstrate natural strengths while remaining attentive to emerging fields with strong growth trajectories. It's about finding your cybersecurity sweet spot, where passion meets opportunity.

Expand your professional presence beyond organisational boundaries by speaking at regional conferences, contributing to industry publications or participating in professional communities. These activities establish your expertise while creating valuable professional networks. Simultaneously, develop adjacent skills that complement your technical capabilities. Project management, financial acumen and communication skills dramatically enhance your effectiveness and advancement potential. Don't underestimate these 'soft skills'; they often separate good cybersecurity professionals from great ones.

In addition to self-study, there is no substitute for formal education and certifications in a cybersecurity professional's development. While university degrees are a great start to a professional's career, certifications validate the practical application of knowledge and skills. They will set you apart, enabling you to be more competitive in the job market and improve your earning potential. Programs such as Certified Information Systems Security Professional (CISSP), Certified Ethical Hacker (CEH), Offensive Security Certified Professional (OSCP) and Certified Cloud Security Professional (CCSP) can equip you with practical, in-depth knowledge across various cybersecurity domains. These certifications are respected industry standards and are instrumental in career advancement.

Beyond technical certifications, gaining expertise in governance and risk management through credentials like Certified Information Security Manager (CISM) or Certified Information Systems Auditor (CISA) can be invaluable. These certifications ensure you understand cybersecurity's technical and business aspects, which is necessary for aligning security strategies with broader organisational goals.

Obtaining these credentials can open doors to advanced roles within the field, such as becoming an Infosec Registered Assessors Program (IRAP) assessor as administered by the ASD. IRAP Assessors are ASD-endorsed ICT professionals from across Australia who combine extensive experience in ICT security assessment and risk management with comprehensive knowledge of ASD's ISM. These specialists help organisations understand, implement and maintain security controls to protect their systems and data.

Technical vs. Management Career Paths

The mid-career transition often presents a critical decision point between technical depth and management breadth. The technical specialist path involves developing elite expertise in specific security domains, becoming the recognised authority organisations turn to for complex challenges. This track requires continuous technical education, research contributions and potentially specialised certifications like OSCP for penetration testing or CISSP-ISSAP for architecture. Technical leaders derive satisfaction from solving intricate security problems and maintaining hands-on involvement with technologies.

Conversely, the management track shifts focus on strategic security leadership, team development and business alignment. Security managers translate technical requirements into business language while championing executive-level security initiatives. This path requires developing budgeting, resource allocation, policy development and team leadership competencies. Management-focused certifications like CISM or CISSP concentrate on Governance, Risk and Compliance (GRC) rather than technical implementation details. Neither path offers inherently superior rewards; the optimal choice depends on individual strengths, values and career objectives.

Realistic Timeframes for Career Progression

Career advancement in cybersecurity generally follows patterns influenced by individual aptitude and market conditions. Early-career professionals typically remain in foundational roles for 1-3 years before advancing to specialised positions with greater autonomy. Mid-level specialist roles generally extend across 3-5 years as professionals build domain expertise and professional recognition. Senior technical specialist or management positions typically require 7-10 years of progressive experience, with executive security leadership roles demanding 10+ years of comprehensive security experience.

This progression timeline can accelerate through demonstrated exceptional performance, strategic certification acquisition and effective professional networking. However, rushing advancement without building adequate technical foundations risks creating knowledge gaps that undermine effectiveness at higher levels. Career advancement represents not merely title changes but expanding impact scope, from securing individual systems to protecting enterprise environments to influencing industry security practices. Each progression stage requires demonstrating tangible security improvements aligned with business objectives. Stay patient, stay focused and remember that building a cybersecurity career is like constructing a fortress—it requires solid foundations, careful planning and persistent effort.

Continuous Learning: Building Your Knowledge Arsenal

As you have probably gathered by now, cybersecurity is not a field where mastery is achieved and left behind. It is a dynamic, ever-evolving domain that requires constant attention, learning and adaptation.

Threats change rapidly, technologies evolve and attackers continue to innovate, making it necessary for cybersecurity professionals to maintain a growth mindset. You must remain informed about the latest developments to safeguard systems and data effectively.

Moore's Law (the principle of doubling transistors on a microchip approximately every two years) no longer applies as it once did. While technological advancement has accelerated in areas like AI and quantum computing, it's slowed in others. What remains certain is that technology is rapidly evolving, making today's cybersecurity environment vastly different from just a few years ago.

New attack vectors have emerged, particularly those exploiting AI, ML and the IoT. Adversaries now use advanced tools like AI-driven malware and deepfake technologies. As a professional in this field, you must embrace the mindset that learning is never-ending. Staying static means falling behind, and in cybersecurity, falling behind can be catastrophic.

Self-Education: Your First Line of Defence

Learning can take many forms, but continuous self-education remains one of the most effective and probably cheapest ways. This involves engaging with cybersecurity publications, blogs and reports that provide timely information on emerging threats.

Websites such as Threatpost and The Hacker News, government resources like cyber.gov.au and podcasts such as Risky.Biz offer a wealth of up-to-date information on trends, vulnerabilities and real-world breaches. Following these sources helps ensure you are always aware of the latest developments and best practices.

The cybersecurity community thrives on collaboration, and attending conferences is one of the most effective ways to engage with this vibrant network of professionals. Events like DEF CON, Black Hat, BSides, Australian Cyber Conference and RSA Conference offer the chance to learn from leading experts, share ideas and discuss the latest research.

These events provide exposure to cutting-edge technologies and strategies while fostering relationships that can lead to mentorship, partnerships and collaboration on research and innovations. Virtual conferences and webinars have also become increasingly popular, providing the flexibility to learn and network from anywhere in the world. Many industry leaders share their knowledge through these platforms, making them excellent resources for continuing education.

Cybersecurity professionals must also cultivate an adaptive mindset. The threats we'll face tomorrow vastly differ from those we'll face today, and adaptability is key to maintaining an effective defence. Understanding emerging technologies like quantum computing, ML, generative AI, securing industrial control systems and integrating DevSecOps methodologies into your work can help you stay agile.

Moreover, participation in Capture The Flag (CTF) challenges or bug bounty programs provides hands-on experience in identifying and mitigating real-world vulnerabilities. These platforms hone your technical skills and sharpen your problem-solving abilities, preparing you for the unpredictable nature of cybersecurity threats.

Strategic Approach to Certifications by Career Goal

Certification strategies should align with your career trajectory rather than following generic recommendations. Consider certifications as building blocks; each one should logically support your overall career architecture.

Technical specialists benefit from progressive certification paths that build upon foundational knowledge. Those pursuing penetration testing careers might progress from CompTIA Security+ to GIAC Penetration Tester (GPEN) to Offensive Security Certified Professional (OSCP), developing increasingly specialised ethical hacking competencies. Security architects should pursue certifications like CISSP-ISSAP or TOGAF that emphasise secure system design principles across enterprise environments.

Security management professionals require credentials demonstrating governance expertise and business alignment. The CISM certification equips security managers with frameworks for developing and managing enterprise security programs. At the same time, ISACA's Certified in Risk and Information Systems Control (CRISC) addresses risk identification and mitigation strategies. For GRC specialists, certifications like CRISC or Certified Compliance and Ethics Professional (CCEP) prove particularly valuable, emphasising regulatory frameworks and compliance mechanisms.

Rather than accumulating certifications indiscriminately, strategically sequence credentials to validate your evolving expertise. Prioritise certifications required in your geographical region or industry vertical. For example, financial services professionals might prioritise CISA certification due to its recognition among banking regulators, while government contractors benefit from pursuing Federal Risk and Authorization Management Program (FedRAMP) or Cybersecurity Maturity Model Certification (CMMC) credentials.

Creating Personal Learning Plans with Templates

Effective learning requires structure rather than haphazard consumption of security content. Develop a personal cybersecurity curriculum by creating a structured learning plan template encompassing multiple knowledge domains. Begin by conducting a skills gap analysis comparing your current capabilities against your target role's requirements. This assessment reveals priority learning areas deserving immediate attention.

Your learning plan template should include four core components: knowledge objectives (specific concepts to master), practical exercises (hands-on activities demonstrating competency), resources (books, courses or mentors) and verification methods (how you'll validate knowledge acquisition). Establish realistic timeframes for each learning module while building accountability through weekly learning reviews and quarterly deep progress assessments.

For implementation, dedicate consistent time blocks, whether daily micro-learning sessions of 20-30 minutes or deeper weekend study periods, to maintain momentum. Technology tools like learning management systems or simple project management applications help track progress through your personalised curriculum, transforming abstract learning goals into measurable achievements.

Measuring Learning Effectiveness

Evaluating learning progress requires objective measurement beyond subjective feelings of competence. Implement practical assessment mechanisms that verify knowledge transformation into applicable skills. Technical challenges like vulnerable virtual machines from platforms such as HackTheBox or VulnHub provide objective validation of offensive security knowledge, while building secure applications or systems demonstrates defensive capabilities.

Peer review represents another powerful validation mechanism. Present your security implementations or research findings to experienced professionals who can identify conceptual gaps or implementation flaws. Participation in professional security communities provides opportunities for such feedback while expanding your professional network.

The ultimate measurement remains your increasing capacity to handle progressively complex security challenges with greater autonomy and effectiveness. When you can confidently tackle problems that once seemed insurmountable, your learning efforts will have paid dividends.

Building Influence: Becoming a Security Leader

To drive meaningful change within your organisation, become a cybersecurity advocate. Educate non-technical employees about security practices, organise training sessions and promote awareness campaigns. Make cybersecurity a business priority by demonstrating how it is not a cost centre but a value multiplier. Show how security mitigates risks, protects assets and enhances customer trust.

Mentorship is another way to make a difference. Sharing your knowledge with newcomers will build the next generation of cybersecurity professionals. This can take many forms, from one-on-one guidance to speaking at industry conferences. Your voice can inspire others to act with purpose.

Beyond advocating for security awareness and mentoring, lead by example in identifying potential risks before they become incidents. Proactively research security tools, study frameworks and standards, and encourage protocol improvements. Take an active role in fortifying your organisation's cybersecurity posture to reduce breach risks and ensure resilience against emerging threats.

Effective Communication of Security Concepts to Non-Technical Stakeholders

Security leaders translate complex technical concepts into business language that resonates with diverse stakeholders. They avoid jargon when communicating with executives. They frame security discussions around business objectives and customer trust. They transform abstract vulnerabilities into concrete scenarios and explain how "customer data could be stolen, resulting in regulatory penalties and brand damage".

Use visual communication when explaining complex concepts. Simple diagrams and metaphors bridge understanding gaps. For example, describe ransomware as 'digital kidnapping' or authentication as a 'digital passport system'. Tailor your approach to your audience's concerns. Executives respond to financial implications, while operational teams focus on workflow impacts. This skill transforms security from a technical domain into a business priority.

Techniques for Security Advocacy Within Organisations

Build a network of security champions across departments by identifying influential allies who can amplify security messages. These champions become your force multipliers, spreading awareness more effectively than anyone could achieve alone.

Connect security initiatives to the company's mission and objectives. Frame security as an enabler supporting innovation while protecting assets. When resistance emerges, redirect toward collaborative solutions. Successful advocacy shifts from fear-based messaging to empowerment, helping colleagues work confidently amid complex threats.

Building Business Cases for Security Investments

Security leaders are adept at constructing compelling business cases that translate security investments into business value. Quantify risks financially and calculate potential breach costs, including penalties, liabilities and operational disruptions.

Research industry benchmarks to establish baseline investment levels for your sector. Structure proposals using return on security investment frameworks demonstrate risk reduction and operational benefits. Identify secondary benefits like improved efficiency or enhanced customer trust. Present tiered implementation options, allowing executives to select approaches aligned with their risk tolerance.

Present tiered implementation options, allowing executives to select approaches aligned with their risk tolerance. Support justifications with evidence from incidents affecting peer organisations, with real-world examples carrying more weight than theoretical scenarios.

Cross-Functional Collaboration Strategies

To be effective as a security leader, you need to establish regular touchpoints with legal, compliance and risk management teams. Participate in early planning phases of initiatives, ensuring security requirements integrate seamlessly. Create joint working groups addressing shared challenges like third-party risk management.

When departmental objectives conflict with security requirements, focus on shared organisational outcomes. Demonstrate flexibility by offering multiple approaches to achieve equivalent protection while accommodating operational constraints.

Successful security leaders transform relationships from compliance checks into strategic partnerships. Such a collaborative approach positions security as an indispensable business enabler that advances organisational goals while protecting critical assets.

Navigating Tomorrow: Emerging Challenges and Opportunities

As we look to the future of cybersecurity, it's clear that the landscape will evolve at an extraordinary pace, presenting challenges and opportunities. The relentless growth of technology, driven by advancements like AI, quantum computing and the IoT, will reshape the field in exciting and daunting ways.

While these technologies offer incredible potential, they also introduce new vulnerabilities that require innovative solutions. For cybersecurity professionals, the future will be defined by our ability to anticipate and adapt to these changes, maintaining resilience in an ever-expanding threat environment.

Alongside these challenges come exciting opportunities. As new technologies emerge, so too does the need for cybersecurity innovation. Cyber professionals have the chance to shape these technologies, building security into their foundations. AI can be trained to anticipate attacks before they occur, and quantum computing may lead to new forms of unbreakable encryption. The possibilities are vast.

Additionally, cybersecurity is becoming integrated into every aspect of life. The emphasis on privacy, data protection and ethical technology use creates opportunities for experts to lead in shaping not just security but digital ethics and governance. This role will be indispensable as technology becomes further entwined with critical infrastructure, finance, healthcare and personal data management.

The future of cybersecurity is bright for those willing to engage with its complexities. As technology evolves, so will the threats and solutions that define this field. The key to success in this dynamic environment is continuous learning, adaptability and a commitment to contributing to the broader cybersecurity community. The trends we see now are just the beginning of a fascinating, ever-evolving landscape.

New Career Opportunities in Emerging Security Domains

The cybersecurity career landscape is expanding beyond traditional roles into specialised domains, blending security with emerging technologies. It's like watching new species evolve in the cybersecurity ecosystem—exciting and full of potential.

Quantum security specialists, who understand both quantum computing principles and cryptographic implications, will become increasingly valuable as quantum computing threatens existing encryption. Organisations already seek professionals who can implement quantum-resistant algorithms and develop post-quantum security strategies.

Autonomous system security represents another burgeoning field. As self-driving vehicles, robots and drones proliferate, professionals skilled in securing these systems face unprecedented demand. This specialisation requires understanding control systems and sensor technologies while developing safeguards against physical and digital attacks.

The healthcare cybersecurity sector continues rapid growth, with specialised roles protecting medical devices and patient data. Medical device security engineers combine biomedical knowledge with cybersecurity principles to secure life-critical technologies.

Meanwhile, AI ethics and security auditors evaluate ML systems for bias and manipulation vulnerabilities. These multidisciplinary roles often command premium compensation whilst addressing meaningful societal challenges. If you're looking for career paths that matter deeply to society's future, these emerging domains offer incredible opportunities. And I haven't even mentioned the cybersecurity of space and robotic systems—look out for them!

Your Security Impact: From Protection to Innovation

Reframing Security from Defensive to Enabling Function

The cybersecurity profession demands a fundamental shift from a defensive barrier to a strategic business enabler. Progressive organisations now recognise security professionals as facilitators of innovation rather than gatekeepers. Security teams propel business initiatives forward by integrating security into development processes, automating compliance procedures and refining risk assessments while preserving essential safeguards. This strategic perspective transforms security from a perceived obstacle into a competitive advantage that confidently empowers digital initiatives.

In your hands lies the ability to influence positive change. You protect critical information, ensure operational continuity and fortify trust throughout our digital ecosystem. Your expertise enables decisive judgments when breach consequences become increasingly significant (and they're growing more serious by the day). Your responsibility transcends your immediate employer, as cyber resilience directly shapes privacy, security and public trust.

The rewards extend beyond the challenges you'll face. The satisfaction from preventing attacks, securing vital systems and educating stakeholders proves profoundly fulfilling. This field delivers professional advancement and measurable impact, plus knowing you're making a real difference.

By embracing this professional path, you join specialists dedicated to safeguarding digital infrastructures worldwide. The insights gained from this book enhance your capacity to contribute meaningfully to this significant mission.

Setting Personal Security Mission Statements

A personal security mission statement captures your distinct values and intended contribution, like having a North Star guide your career decisions. Consider principles that complement your capabilities, establish measurable objectives and review them as your career develops.

This elevates security from merely an occupation to a meaningful vocation, assisting you to stay motivated through challenges while directing expertise toward maximum effectiveness. Your mission statement becomes your professional compass. See the following general example to help get you started.

Example Personal Security Mission Statement

"I am committed to safeguarding digital assets and privacy by creating accessible, innovative security solutions. My mission is to grow technical expertise, share knowledge through education and mentorship, and always uphold user dignity and accessibility."

How This Mission Guides Professional Choices

This foundation prioritises privacy, ethical responsibility, and user-first design, providing clarity when navigating complex decisions.

Career Decisions:

- **Job Choice**—When choosing between ethically questionable roles versus developing privacy tools for healthcare systems, this mission points towards the latter, valuing dignity over profit.
- **Professional Growth**—Selecting conference speaking over another certification advances the education and mentorship commitment
- **Project Direction**—Advocating for intuitive security interfaces over quick patches reflects the user dignity principle

Setting Measurable Goals

Clear objectives make abstract values actionable:

- Earn one or two technical certifications per year
- Contribute to one open-source security project annually
- Mentor one to two professionals
- Maintain over 90% user satisfaction in implementations

- Complete yearly ethics training and support privacy advocacy

Career Evolution Through Your Mission

- **Early Career**—Master technical fundamentals and security principles
- **Mid Career**—Lead innovation projects while developing others
- **Senior Career**—Shape industry standards and educational frameworks

Your mission evolves from individual contribution to broad influence, anchored by consistent core values that define your professional legacy.

Conclusion: Your Cybersecurity Journey Starts Now

As you move forward, embrace the future with optimism. Use what you've learned to empower yourself and others, create lasting change and face the cyber landscape confidently and purposefully. The future is yours to shape.

Cybersecurity needs your skills, insight and passion right now. Whether you're securing networks, developing technologies or educating others about risks, your contributions matter. Stay engaged, informed and curious. The future isn't just something to anticipate; it's something you'll help create.

Cybersecurity isn't a spectator sport. It requires active participation from everyone, whether safeguarding personal data or defending critical infrastructure. Now's the time to step up, enhance your expertise and take ownership of your role.

Whether strengthening personal cyber hygiene, contributing to the professional community or leading initiatives within your organisation, your actions make a tangible difference. Every secure configuration, every awareness session and every vulnerability you patch contribute to our collective digital resilience.

The Payoff: Where Principles Meet Practice

As we conclude this journey through cybersecurity fundamentals, let's revisit the core message: Cybersecurity effectiveness depends not on technological solutions alone, but on understanding and applying fundamental principles that remain constant despite rapidly evolving threats.

Throughout this book, we've demonstrated how this concept holds true across every aspect of cybersecurity:

Part I: Foundations established the timeless principles that anchor all security practices. The CIA Triad provides an unchanging framework for evaluating security needs, whether protecting 1970s mainframes or future quantum computers. The relationship between threats, vulnerabilities and risks remains constant even as specific attack methods evolve. While attack tools change daily, the fundamental nature of what we protect and why endures.

Part II: Implementation demonstrated how principles translate into action through the six-function cybersecurity lifecycle. Rather than prescribing specific tools that become obsolete, we focused on enduring processes: Governance, Identification, Protection, Detection, Response and Recovery. Whether implementing these functions with today's SIEM platforms or tomorrow's AI-driven security orchestration, the underlying principles remain valid. This approach ensures your security knowledge transcends vendor products and technology cycles.

Part III: Human Factors reinforced that even in 'soft skills' (legal compliance, metrics and human factors), principle-based understanding proves more valuable than memorising current regulations or training programs. Laws change, metrics evolve and training methods advance, but the principles of ethical security practice, meaningful measurement and human-centred security design remain constant.

By grounding yourself in these principles rather than chasing the latest security tools, you've developed what truly matters: the ability to adapt to whatever cybersecurity challenges emerge next.

Core Principles: Your Cybersecurity Foundation

This book introduced numerous principles that form the foundation of effective cybersecurity. Here's a consolidated reference of these enduring concepts that will guide your practice regardless of technological change:

Fundamental Security Principles

- CIA Triad: Confidentiality, Integrity and Availability, the three pillars that define security objectives.

- Defence-in-Depth: Multiple layers of security controls ensure resilience when individual measures fail.

- Least Privilege: Users and systems should have only the minimum access necessary for their functions.

- Separation of Duties: Critical tasks should require multiple people to complete, reducing insider risks.
- Zero Trust: "Never trust, always verify", assume no user or system is inherently trustworthy.

Risk Management Principles

- Risk = Threats × Vulnerabilities: The fundamental equation for understanding security exposure.
- Risk-Based Decision-Making: Security investments should align with actual risk levels, not fear or assumptions.
- Continuous Risk Assessment: Threat landscapes evolve; risk evaluations must be ongoing, not one-time events.
- Residual Risk Acceptance: Perfect security is impossible; organisations must consciously accept some level of risk.

Operational Security Principles

- Continuous Monitoring: Security requires constant vigilance, not periodic checks.
- Incident Response Readiness: Assume breaches will occur and prepare accordingly.
- Evidence-Based Security: Decisions should rely on metrics and data, not intuition.
- Security as Business Enabler: Effective security facilitates business objectives rather than obstructing them.

Human-Centered Principles

- Security Culture Over Controls: Technical measures fail without human buy-in and participation.
- Shared Responsibility: Security is everyone's job, not just the IT department's.
- Transparency and Disclosure: Ethical security practice requires honest communication about risks and breaches.
- Continuous Learning: Cybersecurity expertise requires ongoing education and adaptation.

Governance and Compliance Principles

- Accountability: Clear ownership and responsibility for security decisions and outcomes.
- Regulatory Alignment: Security practices must meet legal requirements while serving business needs.
- Ethical Practice: Security work must balance organisational needs with societal responsibilities.
- Documentation and Audit: Security practices must be documented, measurable and verifiable.

Technical Design Principles

- Secure by Design: Security should be built into systems from inception, not added later.
- Fail Secure: When systems fail, they should default to a secure state.

- Minimise Attack Surface: Reduce the number of potential entry points for attackers.
- Encryption as Default: Protect data in all states; at rest, in transit and in use.

Lifecycle Management Principles

- Asset Lifecycle Security: Security considerations from acquisition through disposal.
- Patch Management: Timely updates are essential for maintaining security posture.
- Change Management: All changes should be evaluated for security impact.
- Continuous Improvement: Security programs must evolve based on lessons learned.

These principles transcend specific technologies, vendors or threats. Whether you're securing traditional IT infrastructure, cloud environments, IoT devices or whatever emerges next, these foundational concepts will guide effective security decisions. They represent the *why* behind security practices—knowledge that remains valuable even as the *how* continuously evolves.

Remember: Tools, technologies and threats will change, but these principles endure. Master them, and you'll be prepared for whatever cybersecurity challenges the future brings.

Your cybersecurity journey doesn't end here—it begins now.

Revision Questions

1. What are the three interconnected elements required for understanding cybersecurity principles?

2. What are the three components of the CIA Triad and what does each protect?

3. What are the six essential principles that form the cybersecurity lifecycle framework?

4. What are the two main career paths available to mid-career cybersecurity professionals?

5. Why is continuous learning essential in cybersecurity?

6. What are effective ways to engage with the cybersecurity community for professional growth?

7. How should a personal cybersecurity learning plan be structured?

8. How can security professionals effectively communicate with non-technical stakeholders?

9. What is the recommended approach for building business cases for security investments?

10. What emerging specialised cybersecurity domains are creating new career opportunities?

Cybersecurity Hack

The Final Hack: Essential Eight Assessment

If you have applied the cybersecurity hacks described throughout this book, you and your organisation will now have an effective security posture. To test this, use the following assessment tool to determine your organisation's cybersecurity posture against the Australian Signals Directorate (ASD) Essential Eight Mitigation Strategies. Methodically work through the strategies to identify security vulnerabilities, prioritise remediation efforts and develop a structured approach to enhancing cyber resilience.

Refer to the ASD Essential Eight for a detailed explanation of each mitigation strategy: https://www.cyber.gov.au/resources-business-and-government/essential-cyber-security/essential-eight.

Patch Applications: Ensuring that applications are regularly updated to protect against known vulnerabilities.

- Implement an automated system to periodically scan for and apply patches to applications and operating systems.
- Verify that patches are applied within the recommended timeframes (48 hours), especially for critical vulnerabilities.

Regular Backups: Regularly back up important data to ensure it can be restored in the event of a cyber incident.

- Perform regular backups of critical data and verify that backups are complete and restorable.
- Test backup restoration processes periodically to ensure data can be recovered in the event of a cyber incident.

Restrict Administrative Privileges: Limiting administrative privileges to only those users who need them for their job functions.

- Limit administrative privileges to only those users who require them for their job functions.
- Monitor and audit the use of administrative privileges to detect any unauthorised access or misuse.

Multi-Factor Authentication (MFA): Implementing MFA to add an extra layer of security for accessing sensitive systems and data.

- Implement MFA for all user accounts, especially for remote access and administrative accounts.
- Regularly review and update MFA policies to ensure they are effective and up to date.

User Application Hardening: Disabling unnecessary application features and services to reduce the attack surface.

- Apply the latest security updates and configurations to all user applications.
- Regularly review and update user application settings to align with the latest security guidelines.

Patch Operating Systems: Keeping operating systems updated with the latest security patches.

- Ensure that operating systems are regularly scanned for vulnerabilities and that patches are applied promptly.
- Validate that patches are tested for compatibility before deployment to avoid system disruptions.

Microsoft Office Macro Settings: Limiting the use of macros to only those that are digitally signed by trusted authors.

- Disable macros by default for all users unless there is a demonstrated business need.
- Ensure that macros in files from the internet are blocked and that macro antivirus scanning is enabled.

Application Control: Restricting the execution of executables, scripts and installers to only authorised ones.

- Ensure that only approved applications are allowed to execute by regularly reviewing and updating the application control rules.
- Validate that application control rules are enforced correctly by conducting periodic audits and generating event logs for allowed and blocked executions.

Chapter 20: Answers to Revision Questions

Part I

Chapter 1 - Foundations of Cybersecurity

Question 1. What is the CIA Triad, and why is it considered fundamental to cybersecurity?

The CIA Triad consists of three core principles that form the foundation of information security:

- Confidentiality—Ensures sensitive information is only accessible to authorised users.
- Integrity—Maintains the accuracy and completeness of data, preventing unauthorised alterations.
- Availability—Ensures information and resources are accessible to authorised users when needed.

These principles guide the development of security policies, strategies, architectures and controls worldwide.

Question 2. Explain the relationship between threats, vulnerabilities and risks in cybersecurity.

The relationship is often expressed as Risk = Vulnerabilities × Threats, where:

- Threats are potential sources of harm (natural, human or environmental).

- Vulnerabilities are exploitable weaknesses in systems (software flaws, misconfigurations, poor security practices).
- Risks represent the measurable potential for loss or damage when threats exploit vulnerabilities.

This relationship helps security professionals evaluate potential losses and prioritise security controls.

Question 3. What are the three types of security controls, and how do they work together?

The three types of security controls are:

- Preventive Controls—Stop threats before they occur (firewalls, encryption, access controls).
- Detective Controls—Identify and respond to incidents (IDS, SIEM systems, security audits).
- Corrective Controls—Mitigate damage and restore functionality after an incident (IRPs, system restoration procedures).
- These controls work together to provide comprehensive protection through prevention, detection and recovery.

Question 4. What are some common motivations behind cybercriminal activities?

Cybercriminal motivations include:

- Financial Gain—Using ransomware, fraud and identity theft.
- Political/Social Activism—Hacktivists promoting agendas.
- Espionage—Nation-states gathering intelligence.

- Revenge—Insider threats acting against employers.
- Challenge—Breaking systems out of curiosity.

Question 5. How has the cybersecurity landscape evolved from the 1960s to the present day?

The evolution includes:

- 1960s-70s—Focus on physical security of computer rooms and hardware.
- 1980s-90s—Emergence of network security with internet proliferation.
- 2000s—Rise of organised cybercrime and development of security frameworks.
- Present—Complex threats involving AI, social engineering and sophisticated attack strategies.
- Modern challenges include cloud computing, IoT devices and remote work security.

Question 6. What is defence-in-depth, and why is it important in cybersecurity?

Defence-in-depth is a multi-layered security approach that:

- Creates multiple lines of defence to increase difficulty for attackers
- Implements overlapping security controls at different system levels (network, application, data)
- Ensures that if one security measure fails, others remain to protect systems

- Combines various tools like firewalls, IPS, antivirus, access controls and encryption.

This approach is crucial because no single security measure is completely foolproof.

Question 7. Why is patch management critical to cybersecurity, and how should it be implemented?

Patch management is vital because:

- It addresses vulnerabilities and enhances functionality.
- It prevents exploitation of known weaknesses.

Patch management should be implemented by:

- Identifying and prioritising critical systems
- Conducting vulnerability scans
- Rolling out patches to critical systems first
- Applying updates to less critical systems
- Using automated tools for efficiency.

Question 8. What are the four primary categories of cyber threats?

The four primary categories are:

- Technology-focused threats—Direct attacks on digital infrastructure (malware, DoS attacks).
- People-focused threats—Social engineering tactics (phishing, identity theft).

- Adversarial threats—Deliberate attacks from nation-states, criminals or hacktivists.
- Non-adversarial threats—Unintentional security risks from human error or misconfigurations.

Question 9. What makes Multi-Factor Authentication (MFA) an effective security measure?

MFA is effective because it requires multiple verification methods:

- Something you know (password/PIN)
- Something you have (smartphone/security token)
- Something you are (biometric data).

This layered approach significantly strengthens access control compared to single-password systems.

Question 10. What are the key characteristics of the modern cyber threat landscape?

The modern cyber threat landscape is characterised by:

- Rapid Evolution—Threats advance alongside technological innovation.
- Expansive Scope—Affects all levels from infrastructure to individuals.
- Adaptive Nature—Attackers constantly evolve tactics to bypass security.
- Combination of technical and psychological techniques.

- Requires continuous monitoring and adaptation of defence strategies.

Chapter 2 - The Nature of Threats

Question 1. What is the foundational principle of cybersecurity discussed in the chapter, and why is it important to understand that complete identification and elimination of threats is impossible?

Complete threat elimination is impossible due to rapidly evolving technology and adversary tactics. This emphasises the need for continuous vigilance and adaptation in cybersecurity.

Question 2. How does human nature impact cybersecurity, and what role does fostering a culture of security awareness play?

Human error can compromise technical security measures. Creating a security-aware culture helps ensure staff follow protocols, reducing breach risks.

Question 3. Explain the difference between adversarial and non-adversarial threats with examples from the chapter.

Adversarial threats involve malicious actors using tactics like APTs and malware. Non-adversarial threats come from non-malicious sources like system failures or human errors.

Question 4. What are Advanced Persistent Threats (APTs), and how do they differ from other types of cyber threats?

APTs are sophisticated, state-sponsored attacks focused on maintaining covert network access for espionage. They prioritise stealth over immediate disruption.

Question 5. Describe the impact and significance of the MOVEit attack discussed in the chapter.

The MOVEit attack by the Clop ransomware group in 2023 had massive impact, compromising 2,667 organisations and affecting 84 million individuals worldwide. Its significance lies in how Clop shifted from traditional encryption-based ransomware to a data theft and extortion strategy. This attack demonstrates the evolution of cyber threats and reinforces crucial security principles including rapid vulnerability patching, continuous monitoring and comprehensive incident response planning.

Question 6. What are the key components of a practical approach to understanding and managing cyber threats, as outlined in the chapter?

Key components include threat categorisation, assessment and intelligence gathering to efficiently allocate resources and implement countermeasures.

Question 7. How do insider threats pose a risk to organisations, and what measures can be taken to mitigate these threats?

Insider threats come from trusted individuals misusing access. Mitigation includes access controls, vetting, training and monitoring systems.

Question 8. What are some examples of physical and environmental threats, and why is it important to integrate physical security measures with cybersecurity protocols?

Physical threats include weather, disasters, sabotage and theft. Integration with cybersecurity ensures comprehensive protection against all threats.

Question 9. Explain the process of threat modelling and its significance in the cybersecurity lifecycle.

Threat modelling identifies and mitigates potential system risks through design, break, fix and verify phases, ensuring comprehensive security throughout the lifecycle.

Question 10. How do threat assessments help in managing cyber threats, and what steps are involved in conducting one?

Threat assessments identify and prioritise threats based on likelihood and impact. Steps include asset identification, threat recognition, impact evaluation and countermeasure implementation.

Chapter 3 - Vulnerabilities

Question 1. What are the three interconnected pillars of cybersecurity discussed in the chapter?

The three pillars are threats, vulnerabilities and risks. Threats are malicious actions aimed at stealing, corrupting or disrupting systems; vulnerabilities are weaknesses that can be exploited; and risks represent the potential for loss when threats successfully exploit vulnerabilities.

Question 2. What makes zero-day vulnerabilities particularly dangerous?

Zero-day vulnerabilities are especially dangerous because they are previously unknown weaknesses that attackers can exploit before developers know they exist, creating a race against time. The Heartbleed bug is a notable example that affected OpenSSL, allowing attackers to steal sensitive information before a patch was developed.

Question 3. Name and explain three categories of software vulnerabilities discussed in the chapter.

Three categories include:

- Injection Vulnerabilities—Where malicious code is inserted into application inputs.
- Authentication and Session Management Weaknesses—Vulnerabilities in systems managing digital identity verification.
- Cross-Site Scripting (XSS) —Where malicious scripts are injected into trusted websites.

Question 4. Why are hardware vulnerabilities more challenging to address than software vulnerabilities?

Hardware vulnerabilities are more difficult to address because they often require physical replacement or firmware updates, which are costly and complex to deploy. Unlike software patches that can be quickly deployed, hardware fixes may require extensive testing and can potentially disrupt existing functionality.

Question 5. What are the main phases of the System Development Lifecycle (SDLC), and why is it essential for security?

The main phases are Requirements Analysis, Design, Implementation, Testing, Deployment and Maintenance. The SDLC is important because it provides a structured framework for creating and maintaining secure applications, with each phase presenting unique opportunities to address potential vulnerabilities.

Question 6. What is the difference between static code analysis and Dynamic Application Security Testing (DAST)?

Static code analysis examines source code for vulnerabilities without executing the program, while DAST involves testing the application while it's running to identify vulnerabilities that manifest during execution.

Question 7. What role do configuration vulnerabilities play in system security?

Configuration vulnerabilities arise from improper system settings, such as default passwords and unnecessary pre-enabled services. They are one of the most common causes of security breaches and require careful attention to system hardening practices and configuration management.

Question 8. What is system hardening and why is it important?

System hardening is the process of reducing the attack surface by configuring systems to minimise vulnerabilities. This includes disabling unnecessary services, changing default settings and applying regular patches and updates.

Question 9. Name three key vulnerability scanning tools mentioned in the chapter and their main features.

The tools are:

- Nessus—Known for comprehensive scanning capabilities and configuration audits.
- OpenVAS—An open-source solution offering extensive scanning and management features.
- Qualys—Provides cloud-based scanning solutions with seamless security platform integration.

Question 10. What is the purpose of responsible disclosure programs?

Responsible disclosure programs encourage researchers and ethical hackers to report vulnerabilities to affected organisations before making them public, allowing time for fixes to be developed. They often include clear reporting guidelines and may offer rewards through bug bounty programs.

Chapter 4 - Risk Management Essentials

Question 1. What is the relationship between threats, vulnerabilities and risk in cybersecurity?

Risk is calculated as Risk = Threats × Vulnerabilities, where threats are potential harmful events and vulnerabilities are exploitable weaknesses. When threats exploit vulnerabilities, they create risk—the potential for loss or damage to organisational assets.

Question 2. Describe the four phases of the risk management process and their key activities.

The four phases are:

- **Risk Identification**—Using threat modelling, vulnerability assessments and trend analysis.
- **Risk Assessment**—Evaluating likelihood and impact using qualitative/quantitative methods.
- **Risk Treatment**—Implementing avoidance, reduction, transfer or acceptance strategies.
- **Continuous Monitoring**—Ongoing evaluation of control effectiveness and emerging threats.

Question 3. What are the six core functions of the NIST Cybersecurity Framework (CSF)?

The six functions are:

- **Govern**—Aligning cybersecurity with business goals.
- **Identify**—Understanding assets, risks and vulnerabilities.
- **Protect**—Implementing safeguards and controls.
- **Detect**—Identifying cybersecurity incidents.
- **Respond**—Taking action on detected incidents.
- **Recover**—Maintaining resilience and restoring capabilities.

Question 4. Compare ISO/IEC 27001 and the ASD-ACSC Essential Eight frameworks in terms of their approach and focus.

ISO/IEC 27001 establishes a comprehensive Information Security Management System (ISMS) using a Plan-Do-Check-Act (PDCA) cycle, requiring extensive documentation and formal audits. It's globally recognised but resource intensive. The ASD-ACSC Essential Eight focuses on eight specific security controls (including application control, patching and MFA) that effectively mitigate common threats. It offers a pragmatic approach for organisations seeking significant security improvements with minimal investment but has a narrower scope than ISO/IEC 27001.

Question 5. What are the Essential Eight controls recommended by the ASD-ACSC?

The Essential Eight controls are:

1. Application Control.

2. Application Patching.

3. Microsoft Office Macro Restrictions.

4. User Application Hardening.

5. Restricted Administrative Privileges.

6. Operating System Patching.

7. Multi-Factor Authentication.

8. Regular Backups.

Question 6. What key lessons were learned from the 2023 MGM Resorts cyberattack?

Key lessons included:

- Importance of Zero Trust Architecture (ZTA)
- Critical role of human behaviour in security
- Need for robust help desk verification protocols
- Value of network segmentation
- Necessity of tested offline backup systems
- Importance of comprehensive IRPs.

Question 7. How does AI/ML impact modern cybersecurity risk management?

AI/ML enhances threat detection and response capabilities by processing vast amounts of security data and identifying patterns and anomalies.

However, it also introduces new risks through algorithmic biases and potential manipulation of AI models by adversaries.

Question 8. What are the four main risk treatment strategies and when should each be used?

The strategies are:

- Avoidance—Eliminating high-risk activities.
- Reduction—Implementing controls to decrease likelihood or impact.
- Transfer—Shifting risk to third parties (e.g., cyber insurance).
- Acceptance—Taking on low-severity risks when mitigation costs exceed benefits.

Question 9. How does Zero Trust Architecture (ZTA) differ from traditional security approaches?

Zero Trust replaces traditional perimeter-based "trust but verify" approaches with a "never trust, always verify" model, requiring continuous validation of every user and device regardless of location or previous authentication.

Question 10. What core practices should organisations adopt for ongoing risk management success?

Key practices include:

- Establishing a Risk Management Committee
- Implementing continuous feedback loops
- Leveraging automation where appropriate

- Staying informed of emerging threats
- Investing in ongoing training.

Chapter 5 - Risk Management Essentials Part 2—Frameworks and Standards

Question 1. What is the fundamental difference between cybersecurity frameworks and standards, and what roles do they play in risk management?

Frameworks provide flexible guidelines that help organisations systematically identify, protect, detect, respond to and recover from cyber threats, while standards define specific technical requirements and measurable criteria for security controls. Together, they create a foundation for strengthening security postures, establishing a common language and enabling risk management.

Question 2. What are the six essential functions of the NIST Cybersecurity Framework (CSF), and how do they work together to create a resilient security strategy?

The NIST CSF's six essential functions, Govern, Identify, Protect, Detect, Respond and Recover, work together to create a resilient security strategy by establishing an interconnected approach that aligns cybersecurity with organisational goals while managing risks throughout the security lifecycle.

Question 3. What are the seven steps of the NIST Risk Management Framework (RMF), and how does it differ from the CSF?

The NIST RMF's seven steps are, prepare, categorise, select, implement, assess, authorise and monitor. They differ from the CSF by providing a more structured, compliance-focused process specifically designed for System

Development Lifecycle (SDLC), while the CSF offers a flexible, outcome-driven approach applicable to organisations at any maturity level.

Question 4. What are the main components of the Australian ASD-ACSC Information Security Manual (ISM), and what advantages does it offer Australian organisations?

The ISM consists of seven components: risk management framework, security controls, compliance and assurance, incident response and management, governance and accountability, security awareness program and continuous improvement process. Its advantages include being government-endorsed and aligned with Australian standards.

Question 5. What are the eight key strategies of the Australian Essential Eight framework, and why is it particularly beneficial for organisations with limited resources?

The eight strategies are: Application Control, Application Patching, Restrict Microsoft Office Macros, User Application Hardening, Restrict Administrative Privileges, Patch Operating Systems, Multi-Factor Authentication and Regular Backups. It's beneficial for resource-limited organisations because it provides a practical and effective cybersecurity baseline with a clear maturity model.

Question 6. How does COBIT differ from other cybersecurity frameworks, and what are its key components for effective IT governance?

COBIT differs by focusing specifically on IT governance and aligning technology with business objectives. Its key components are: Governance and Management Framework, Process Model, Business-IT Alignment and Performance Measurement. This approach bridges the gap between business needs and IT capabilities.

Question 7. What are the core principles of the CIS Controls framework, and how does its Implementation Group (IG) structure benefit organisations of varying sizes?

The core principles are prioritisation of security measures, defence-in-depth approach and continuous improvement. The framework's Implementation Groups (IGs) allow organisations to adopt security measures progressively based on their size, resources and risk exposure, ensuring effective defences regardless of maturity level.

Question 8. What are the five interconnected components of ISO 31000, and how does it differ from ISO/IEC 27001?

ISO 31000's components are context establishment, risk identification, risk assessment, risk treatment, and monitoring and review. While ISO 31000 provides a general risk management framework applicable across all types of risks, ISO/IEC 27001 specifically focuses on information security management systems (ISMS).

Question 9. What key factors should organisations consider when selecting a cybersecurity framework, particularly in the Australian context?

Organisations should consider: Industry and regulatory requirements, alignment with ACSC Guidance (incorporating Essential Eight and other ACSC strategies), organisational size and resources, and risk profile.

Question 10. What are the essential steps for successfully implementing a cybersecurity framework, and why is each step important?

The essential steps are conduct risk assessments, develop a cybersecurity, identify gaps and prioritise actions, implement controls, train staff and conduct regular monitoring. Each step is important because it builds upon the previous one to create a comprehensive, adaptable security program that addresses specific organisational risks and regulatory requirements.

Chapter 6 - Confidentiality

Question 1. What is the primary role of confidentiality in cybersecurity?

The primary role of confidentiality in cybersecurity is to ensure that sensitive information is protected from unauthorised access and disclosure. This protection is crucial for maintaining privacy and safeguarding personal data, intellectual property and proprietary business information.

Question 2. Explain the legal implications of data breaches under *The General Data Protection Regulation* (GDPR) and *The Health Insurance Portability and Accountability Act* (HIPAA).

Under the GDPR, organisations face fines of up to €20 million or 4% of annual global turnover for data breaches. HIPAA imposes strict penalties for healthcare data breaches in the United States.

Question 3. Describe the CIA Triad and the interdependence between its three components: Confidentiality, Integrity and Availability.

The CIA Triad consists of Confidentiality, Integrity and Availability. Confidentiality ensures authorised access only; Integrity guarantees data accuracy; and Availability ensures authorised users can access resources when needed. These components are interdependent, as a breach in one often affects the others.

Question 4. Provide real-world examples of data breaches and discuss their impacts on organisations and individuals.

The Equifax breach (2017) exposed 147 million individuals' personal information, causing financial losses and reputational damage. The Yahoo breaches affected three billion accounts, leading to lost user trust and decreased market value.

Question 5. Why is data confidentiality critical for both individuals and organisations?

For individuals, it protects against identity theft and financial loss. For organisations, it safeguards trade secrets, intellectual property and customer data, preventing financial losses and reputational damage.

Question 6. Identify and describe the various types of threats to data confidentiality, including unauthorised access, interception, social engineering and physical security breaches.

- Unauthorised access—Hackers and malware gaining unauthorised system entry.
- Interception—Data transmission eavesdropping through Man-In-The-Middle (MITM) attacks.
- Social engineering—Manipulating individuals through phishing and pretexting.
- Physical security breaches—Device theft or unauthorised physical access.

Question 7. What role does human error play in compromising data confidentiality, and how can it be mitigated?

Human error contributes through mishandling sensitive information and weak passwords. Mitigation involves security awareness training and implementing strong password policies.

Question 8. Discuss the different methods for protecting data confidentiality, including access control, data classification and encryption.

- Access control—Using RBAC and MFA.
- Data classification—Prioritising protection based on sensitivity.
- Encryption—Securing data during storage and transmission.
- Data minimisation—Collecting only necessary data.

Question 9. What are the key legal and regulatory frameworks governing data confidentiality, and what are their main requirements?

- APPs—13 principles in *The Privacy Act 1988* for handling personal information.
- GDPR—Requires data minimisation and breach notification within 72 hours.
- HIPAA—Mandates protection standards for healthcare information.

Question 10. What are the essential components of a comprehensive confidentiality program and why is continuous improvement necessary?

A comprehensive confidentiality program comprises six essential components:

- Security Policy—Clear data handling procedures and access controls with dual-authorisation requirements.
- Employee Training—Regular security awareness education covering phishing recognition and data handling best practices.
- Incident Response Plan (IRP)—Structured procedures for detecting, containing and recovering from security incidents.
- Data Loss Prevention (DLP)—Monitoring and blocking unauthorised data transfers through network scanning and policy enforcement.
- Auditing and Monitoring—Continuous oversight using SIEM systems to identify vulnerabilities and suspicious activities.
- Continuous Improvement—Ongoing assessment addressing emerging threats and evolving security technologies.

Continuous improvement is necessary because security is an ongoing process. Organisations must adapt to evolving cyber threats and invest in new technologies to maintain effective protection.

Chapter 7 - Integrity

Question 1. What is data integrity, and why is it a critical component of the CIA Triad?

Data integrity ensures information remains accurate, consistent and unaltered within the CIA Triad.

Question 2. What are the primary consequences of compromised data integrity?

Compromised data integrity can lead to financial losses, operational disruptions and reputational damage.

Question 3. Can you provide a real-world example where data integrity was compromised and describe the impact?

The NotPetya attack caused widespread data corruption and operational disruptions, resulting in billions in damages.

Question 4. What are some common threats to data integrity, and how do they differ?

Common threats include cyberattacks, hardware failures, software bugs, human error and insider threats, each requiring specific mitigation approaches.

Question 5. How does hashing help in maintaining data integrity?

Hashing generates a unique data fingerprint to detect unauthorised changes through hash value comparison.

Question 6. What role do digital signatures play in ensuring data integrity?

Digital signatures verify data authenticity and origin, providing non-repudiation and tamper detection.

Question 7. Why are checksums necessary for data integrity, and what are their limitations?

Checksums detect data errors through redundant values but are less secure against deliberate tampering than hashing.

Question 8. What is the significance of data logging and auditing in maintaining data integrity?

Data logging and auditing track changes and activity to monitor and detect anomalies.

Question 9. How do access controls contribute to data integrity?

Access controls prevent unauthorised modifications by restricting data access based on predefined policies.

Question 10. What are the legal and regulatory requirements related to data integrity, and why are they important?

Regulations like APPs, HIPAA and PCI-DSS mandate data integrity controls to protect information and ensure legal compliance.

Chapter 8 - Availability

Question 1. What is availability, and why is it considered a core component of the CIA Triad in cybersecurity?

Availability ensures that systems, data and services are accessible when needed, which is crucial for maintaining business operations and continuous access to resources.

Question 2. How can downtime impact businesses financially and reputationally? Provide examples.

Downtime can cause financial losses through halted operations and lost revenue while damaging reputation and customer trust. Examples include the 2016 Dyn DDoS attack and the 2017 WannaCry ransomware attack.

Question 3. Explain the difference between High Availability (HA) and Disaster Recovery (DR). How do they complement each other?

HA ensures uninterrupted operation through redundancy and load balancing. DR enables quick recovery from disruptions through backups and recovery plans. Together, they provide protection and rapid recovery capabilities.

Question 4. Identify and describe the main types of threats to availability.

Main threats include cyberattacks (DoS, ransomware), hardware failures (equipment malfunctions, power outages), software bugs (system crashes) and human error (misconfigurations).

Question 5. Why is it essential to have a multi-layered defence approach to address the increasing complexity of threats to availability?

A multi-layered approach is necessary because modern threats are sophisticated and often combine different tactics, requiring comprehensive protection measures.

Question 6. What role do insider threats play in disrupting availability, and how can organisations mitigate these threats?

Insider threats can significantly disrupt availability through malicious or unintentional actions. Mitigation includes access controls, monitoring and employee training.

Question 7. Discuss the significance of network security controls like firewalls, IDS and IPS in maintaining network availability.

These controls protect against unauthorised access and attacks by filtering traffic, monitoring suspicious activity and blocking threats.

Question 8. How does continuous network monitoring contribute to maintaining availability?

Continuous monitoring provides real-time visibility into network performance and detects anomalies before they escalate into major issues.

Question 9. Explain the importance of Service Level Agreements (SLAs) in cloud availability and what factors should be considered when evaluating SLAs.

SLAs define service terms between providers and organisations, including uptime guarantees and recovery times. Consider provider history and financial implications when evaluating.

Question 10. How does the shared responsibility model work in cloud environments, and why is understanding this model critical for ensuring availability?

The shared responsibility model defines which security and availability tasks fall to the cloud provider versus the customer. Providers typically secure the underlying infrastructure (hardware, networking, data centres), while organisations remain responsible for their data, applications and access controls. The division varies by service model (SaaS, PaaS, IaaS).

Chapter 9 - CTI

Question 1. What is Cyber Threat Intelligence (CTI), and why is it critical in modern cybersecurity?

CTI involves collecting and analysing information about cyber threats to anticipate and respond to incidents effectively.

Question 2. Explain the shift from reactive to proactive defence strategies in cybersecurity.

Reactive strategies respond after incidents occur, while proactive strategies use CTI to anticipate and prevent threats.

Question 3. Describe the stages of the intelligence cycle adapted for CTI.

The intelligence cycle includes Planning, Collection, Processing and Analysis, Dissemination and Feedback.

Question 4. What are the different types of CTI based on source and focus?

Types include Strategic, Tactical, Technical, Open-Source and Commercial Threat Intelligence.

Question 5. How can organisations use CTI in threat hunting?

CTI provides patterns and indicators to help proactively search for potential attacks within networks.

Question 6. In what ways does CTI enhance incident response efforts?

CTI provides information about attackers and methods, enabling more effective incident identification and mitigation.

Question 7. What role does CTI play in security awareness training?

CTI informs awareness programs with current threat information to help employees recognise risks.

Question 8. Discuss how CTI is integrated into vulnerability management.

CTI helps prioritise patching by identifying actively exploited vulnerabilities.

Question 9. Identify and describe two major challenges organisations face when implementing CTI.

Major challenges include data overload and converting raw data into actionable insights.

Question 10. Analyse how the SolarWinds cyberattack of 2020 demonstrates the importance of CTI in detecting and responding to sophisticated threats.

The SolarWinds attack demonstrated CTI's importance through enabling detection, facilitating rapid response coordination through shared IOCs and

TTPs, revealing supply chain vulnerabilities and informing defensive improvements for future protection.

Part II

Chapter 10 - Governance

Question 1. What is cybersecurity governance, and why is it critical in managing cyber risks for organisations?

Cybersecurity governance includes policies and structures guiding security management, ensuring risks are identified and mitigated while aligning with business objectives.

Question 2. How has the cybersecurity landscape shifted in recent years, and what impact does this have on governance?

Technological advances like IoT and cloud have increased attack sophistication and expanded attack surfaces, requiring more robust governance.

Question 3. Explain the core principles for establishing effective cybersecurity governance.

Core principles include executive commitment, risk management framework, clear structure, defined accountability and regulatory compliance.

Question 4. What are the benefits of having a cybersecurity governance framework?

A framework provides structured security management, facilitates communication and ensures consistent practices throughout the organisation.

Question 5. What are the key components of an effective cybersecurity governance model, and how do they interact?

A governance model includes structure (roles and committees), accountability mechanisms, cultural dynamics and infrastructure (policies and procedures) that together enable effective security oversight and risk management.

Question 6. How do cybersecurity policies translate governance principles into operational practices?

Policies provide guidelines for IT resource use, incident reporting and response protocols for consistent governance application.

Question 7. What types of cybersecurity policies are crucial for organisations, and what do they entail?

Crucial policies include Acceptable Use, Password, Incident Response and Data Breach Notification policies.

Question 8. Why is fostering a culture of cybersecurity awareness important, and what are its key elements?

Awareness empowers employees to handle threats effectively through leadership commitment, training and shared responsibility.

Question 9. What steps are involved in the effective implementation and ongoing maintenance of cybersecurity governance?

Steps include framework dissemination, training, compliance monitoring and continuous improvement.

Question 10. Why is continuous improvement essential in cybersecurity governance, and how can organisations achieve it?

Continuous improvement addresses evolving threats through regular framework updates and incorporation of lessons learned.

Chapter 11 - Identification

Question 1. What does the Identify principle refer to in cybersecurity, and why is it fundamental?

Identification refers to the systematic process of recognising and cataloguing all assets within an organisation's digital ecosystem, forming the foundation for all subsequent security measures.

Question 2. What are the four essential components of the Identify principle?

The four components are asset management, risk assessment, governance and supply chain security.

Question 3. What was the WannaCry ransomware outbreak of 2017 an example of?

WannaCry exemplifies the consequences of identification failures, where organisations failed to identify unpatched Windows systems within their networks, resulting in billions in damages.

Question 4. What major categories should be included in a comprehensive asset inventory?

A comprehensive inventory should include hardware assets, software assets, cloud services, data assets and personnel.

Question 5. What is the CIA Triad used for in asset classification?

The CIA Triad (Confidentiality, Integrity, Availability) is used to evaluate and classify assets based on their importance to organisational operations and potential impact if compromised.

Question 6. What roles do different teams play in asset management?

IT teams handle provisioning and decommissioning, security teams identify vulnerabilities and enforce controls, while business units align assets with operational objectives.

Question 7. What is shadow IT and why is it a significant risk?

Shadow IT refers to unauthorised applications, devices or cloud services introduced without security oversight, creating dangerous blind spots that increase breach risks.

Question 8. How does risk identification differ from risk assessment?

Risk identification focuses on systematically discovering potential risks before they escalate, while risk assessment evaluates the likelihood and impact of those threats.

Question 9. What major cybersecurity incidents demonstrate the importance of supply chain security?

The 2020 SolarWinds attack and 2021 Kaseya VSA attack demonstrate how compromises in the supply chain can allow threat actors to infiltrate numerous organisations simultaneously.

Question 10. How are Zero Trust and AI transforming asset discovery?

Zero Trust redefines asset identification through continuous verification of users, devices and workloads, while AI-driven tools automatically detect rogue devices, anomalous connections and shadow IT without manual intervention.

Chapter 12 - Protection

Question 1. What is the Protection principle in cybersecurity frameworks, and how does it function?

Protection transforms abstract security strategies into concrete defences, implementing safeguards for critical infrastructure while working alongside other framework components to create a comprehensive security posture.

Question 2. What is the CIA Triad, and why is it important to the Protection principle?

The CIA Triad—Confidentiality, Integrity and Availability—provides the essential framework for evaluating security controls, guiding organisations in prioritising protection measures and determining how they should be balanced.

Question 3. How has the Protection strategy evolved from traditional approaches?

Protection has evolved from traditional perimeter defences toward Zero Trust Architecture (ZTA), enforcing "never trust, always verify" principles that require continuous validation for every user and device attempting to access resources.

Question 4. What is Identity and Access and Privileged Access Management (IAM and PAM), and how does it enhance security?

IAM and PAM ensure only authorised individuals access specific resources using the Authentication, Authorisation and Accounting (AAA) model while implementing least privilege principles to reduce the attack surface.

Question 5. What are the three states in which data requires encryption protection?

Data requires encryption across three distinct states: at rest (stored data), in transit (data traversing networks) and in use (data being processed in memory).

Question 6. Why is network segmentation important in security architecture?

Network segmentation divides large networks into smaller, isolated sections to contain breaches and restrict lateral movement, preventing attackers from easily traversing the network after initial compromise.

Question 7. What primary security failure led to the 2022 Optus data breach?

The Optus breach resulted from inadequate API security controls, specifically an unauthenticated API endpoint that allowed attackers to bypass perimeter defences and access customer data directly.

Question 8. How does system hardening contribute to security?

System hardening transforms vulnerable default configurations into strengthened defensive positions by reducing the attack surface through elimination of unnecessary services, removal of default credentials and disabling unused features.

Question 9. What role do humans play in the Protection principle?

Humans represent both the greatest vulnerability and most adaptable defence mechanism; cybersecurity awareness training transforms employees from potential liabilities into vigilant human firewalls capable of identifying threats.

Question 10. Why is secure hardware lifecycle management necessary?

Secure hardware lifecycle management ensures appropriate handling of hardware from acquisition through disposal, preventing data leakage and malware introduction that could establish footholds in ICT infrastructure.

Chapter 13 - Detection

Question 1: What is the Detection principle in cybersecurity and why is it important?

Detection identifies potential threats before they escalate, acknowledging that some threats will bypass protective measures. Early identification dramatically reduces potential damage and transforms cybersecurity into a dynamic, responsive function.

Question 2: What are the three fundamental pillars of effective cybersecurity detection?

The three pillars are monitoring (continuous surveillance), analysis (converting data into actionable insights) and alerting (ensuring identified threats receive appropriate attention).

Question 3: According to IBM security reports, how long do organisations typically take to identify breaches?

Organisations take an average of 204 days to identify breaches.

Question 4: What is Network Traffic Analysis (NTA) and what techniques does it use?

NTA provides insights into threat activities through Deep Packet Inspection (DPI) (examining packet contents) and flow analysis (examining traffic patterns, communication relationships and data volume changes).

Question 5: What are Indicators of Compromise (IoCs) and what categories do they fall into?

IoCs are telltale signs of malicious activity that include file-based (hashes, names), network-based (IP addresses, domains), behavioural (unusual activity) and artifact-based (registry keys, log modifications) indicators.

Question 6: What detection challenge did the 2013 Target data breach illustrate?

The Target breach illustrated 'alert fatigue', where multiple automated alerts warning of a potential intrusion were missed due to the overwhelming volume of false alarms.

Question 7: What is Living off the Land (LotL) and why is it challenging for detection systems?

LotL attacks weaponize legitimate system tools like PowerShell or Windows Management Instrumentation (WMI) to execute malicious actions without triggering signature-based alerts.

Question 8: How does User and Behavioural Analytics (UBA) enhance detection capabilities?

UBA establishes normal activity patterns, detects anomalies and intercepts potential threats by focusing on human behaviour rather than relying solely on static rules.

Question 9: Why is physical environment monitoring important for cybersecurity?

Physical monitoring ensures unauthorised attempts to access physical assets like servers and network equipment are promptly detected, preventing severe breaches.

Question 10: What approach to supply chain security assumes breaches will occur and requires verification for all third-party interactions?

The Zero Trust approach assumes breaches will occur and requires verification for all third-party interactions, never automatically trusting any vendor.

Chapter 14 - Response

Question 1. What is the Respond principle in cybersecurity and why is it important?

The Respond principle focuses on coordinated actions taken after incident identification, including containment, eradication and recovery strategies that minimise impact and ensure rapid restoration of normal operations.

Question 2. What are the key components of an effective incident management framework?

Key components include defined lifecycle processes, categorisation and prioritisation systems, clear stakeholder roles, response playbooks, automation capabilities and measurement mechanisms like MTTD and MTTR.

Question 3. What essential components should be included in an Incident Response Plan (IRP)?

An IRP should include clearly defined roles and responsibilities, communication protocols, tiered escalation procedures and detailed recovery steps for system restoration and forensic analysis.

Question 4. How should organisations approach incident triage?

Organisations should identify potential incidents through monitoring systems, prioritise incidents based on data sensitivity and business impact, validate threats to eliminate false positives and preserve evidence while notifying key stakeholders.

Question 5. What factors should be considered when categorising and prioritising incidents?

Organisations should consider impacts on confidentiality, integrity and availability; use frameworks like NIST SP 800-61; employ quantitative models like CVSS; and account for regulatory requirements and organisational risk.

Question 6. What containment strategies can be employed during cyber incidents?

Containment strategies include immediate isolation versus delayed containment for intelligence gathering, network-level controls like firewall rules, endpoint-level actions like account disabling and targeted restrictions that maintain operational continuity.

Question 7. How should digital evidence be collected and preserved during incident response?

Evidence should be collected based on volatility (RAM contents first, then disk images), in compliance with legal frameworks, maintaining chain of custody documentation and using secure handling with cryptographic hashing and write-blocking.

Question 8. What analytical approaches help identify the root causes of cyber incidents?

Root Cause Analysis (RCA) frameworks (Five Whys, Fishbone Diagrams), event correlation, timeline construction, behavioural analysis and memory/disk forensics help identify initial compromise points and attack patterns.

Question 9. What internal communication strategies are essential during a cybersecurity incident?

Internal communication requires clear protocols with regular executive briefings, cross-functional team coordination with defined roles and employee communications that provide actionable guidance and dedicated support channels.

Question 10. What external communication responsibilities do organisations have during cyber incidents?

Organisations must comply with regulatory reporting requirements like the GDPR and NDB, communicate transparently with affected customers, manage media relations with a prepared spokesperson and highlight recovery efforts to rebuild confidence.

Chapter 15 - Recover

Question 1. What is the Recover principle in cybersecurity, and how does it differ from disaster recovery?

Recover is the process of restoring systems after a security incident, extending beyond technical restoration to rebuilding trust and strengthening security posture. Unlike disaster recovery, it specifically addresses malicious activities, requiring evidence preservation and consideration of attacker persistence.

Question 2. What are RTOs and RPOs, and why are they important for recovery planning?

Recovery Time Objectives (RTOs) specify maximum acceptable downtime, while Recovery Point Objectives (RPOs) define acceptable data loss. These metrics guide investment decisions, set stakeholder expectations and provide benchmarks for evaluating recovery performance.

Question 3. What is a Business Impact Analysis (BIA), and what role does it play in recovery planning?

BIA identifies critical business functions and evaluates potential financial, operational and reputational consequences of disruptions, forming the foundation for prioritising recovery efforts and resource allocation during crises.

Question 4. Explain the 3-2-1 Backup Strategy and why it's effective for cybersecurity recovery?

The 3-2-1 Strategy maintains three copies of data on two different storage media with one copy stored offsite, creating multiple recovery options and protecting against various failure scenarios including ransomware attacks.

Question 5. What are the differences between hot, warm and cold recovery sites?

Hot sites maintain fully configured environments with real-time replication for near-immediate failover; warm sites require only data restoration and final configuration; cold sites provide basic infrastructure without pre-installed systems for more economical recovery when longer timeframes are acceptable.

Question 6. How do cloud-based recovery environments compare to traditional on-premises recovery approaches?

Cloud-based recovery offers flexibility and scalability without procurement delays, allowing dynamic resource provisioning compared to traditional approaches that require extensive time for physical infrastructure rebuilding and configuration.

Question 7. Why is communication critical during recovery, and what phases should it address?

Communication maintains coordination and stakeholder trust, spanning three phases: pre-incident planning with defined roles and templates, real-time updates during incidents and post-incident trust rebuilding through transparent reporting.

Question 8. What is an After-Action Review (AAR), and how does it contribute to organisational resilience?

An AAR is a structured analysis examining what happened, why it happened, how teams responded and what should change, transforming incidents into learning opportunities that systematically strengthen security programs through improvement implementation.

Question 9. What metrics can be used to evaluate recovery effectiveness?

Recovery effectiveness metrics include capability metrics (percentage of systems with documented procedures), performance metrics (actual recovery times compared to RTOs) and stakeholder-focused metrics (customer satisfaction following recovery).

Question 10. What challenges do modern technology environments present for recovery efforts?

Modern environments create challenges through architectural complexity, tight interdependencies between systems, cloud-native applications with third-party dependencies and microservice architectures requiring complex recovery sequencing.

Part III

Chapter 16 - Legal and Ethical

Question 1: Why is understanding the relationship between cybersecurity, legal frameworks and ethical considerations important?

Legal and ethical frameworks establish operational boundaries for security measures, ensuring they protect systems without infringing on individual rights or violating ethical standards.

Question 2: What are the three core ethical principles in cybersecurity discussed in the chapter?

The three core ethical principles are confidentiality (protecting sensitive data from unauthorised access), privacy (balancing security with respect for individual privacy rights) and transparency/disclosure (honestly informing stakeholders about breaches).

Question 3: What does Australia's Notifiable Data Breaches (NDB) scheme require?

The NDB scheme requires organisations to report data breaches to the Office of the Australian Information Commissioner and affected individuals when a breach could result in serious harm, as soon as practicable after discovery.

Question 4: How has breach notification law impacted cybersecurity practices?

Breach notification requirements have motivated organisations to implement stronger security measures, enhance incident response capabilities and adopt proactive approaches to prevent public scrutiny and reputational damage.

Question 5: What distinguishes ethical hacking from criminal hacking activities?

Ethical hacking operates within specific parameters including explicit authorisation, defined scope and formal agreements, converting potentially illegal activities into valuable security services that enhance organisational defences.

Question 6: What key Australian legislation governs cybersecurity practices?

Key Australian legislation includes *The Privacy Act 1988, The Cyber Security Act (CSA) 2024, The Security of Critical Infrastructure (SOCI) Act 2018, The Telecommunications Act 1997* and *The Criminal Code Act 1995*.

Question 7: What is the *Budapest Convention on Cybercrime* and why is it significant?

The *Budapest Convention* is the most significant international treaty that establishes common definitions of cybercrime offenses, harmonises national laws and creates mechanisms for international cooperation, though its effectiveness is limited by the absence of major cyber powers.

Question 8: What were the key components of Toll Group's response to their 2020 ransomware attack?

Toll Group contained the incident by disconnecting affected systems, refused to pay the ransom, restored systems from secured backups, enlisted specialised cybersecurity firms, notified authorities, maintained operations through manual processes and provided regular customer updates.

Question 9: What elements should be included in a cybersecurity professional's legal and ethical toolkit?

A comprehensive toolkit includes documentation and evidence preservation protocols, effective collaboration with legal counsel, ethical decision-making frameworks, current legal knowledge maintenance and a comprehensive legal and ethical decision checklist.

Question 10: What challenges exist in international cybersecurity law enforcement?

Challenges include the borderless nature of cyberspace versus geographical jurisdictions, differing legal definitions and procedures across countries, extradition difficulties with 'safe haven' nations, and cultural and jurisdictional differences in approaches to cybersecurity governance.

Chapter 17 - Metrics That Matter

Question 1. What is the difference between metrics and KPIs in cybersecurity?

Metrics are quantifiable measures tracking specific security processes, while KPIs operate at a strategic level aligning with broader organisational goals and indicating meaningful progress.

Question 2. Why is data-driven decision-making important in cybersecurity assessment?

Data-driven decision-making provides an objective foundation for accurately understanding security status, avoiding cognitive biases of intuition-based approaches and identifying specific areas for improvement.

Question 3. What are the three critical areas covered by core cybersecurity metrics?

Core cybersecurity metrics cover security posture (preparedness against threats), operations (efficiency of security processes) and business impact (connection to organisational outcomes).

Question 4. Name three key security posture metrics discussed in the chapter.

Important security posture metrics include vulnerability identification and remediation rates, patching frequency and unauthorised access attempts.

Question 5. What do the acronyms MTTD, MTTR and MTTC represent and why are they important?

They represent Mean Time to Detect, Mean Time to Respond and Mean Time to Contain—critical indicators of an organisation's incident response capabilities and efficiency.

Question 6. How can organisations measure the financial impact of security incidents?

Organisations can estimate direct costs (recovery, legal fees, fines) and indirect costs (revenue loss, reputational damage) and calculate potential ROI for security investments.

Question 7. What criteria should be considered when selecting appropriate KPIs?

KPIs should align with security goals and risk profiles, be limited to a manageable set, and be measurable, actionable and relevant to stakeholders.

Question 8. What is the process for using KPIs to drive continuous improvement?

The process involves establishing baseline metrics, setting realistic targets, tracking trends regularly and benchmarking against industry standards.

Question 9. What challenges do organisations face in cybersecurity data collection and analysis?

Challenges include managing overwhelming data volumes, ensuring data quality, implementing advanced analytics techniques and delivering real-time analysis.

Question 10. How can organisations balance security and usability effectively?

Organisations can implement a risk-based approach with proportionate security measures aligned to identified risk levels and use metrics like Key Risk Indicators (KRIs) to measure risk tolerance.

Chapter 18 - The Human Firewall

Question 1. What is the 'human firewall' and why is it important in cybersecurity defence?

The 'human firewall' refers to people as the first line of defence against cyber threats; when employees are trained to recognise and respond to security risks, they transform from potential vulnerabilities into active defenders of an organisation's digital infrastructure.

Question 2. How does security awareness training impact an organisation's vulnerability to phishing attacks?

According to Proofpoint data, organisations that increased phishing training by 52% saw successful spear-phishing attacks drop from 88% in 2022 to 56% in 2023, demonstrating the measurable impact of consistent training.

Question 3. What is the first step in creating an effective security awareness training program?

Begin with a thorough needs assessment to identify your organisation's unique security vulnerabilities, which reveals knowledge gaps, highlights departmental risk variations and ensures materials directly apply to employees' specific roles.

Question 4. What delivery methods are recommended for effective security awareness training?

Effective delivery combines instructor-led workshops for complex topics with self-paced digital modules, concise microlearning units and simulation platforms that allow employees to safely encounter and respond to simulated threats.

Question 5. What are the five key topics that should be covered in comprehensive security training?

The five key topics are: phishing and social engineering, password security and authentication, mobile device security, cloud security, and insider threats and data protection.

Question 6. How can organisations measure the effectiveness of their security awareness training?

Effectiveness can be measured through phishing simulation results, security incident metrics, knowledge assessments and behavioural change indicators that together provide both quantitative and qualitative insights.

Question 7. Why is it important to provide targeted training based on audience characteristics?

Employees have different roles, access levels and learning styles; personalised training ensures content remains relevant and practical for each group, with appropriate depth based on their specific security responsibilities.

Question 8. What engagement methods help ensure knowledge retention in security training?

Effective knowledge retention methods include scenario-based learning, phishing simulations, role-playing exercises, microlearning with short videos and gamification with points, badges and leaderboards.

Question 9. What are the essential content pillars of security awareness training?

Essential content pillars include password hygiene, phishing recognition, secure browsing habits, data security principles, BYOD policies, incident reporting and security policies.

Question 10. Why is continuous reinforcement important in security awareness programs?

Cybersecurity threats constantly evolve, making continuous education vital; regular security awareness campaigns, training refreshers and varied communication formats help keep employees updated on the latest threats and maintain security as a priority despite competing demands.

Chapter 19 – From Learner to Leader

Question 1. What are the three interconnected elements required for understanding cybersecurity principles?

Threats (malicious actors and techniques), vulnerabilities (weaknesses in systems) and risks (potential for harm when threats exploit vulnerabilities).

Question 2. What are the three components of the CIA Triad and what does each protect?

Confidentiality (ensures sensitive information is only accessible to authorised persons), Integrity (maintains accuracy and trustworthiness of data) and Availability (guarantees information is accessible when needed).

Question 3. What are the six essential principles that form the cybersecurity lifecycle framework?

Governance, Identification, Protection, Detection, Response and Recovery.

Question 4. What are the two main career paths available to mid-career cybersecurity professionals?

Technical specialist path (developing elite expertise in specific security domains) and management track (focusing on strategic security leadership and business alignment).

Question 5. Why is continuous learning essential in cybersecurity?

Because threats change rapidly, technologies evolve and attackers continue to innovate, making it necessary to maintain a growth mindset and stay informed about latest developments.

Question 6. What are effective ways to engage with the cybersecurity community for professional growth?

Attending conferences (like DEF CON, Black Hat, BSides), participating in virtual conferences/webinars and engaging in Capture The Flag (CTF) challenges or bug bounty programs.

Question 7. How should a personal cybersecurity learning plan be structured?

With knowledge objectives, practical exercises, resources, verification methods and realistic timeframes for each learning module.

Question 8. How can security professionals effectively communicate with non-technical stakeholders?

By avoiding jargon, framing security in business terms, using visual communication/metaphors and tailoring the approach to the audience's specific concerns.

Question 9. What is the recommended approach for building business cases for security investments?

Quantify risks financially, research industry benchmarks, structure proposals using return on security investment frameworks and present tiered implementation options.

Question 10. What emerging specialised cybersecurity domains are creating new career opportunities?

Quantum security, autonomous system security, healthcare cybersecurity and AI ethics and security auditing.

About the Author

Alec Sklepic is a seasoned ICT security professional with over 25 years of experience in the Australian Federal Government, including the Australian Department of Defence and the Australian Signals Directorate (ASD) within the Australian Cyber Security Centre (ACSC). His practical approach to cybersecurity encompasses information assurance, security assessments, risk management and policy development. View Alec's professional profile at www.linkedin.com/in/alec-s-cyber-security.

Today, as Principal Consultant of his cybersecurity business, Alec works directly with government agencies, space service providers and other enterprises to strengthen security frameworks and conduct comprehensive security assessments. His credentials include being an ASD-endorsed Information Security Registered Assessors Program (IRAP) assessor, complemented by professional certifications such as Certified Information Systems Security Professional (CISSP) and ISO/IEC 27001 Lead Auditor.

With a foundation in IT, Alec's career spans system administration, project leadership and business analysis. This book emerges from his frontline cybersecurity experience and is specifically designed to guide beginners and early-career professionals through essential cybersecurity principles. His aim is to transform complex security concepts into accessible and actionable knowledge, equipping you with practical skills you can apply immediately in your professional journey while helping you navigate the increasingly critical field of cybersecurity.

Acknowledgements

When I began this project, I was under the illusion that I could complete it alone. With years of experience in cybersecurity, I believed that combining my knowledge with a bit of extra research would be enough. I quickly learned otherwise. Writing and publishing a book is far from a solo endeavour—it's a collaborative process that requires insight, support, and encouragement from others. It's a team effort in every sense.

First and foremost, to my wife, Penny—thank you for your steady support, patience and belief in me throughout this journey. Your encouragement sparked the beginning of this book and sustained me through to the end. You carried me through moments of doubt and gave me the time and space to focus. I couldn't have done this without you.

Thanks also to Heather Moore, your curiosity and enthusiasm were a source of motivation. Your perspective helped shape the book and clarified its message. You reminded me how vital it is to make complex ideas accessible and engaging.

To the colleagues who offered their time, technical expertise and honest feedback—a special thank you. Your contributions sharpened the content, challenged assumptions and helped ensure this book is credible and practical. I'm especially grateful to Kim Knowles and, in particular, to Justin Baldock, whose detailed feedback went above and beyond. I truly appreciate their support and their generosity.

My sincere thanks to my talented editor, Sarah Chalmers, whose insightful guidance and sharp editorial eye elevated every page of this book. I am also grateful to my cover designer, Mahnoor Nisar, whose creative vision made this book truly distinctive.

Finally, to you, the reader, thank you for investing your time and for your commitment to building a more cyber-resilient and secure world.

A Personal Note from the Author

You just did something amazing.

You read this book from beginning to end. In a world of endless distractions, infinite scrolling, and shrinking attention spans, you chose to spend hours of your precious time understanding cybersecurity principles. For that, I'm deeply grateful.

Your Voice Matters More Than You Think

Here's something most readers don't realise: your opinion carries incredible power.

If this book helped you better understand cybersecurity concepts, challenged your thinking, or gave you a few "aha" moments—I'd really appreciate it if you'd take a moment to leave a review on Amazon. It will help others decide if this book is right for them.

Your review can make a real impact by helping to:

- Inform other readers navigating the same questions you once had.

- Improve discoverability so more people can find trusted cybersecurity guidance.

- Support authors who invest in sharing knowledge that empowers and protects.

Don't overthink it. There's no perfect review. Whether you write two sentences or two paragraphs, your honest feedback makes a genuine difference.

One More Thing...

If you enjoyed this book, I occasionally send updates and exclusive content to readers who want to stay connected. You can find me at cyops.com.au. Don’t forget to claim your copy of the indispensable Incident Response Checklist.

Thank you for reading and for your time. Whether you leave a review or simply pass the book on to a curious friend or colleague, thank you. Helping people build cyber resilience is what this book is all about.

With gratitude,

Alec Sklepic

P.S. Reviews don't have to be lengthy academic essays. "Loved it," "Made me think," or "Great reference" are perfect reviews that take 10 seconds to write but mean the world to future readers.

Notes

Introduction

[1] Tidy, J. (2024). *CrowdStrike IT outage affected 8.5 million Windows devices, Microsoft says*. BBC. https://www.bbc.com/news/articles/cpe3zgznwjno

[2] Petrosyan, A. (2024). *Annual number of data compromises and individuals impacted in the United States from 2005 to 2023*. Statistica. https://www.statista.com/statistics/273550/data-breaches-recorded-in-the-united-states-by-number-of-breaches-and-records-exposed/

Chapter 1

[3] Kost, E. (2024). *How to Avoid a Disaster Like the Optus Breach*. UpGuard. https://www.upguard.com/blog/how-to-avoid-a-disaster-like-the-optus-breach

[4] Australian Broadcasting Corporation (ABC). (2019). *ANU data breach stretching back 19 years detected.* https://www.abc.net.au/news/2019-06-04/anu-data-hack-bank-records-personal-information/11176788

[5] ABC. (2019). *ANU data breach stretching back 19 years detected.* https://www.abc.net.au/news/2019-06-04/anu-data-hack-bank-records-personal-information/11176788

[6] Australian National University. (2019). *Incident report on the breach of the Australian National University's administrative systems.* https://imagedepot.anu.edu.au/scapa/Website/SCAPA190209_Public_report_web_2.pdf

[7] Cybersecurity & Infrastructure Security Agency (CISA). (2023). *Understanding patches and software updates.* https://www.cisa.gov/news-events/news/understanding-patches-and-software-updates

[8] The Federal Bureau of Investigation (FBI). (2018). *The Morris worm.* https://www.fbi.gov/news/stories/morris-worm-30-years-since-first-major-attack-on-internet-110218

Chapter 2

[9] McAfee. (2023). *CLOP Ransomware exploits MOVEit software.* https://www.mcafee.com/blogs/other-blogs/mcafee-labs/clop-ransomware-exploits-moveit-software/

[10] Kaspersky. (2024). *Ransomware: The most high-profile attacks of 2023.* https://www.kaspersky.com/blog/ransowmare-attacks-in-2023/50634/

[11] Microsoft. (2024) *Introduction to threat modeling - Training | Microsoft Learn.* https://learn.microsoft.com/en-us/training/modules/tm-introduction-to-threat-modeling/

[12] Microsoft. *Introduction to threat modeling - Training | Microsoft Learn.* https://learn.microsoft.com/en-us/training/modules/tm-introduction-to-threat-modeling/

[13] Threat Modeling Manifesto Working Group. *Threat modeling manifesto.* https://www.threatmodelingmanifesto.org/

[14] Kidd. C, Raza.M. (2024). *Threat modeling: The ultimate guide.* Splunk Inc. https://www.splunk.com/en_us/blog/learn/threat-modeling.html

[15] Cado Security. (2024). *SOC playbook examples for real-world cyber threats.* https://www.cadosecurity.com/wiki/soc-playbook-examples-for-real-world-cyber-threats

Chapter 3

[16] The MITRE Corporation. (2023) *CWE top 25 most dangerous software weaknesses.* https://cwe.mitre.org/top25/index.html

[17] The Open Web Application Security Project (OWASP). (2023). *OWASP top ten.* https://owasp.org/www-project-top-ten/

[18] OWASP. (2024). *OWASP top 10 for large language model applications.* https://genai.owasp.org/

[19] The MITRE Corporation. (2023). *Common vulnerabilities and exposures (CVE).* https://cve.mitre.org/

[20] Forum of Incident Response and Security Teams, Inc. (FIRST). (2023). *Common vulnerability scoring system (CVSS).* https://www.first.org/cvss/

[21] CISA. *Known exploited vulnerabilities catalog (KEV).* https://www.cisa.gov/known-exploited-vulnerabilities-catalog

[22] National Institute of Standards and Technology (NIST). (2024). *National vulnerability database (NVD).* https://nvd.nist.gov/

[23] Hossain, M. (2023). *Software development life cycle (SDLC) methodologies for information systems project management.* https://www.researchgate.net/publication/373800862_Software_Development_Life_Cycle_SDLC_Methodologies_for_Information_Systems_Project_Management

[24] GitHub Inc. (2024). *Understanding the SDLC: Software development lifecycle explained.* https://resources.github.com/software-development/what-is-sdlc/

[25] OWASP. (2021). *OWASP top ten.* https://owasp.org/www-project-top-ten/

[26] Higgins, K, J. (2025). *Pen testers who got arrested doing their jobs tell all.* Darkreading. https://www.darkreading.com/vulnerabilities-threats/pen-testers-who-got-arrested-doing-their-jobs-tell-all

[27] CISA. (2023). *Continuous diagnostics and mitigation (CDM) program.* https://www.cisa.gov/cdm

[28] OWASP. (2023). *OWASP secure coding practices-quick reference guide.* https://owasp.org/www-project-secure-coding-practices-quick-reference-guide/

[29] IBM. (2025). *What is a zero-day exploit?* https://www.ibm.com/think/topics/zero-day

[30] MeltdownAttack. (2021). *Meltdown and spectre vulnerabilities.* https://meltdownattack.com/

[31] Google Project Zero. (2015). *Exploiting the DRAM rowhammer bug to gain kernel privileges.* https://googleprojectzero.blogspot.com/2015/03/exploiting-dram-rowhammer-bug-to-gain.html

[32] FireCompass. (2024). *Breaking the Trust: How SS7 attacks exploit telecom security gaps.* https://www.firecompass.com/exploiting-ss7-vulnerabilities-sigploit/

Chapter 4

[33] Palatty. J. N. (2025). 90+ cyber crime statistics 2025: Cost, industries & trends. https://www.getastra.com/blog/security-audit/cyber-crime-statistics/

[34] Petrosyan, A. Statista. (2024). Annual cost of cybercrime worldwide 2018-2029. https://www.statista.com/forecasts/1280009/cost-cybercrime-worldwide

[35] MGM Resorts International. (2023). *MGM resorts update on recent cybersecurity issue.* https://investors.mgmresorts.com/investors/news-releases/press-release-details/2023/MGM-RESORTS-UPDATE-ON-RECENT-CYBERSECURITY-ISSUE/default.aspx

Chapter 5

[36] NIST. *Risk Management.* https://www.nist.gov/risk-management

Chapter 6

[37] Reuters. (2017). *Yahoo says all three billion accounts hacked in 2013 data theft.* https://www.reuters.com/article/us-yahoo-cyber/yahoo-says-all-three-billion-accounts-hacked-in-2013-data-theft-idUSKCN1C82O1

[38] GDPR.EU. *What are the GDPR fines?* https://gdpr.eu/fines/

[39] IBM Security. (2023). *Cost of a data breach report 2023.* https://www.ibm.com/security/data-breach

[40] Root, E. Kaspersky. (2024). *A scientific approach to eavesdropping via HDMI.* https://www.kaspersky.com/blog/deep-tempest-side-channel-hdmi/52058/

[41] Verizon. (2024). *Data breach investigations report 2024.* https://www.verizon.com/business/resources/reports/dbir/

[42] DarkReading. (2023). *World security report finds physical security incidents cost companies USD $1T in 2022.* https://www.darkreading.com/cybersecurity-operations/world-security-report-finds-physical-security-incidents-cost-companies-usd-1t-in-2022

[43] IT Governance UK Blog. (2023). *List of data breaches and cyber attacks in 2023 – 8,214,886,660 records breached.* https://www.itgovernance.co.uk/blog/list-of-data-breaches-and-cyber-attacks-in-2023

[44] Office of the Australian Information Commissioner. *Privacy Act 1988.* https://www.oaic.gov.au/privacy/the-privacy-act/

[45] GDPR.EU. *What are the GDPR fines?* https://gdpr.eu/fines/

[46] U.S. Department of Health and Human Services. *Breach Notification Rule.* https://www.hhs.gov/hipaa/for-professionals/breach-notification/index.html

Chapter 7

[47] cyber.uk. *Nation state hackers case study: Bangladesh bank heist.* https://cyber.uk/areas-of-cyber-security/cyber-security-threat-groups-2/nation-state-hackers-case-study-bangladesh-bank-heist/

[48] WatchGuard Technologies. (2023). *The evolution of fileless malware in 2023*. https://www.watchguard.com

[49] Federal Trade Commission. (2019). *Equifax to pay $575 million in settlement*. https://www.ftc.gov/

[50] U.S. House of Representatives. (2020). *The Equifax data breach: A case study on ineffective cybersecurity practices*. Committee on Oversight and Reform. https://oversight.house.gov/

[51] Ponemon Institute. (2021). *Cost of a data breach report*. IBM Security. https://www.ibm.com/security/data-breach

[52] U.S. Securities and Exchange Commission. (2021). *Equifax Inc. financial disclosures and breach-related costs*. https://www.sec.gov/

[53] Wired. (2019). *Inside the Equifax hack*. https://www.wired.com/

[54] Cybersecurity & Infrastructure Security Agency. (2018). *Apache Struts vulnerability exploited in Equifax breach*. https://www.cisa.gov/

[55] Krebs, B. (2019). *The Equifax breach in depth*. Krebs on Security. https://krebsonsecurity.com/

[56] U.S. House of Representatives. (2020). *The Equifax data breach: A case study on ineffective cybersecurity practices*. Committee on Oversight and Reform. https://oversight.house.gov/

[57] PostgreSQL. (2021). *Write-ahead logging (WAL)*. https://www.postgresql.org/docs/current/wal-intro.html

[58] Vadala, D. (2002). *Journaling filesystems*. O'Reilly. https://www.oreilly.com/library/view/managing-raid-on/9780596802035/ch06s01s03.html

[59] Sheldon, R. (2023). *What is a message authentication code (MAC)?* TechTarget. https://www.techtarget.com/searchsecurity/definition/message-authentication-code-MAC

[60] Therapeutic Goods Administration (TGA). (2017). *Data management and data integrity (DMDI)*. https://www.tga.gov.au/data-management-and-data-integrity-dmdi

[61] PCI Security Standards Council. (2018). *PCI DSS quick reference guide*. https://www.pcisecuritystandards.org/documents/PCI_DSS-QRG-v3_2_1.pdf

Chapter 8

[62] BBC. (2016). *Australian census attacked by hackers.* https://www.bbc.com/news/world-australia-37008173

[63] cyberint. (2024). *Ransomware groups report 2024 – Q1.* https://cyberint.com/blog/research/ransomware-trends-2024-report/

[64] Lai, B. *The threat of ransomware*. Foreign Affairs. https://www.aph.gov.au/About_Parliament/Parliamentary_Departments/Parliamentary_Library/pubs/BriefingBook47p/ThreatRansomware

[65] The New York Times. (2019). *How one computer system tangled up several airlines.* https://www.nytimes.com/2019/04/29/business/airlines-computer-glitch-sabre.html

[66] Wikipedia. (2025). *2021 Facebook outage.* https://en.wikipedia.org/wiki/2021_Facebook_outage

[67] Ponemon Institute. (2022). *2022 cost of insider threats: Global report.* https://protectera.com.au/wp-content/uploads/2022/03/The-Cost-of-Insider-Threats-2022-Global-Report.pdf

[68] A10 Networks. (2024). *Five most famous DDoS attacks and then some.* https://www.a10networks.com/blog/5-most-famous-ddos-attacks/

[69] Data Foundry. (2021). *How to calculate the true cost of downtime.* https://www.datafoundry.com/how-to-calculate-the-true-cost-of-downtime/

[70] AWS. *Best practices for security, identity, & compliance.* https://aws.amazon.com/architecture/security-identity-compliance/

[71] Microsoft. (2024). *Service level agreements (SLA) for online services.* https://www.microsoft.com/licensing/docs/view/Service-Level-Agreements-SLA-for-Online-Services

[72] AWS. (2024). *AWS elastic disaster recovery.* https://aws.amazon.com/disaster-recovery/

[73] Microsoft. (2024). *Azure site recovery.* https://azure.microsoft.com/en-us/products/site-recovery

[74] Google Cloud. (2024). *Trusted infrastructure.* https://cloud.google.com/security/infrastructure

Chapter 9

[75] BBC. (2017). *Cyber-attack: Europol says it was unprecedented in scale.* https://www.bbc.com/news/world-europe-39907965

[76] Zetter, K. (2023). *The untold story of the boldest supply-chain hack ever.* Wired. https://www.wired.com/story/the-untold-story-of-solarwinds-the-boldest-supply-chain-hack-ever/

[77] Reuters. (2020). *Hackers used SolarWinds' dominance against it in sprawling spy campaign.* https://www.reuters.com/article/world/india/hackers-used-solarwinds-dominance-against-it-in-sprawling-spy-campaign-idUSKBN28Q07O/

[78] TechTarget. (2023). *SolarWinds hack explained: Everything you need to know.* https://www.techtarget.com/whatis/feature/SolarWinds-hack-explained-Everything-you-need-to-know

[79] FireEye. (2020). *Highly evasive attacker leverages SolarWinds supply chain to compromise multiple global victims with SUNBURST backdoor.* https://cloud.google.com/blog/topics/threat-intelligence/evasive-attacker-leverages-solarwinds-supply-chain-compromises-with-sunburst-backdoor/

[80] Symantec. (2017). *WannaCry ransomware attack: What you need to know.* https://www.symantec.com/blogs/threat-intelligence/wannacry-ransomware-attack

[81] CISA. (2024). *Known exploited vulnerabilities catalog.* https://www.cisa.gov/known-exploited-vulnerabilities-catalog

[82] Splunk. (2023). *SIEM: Security information & event management explained.* https://www.splunk.com/en_us/blog/learn/siem-security-information-event-management.html

[83] CISA. (2022). *Traffic light protocol (TLP) definitions and usage.* https://www.cisa.gov/news-events/news/traffic-light-protocol-tlp-definitions-and-usage

[84] The MITRE Corporation. *Structured threat information eXpression — STIX.* https://makingsecuritymeasurable.mitre.org/docs/stix-intro-handout.pdf

[85] The MITRE Corporation. *Trusted automated exchange of indicator information — TAXII.* https://makingsecuritymeasurable.mitre.org/docs/taxii-intro-handout.pdf

Chapter 10

[86] Industrial Cyber. (2024). *Australia updates cyber security governance principles to address emerging cyber threats, regulatory needs.* https://industrialcyber.co/regulation-standards-and-compliance/australia-

updates-cyber-security-governance-principles-to-address-emerging-cyber-threats-regulatory-needs/

[87] Australian Institute of Company Directors. (2024). *Cyber security governance principles, version 2.* https://www.aicd.com.au/risk-management/framework/cyber-security/cyber-security-governance-principles.html

[88] The Department of Home Affairs. (2023). *2023-2030 Australian cyber security strategy.* https://www.homeaffairs.gov.au/about-us/our-portfolios/cyber-security/strategy/2023-2030-australian-cyber-security-strategy

[89] Hoy, T. (2023). *What is a governance model and why does it matter?* BoardEffect. https://www.boardeffect.com/blog/what-governance-model-why-does-matter/

[90] ASD-ACSC. (2021). *Infosec registered assessors program (IRAP).* https://www.cyber.gov.au/irap

[91] International Organization for Standardization (ISO). *ISO/IEC 27001:2013 Information security management.* https://www.iso.org/standard/27001

Chapter 11

[92] BBC News. (2017). Massive ransomware infection hits computers in 99 countries. https://www.bbc.com/news/technology-39901382

[93] ISO. (2024). *ISO 55000:2024 Asset management — Vocabulary, overview and principles.* https://www.iso.org/standard/83053.html

[94] AWS Insider.net. (2024). *Yet another misconfigured AWS S3 bucket exposes sensitive customer data.* https://awsinsider.net/Articles/2024/12/10/Yet-Another-Misconfigured-AWS-S3-Bucket-Exposes-Sensitive-Customer-Data.aspx

[95] Kost. E. Upguard. (2024). *How did Kaseya get hacked?* https://www.upguard.com/blog/how-did-kaseya-get-hacked

Chapter 12

[96] Office of the Australian Information Commissioner. (2022). *Commissioner launches investigation into Optus breach.* https://www.oaic.gov.au/news/media-centre/oaic-opens-investigation-into-optus-over-data-breach

[97] ACSC. *Optus data breach*. https://www.cyber.gov.au/about-us/alerts/optus-data-breach

[98] Attorney-General's Department. (2022). *Privacy Act review discussion paper*. https://consultations.ag.gov.au/rights-and-protections/privacy-act-review-discussion-paper/

[99] Gartner In. (2025). *Cloud security posture management tools reviews and ratings*. https://www.gartner.com/reviews/market/cloud-security-posture-management-tools

[100] Krebs, B. (2014). *Target hackers broke in via HVAC company*. Krebs on Security. https://krebsonsecurity.com/2014/02/target-hackers-broke-in-via-hvac-company/

[101] Lakshmanan. R. (2024). *Critical flaws found in ConnectWise ScreenConnect software - patch now*. The Hacker News. https://thehackernews.com/2024/02/critical-flaws-found-in-connectwise.html

[102] ASD. (March 2024). *Information security manual (ISM)*. https://www.cyber.gov.au/resources-business-and-government/essential-cyber-security/ism

[103] NIST. (2006). *Guide to computer security log management*. https://csrc.nist.gov/pubs/sp/800/92/final

[104] RedHat. (2023). *What is CI/CD?* https://www.redhat.com/en/topics/devops/what-is-ci-cd

Chapter 13

[105] UpGuard. (2024). *What is the cost of a data breach in 2024?* https://www.upguard.com/blog/cost-of-a-data-breach-2024

[106] Committee On Commerce, Science, And Transportation. Majority Staff Report for Chairman Rockefeller (2014). *A "Kill Chain" Analysis of the 2013 Target Data Breach*. https://www.commerce.senate.gov/services/files/24d3c229-4f2f-405d-b8db-a3a67f183883?utm_source=chatgpt.com

Chapter 14

[107] ASD-ACSC. (2023). *Cyber incident response plan guidance*. https://www.cyber.gov.au/sites/default/files/2023-03/ACSC%20Cyber%20Incident%20Response%20Plan%20Guidance_A4.pdf

Chapter 15

[108] The Economic Times. (2022). *NotPetya: The cyberattack that shook the world.* https://economictimes.indiatimes.com/tech/newsletters/ettech-unwrapped/notpetya-the-cyberattack-that-shook-the-world/articleshow/89997076.cms

[109] INSURICA. *Cyber case study: Colonial pipeline ransomware attack.* https://insurica.com/blog/colonial-pipeline-ransomware-attack/

[110] Federal Trade Commission (FTC). (2019). *Equifax data breach settlement: What you should know.* https://consumer.ftc.gov/consumer-alerts/2019/07/equifax-data-breach-settlement-what-you-should-know

[111] Australian Prudential Regulation Authority. (2019). *Prudential standard CPS 234: Information security.* https://www.apra.gov.au/sites/default/files/cpg_234_information_security_june_2019_0.pdf

Chapter 16

[112] BBC. (2021). *Hacker tried to poison Florida town's water supply.* https://www.bbc.com/news/world-us-canada-55989843

[113] Commonwealth Consolidated Acts. *Privacy Act 1988 - Schedule 1 Australian privacy principles.* https://www.austlii.edu.au/cgi-bin/viewdoc/au/legis/cth/consol_act/pa1988108/sch1.html

[114] Attorney-Generals Department. *Privacy.* https://www.ag.gov.au/rights-and-protections/privacy

[115] European Commission. (2020). *Data protection.* https://ec.europa.eu/info/law/law-topic/data-protection_en

[116] State of California Department of Justice. (2024). *California Consumer Privacy Act (CCPA).* https://oag.ca.gov/privacy/ccpa

[117] U.S. Department of Health and Human Services. (2024). *Health information privacy.* https://www.hhs.gov/hipaa/index.html

[118] Australian Department of Home Affairs. (2024). *Cyber Security Act,* https://www.homeaffairs.gov.au/cyber-security-subsite/Pages/cyber-security-act.aspx

[119] Federal Register of Legislation. *Security of Critical Infrastructure Act 2018.* https://www.legislation.gov.au/C2018A00029/latest/versions

[120] Commonwealth Consolidated Acts. *Telecommunications Act 1997.* https://www8.austlii.edu.au/cgi-bin/viewdb/au/legis/cth/consol_act/ta1997214/

[121] Federal Register of Legislation. *Criminal Code Act 1995.* https://www.legislation.gov.au/C2004A04868/2022-11-10/text

[122] US Department of Justice. *Justice manual. Title 9: Criminal 9-48.000 - Computer Fraud and Abuse Act.* https://www.justice.gov/jm/jm-9-48000-computer-fraud

[123] CISA. *Federal Information Security Modernization Act.* https://www.cisa.gov/topics/cyber-threats-and-advisories/federal-information-security-modernization-act

[124] Council of Europe. *The convention on cybercrime (Budapest Convention, ETS No. 185) and its protocols.* https://www.coe.int/en/web/cybercrime/the-budapest-convention

[125] European Union. (2018). *General Data Protection Regulation (GDPR).* https://gdpr.eu/

[126] Car and Driver. (2019). *Everything you need to know about the VW diesel-emissions scandal.* https://www.caranddriver.com/news/a15339250/everything-you-need-to-know-about-the-vw-diesel-emissions-scandal/

[127] The Guardian. *The Cambridge Analytica files.* https://www.theguardian.com/news/series/cambridge-analytica-files

[128] Federal Trade Commission (FTC). (2019). *Equifax data breach settlement: What you should know.* https://consumer.ftc.gov/consumer-alerts/2019/07/equifax-data-breach-settlement-what-you-should-know

[129] IT News. (2020). *Toll Group 'returns to normal' after Mailto ransomware attack.* https://www.itnews.com.au/news/toll-group-returns-to-normal-after-mailto-ransomware-attack-539507

Chapter 17

[130] ASD-ACSC. (2024). *Information security manual (ISM).* https://www.cyber.gov.au/resources-business-and-government/essential-cyber-security/ism

Chapter 18

[131] Proofpoint. (2024). *2024 State of the phish – Today's cyber threats and phishing protection.* https://www.proofpoint.com/au/resources/threat-reports/state-of-phish

Index

E

Made in United States
Orlando, FL
04 November 2026

90201518R00354